Mathematics and Visualization

Springer
Berlin
Heidelberg
New York
Barcelona
Hong Kong
London
Milan
Paris
Tokyo

Hartmut Prautzsch
Wolfgang Boehm
Marco Paluszny

Bézier and B-Spline Techniques

With 182 Figures

Springer

Hartmut Prautzsch

Universität Karlruhe (TH)
Geometrische Datenverarbeitung
Am Fasanengarten 5
76128 Karlsruhe, Germany

e-mail: prau@ira.uka.de

Wolfgang Boehm

Technische Universität Braunschweig
Angewandte Geometrie und Computergraphik
Pockelsstraße 14
38106 Braunschweig, Germany

e-mail: w.boehm@tu-bs.de

Marco Paluszny

Universidad Central de Venezuela
Escuela de Matematicas, Facultad de Ciencias
Paseo Los Ilustres
20513 Caracas, Venezuela

e-mail: marco@euler.ciens.ucv.ve

Catalog-in-Publication Data applied for
Die Deutsche Bibliothek - CIP-Einheitsaufnahme
Prautzsch, Hartmut:
Bezier and B-spline techniques / Hartmut Prautzsch ; Wolfgang Boehm ; Marco Paluszny. - Berlin ; Heidelberg ; New York ; Barcelona ; Hong Kong ; London ; Milan ; Paris ; Tokyo : Springer, 2002
(Mathematics and visualization)
ISBN 3-540-43761-4

Mathematics Subject Classification (2000): 65D17, 68U07, 65D07, 65D05

ISBN 3-540-43761-4 Springer-Verlag Berlin Heidelberg New York

Springer-Verlag Berlin Heidelberg New York
a member of BertelsmannSpringer Science+Business Media GmbH

http://www.springer.de

Typeset in TEX by the authors
Cover design: *design & production* GmbH, Heidelberg

SPIN 10881822 46/3142db - 5 4 3 2 1 0 – Printed on acid-free paper

To

Paul de Faget de Casteljau

Preface

Computer-aided modeling techniques have been developed since the advent of NC milling machines in the late 40's. Since the early 60's Bézier and B-spline representations evolved as the major tool to handle curves and surfaces. These representations are geometrically intuitive and meaningful and they lead to constructive numerically robust algorithms.

It is the purpose of this book to provide a solid and unified derivation of the various properties of Bézier and B-spline representations and to show the beauty of the underlying rich mathematical structure. The book focuses on the core concepts of Computer-aided Geometric Design (CAGD) with the intent to provide a clear and illustrative presentation of the basic principles as well as a treatment of advanced material, including multivariate splines, some subdivision techniques and constructions of arbitrarily smooth free-form surfaces.

In order to keep the book focused, many further CAGD methods are excluded. In particular, rational Bézier and B-spline techniques are not addressed since a rigorous treatment within the appropriate context of projective geometry would have been beyond the scope of this book.

The book grew out of several courses taught repeatedly at the graduate and intermediate under-graduate levels by the authors at the Rensselaer Polytechnic Institute, USA, the Universities of Braunschweig and Karlsruhe, Germany, and the Universidad Central de Venezuela. These courses were taught as part of the curricula in mathematics and computer sciences, and they were regularly attended also by students from electrical and mechanical engineering, geophysics and other sciences.

For the careful proofreading of parts of the manuscript, we like to thank Stefan Bischoff, Bernhard Garz, Georg Umlauf, Claudia Bangert and Norbert Luscher. Especially, we thank Christoph Pennekamp and Natalie Spinner for preparing the LaTeX file and Bernd Hamann for his critical, thorough and final proofreading.

Wolfenbüttel, Wolfgang Boehm
Caracas, Marco Paluszny
Karlsruhe, Hartmut Prautzsch

Contents

III Multivariate Splines

Part I

Curves

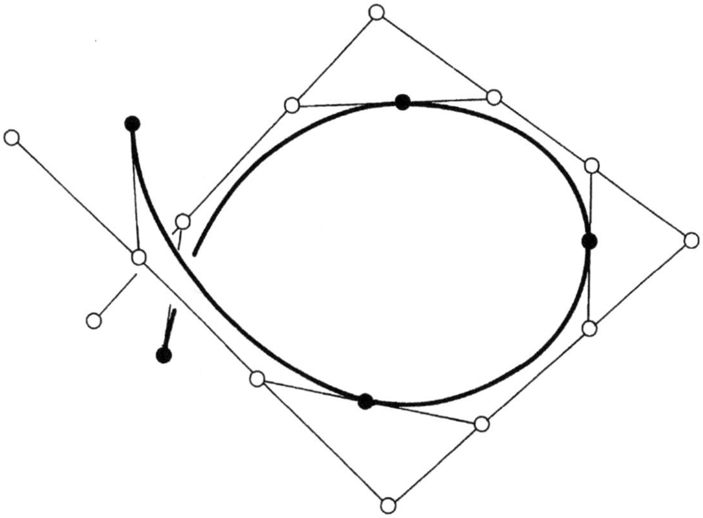

1 Geometric fundamentals

The world can be seen as a space of points while vectors describe the directions and lengths of line segments between pairs of points. The interpretation of the world as a point space, and not as a vector space, has the advantage that any point can may serve as the origin. This fact is also reflected by the symmetry of barycentric coordinates.

Since this book relies heavily on the concept of point or affine spaces, this chapter provides a brief recapitulation of their fundamental properties that are used throughout the book.

1.1 Affine spaces

An **affine space** $\mathcal{A}$ is a point space with an underlying vector space $\mathbf{V}$. Here we consider only finite-dimensional spaces over $\mathbb{R}$, which implies that points as well as vectors can be represented by the elements of some $\mathbb{R}^n$. Thus, any $\mathbf{x} \in \mathbb{R}^n$ represents a point or a vector, depending on the context. Moreover, we only work with such a coordinate representation and, therefore, simply regard $\mathcal{A}$ and $\mathbf{V}$ as some $\mathbb{R}^n$.

Given two points $\mathbf{p}$ and $\mathbf{q}$, the vector pointing from $\mathbf{p}$ to $\mathbf{q}$ is obtained as their difference,

$$\mathbf{v} = \mathbf{q} - \mathbf{p} \ ,$$

as illustrated in Figure 1.1. Note that a vector can be added to a point, but the sum of two points is undefined.

One can distinguish between points and vectors by **extended coordinates** with

$$\mathbb{x} = \begin{bmatrix} \mathbf{x} \\ e \end{bmatrix} \quad \text{representing a} \quad \begin{cases} \text{point} \\ \text{vector} \end{cases} \quad \text{if} \quad e = \begin{cases} 1 \\ 0 \end{cases} \ .$$

The above representation of points and vectors depends on a coordinate system. One can choose any point $\mathbf{p}$ of $\mathcal{A}$ and any n vectors $\mathbf{v}_1, \ldots, \mathbf{v}_n$ forming a basis of $\mathbf{V}$, where n is the dimension of $\mathbf{V}$. Then each point $\mathbf{q}$ of $\mathcal{A}$ has a unique representation $\mathbf{q} = \mathbf{p} + \mathbf{v}_1 x_1 + \cdots + \mathbf{v}_n x_n$, i.e., the coordinate column $\mathbf{x} = [x_1 \ \ldots \ x_n]^t \in \mathbb{R}^n$ represents the point $\mathbf{q}$ with respect to the **affine system** $\mathbf{p}; \mathbf{v}_1, \ldots, \mathbf{v}_n$ The point $\mathbf{p}$ is referred to as the **origin** of the coordinate system and has the coordinate column $\mathbf{x} = \mathbf{o} := [0 \ldots 0]^t$.

The **dimension** of an affine space $\mathcal{A}$ is defined as the dimension of its underlying vector space $\mathbf{V}$.

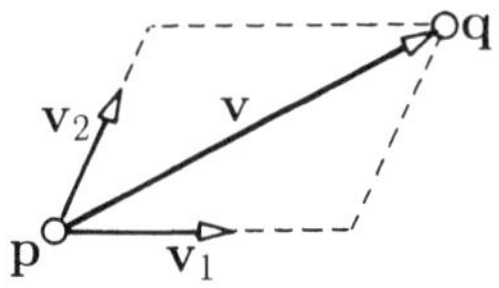

Figure 1.1: Points, vector and affine system.

1.2 Affine combinations

Any point sequence $\mathbf{p}_0, \ldots, \mathbf{p}_m$ of an affine space $\mathcal{A}$ is called **affinely independent** if the vector sequence $\mathbf{p}_1 - \mathbf{p}_0, \ldots, \mathbf{p}_m - \mathbf{p}_0$ is linearly independent. This definition does not depend on the order of the points $\mathbf{p}_i$, see Problem 1. Although independence is a property of sequences, and not of their individual members, we will also speak of independent points when we mean that the sequence formed by these points is affinely independent.

Let n be the dimension of $\mathcal{A}$. Then, any independent sequence $\mathbf{p}_0, \ldots, \mathbf{p}_n$ of $n+1$ points forms a **frame** of $\mathcal{A}$, and each point $\mathbf{q}$ of $\mathcal{A}$ can uniquely be written as

$$\begin{aligned} \mathbf{q} &= \mathbf{p}_0 + (\mathbf{p}_1 - \mathbf{p}_0)x_1 + \cdots + (\mathbf{p}_n - \mathbf{p}_0)x_n \\ &= \mathbf{p}_0 x_0 + \cdots + \mathbf{p}_n x_n \end{aligned}$$

where $1 = x_0 + \cdots + x_n$. The coefficients x_i are called the **barycentric coordinates** of $\mathbf{q}$ with respect to the frame $\mathbf{p}_0 \ldots \mathbf{p}_n$.

Note that $x_0, \ldots, x_{j-1}, x_{j+1}, \ldots, x_n$ are the affine coordinates of $\mathbf{q}$ with respect to the origin $\mathbf{p}_j$ and the n vectors $\mathbf{p}_i - \mathbf{p}_j$, $i \neq j$.

In particular, if $n = 1$, the point $\mathbf{q} = \mathbf{q}(x) = \mathbf{p}_0(1-x) + \mathbf{p}_1 x$ moves on the **linear interpolant** to $\mathbf{p}_0$ and $\mathbf{p}_1$ at the abscissae 0 and 1, respectively. The ratio $x : (1-x)$ is the **ratio** of $\mathbf{q}$ with respect to $\mathbf{p}_0$ and $\mathbf{p}_1$, see Figure 1.2.

Let $\mathbf{a}_1, \ldots, \mathbf{a}_m$ be the affine, extended or barycentric coordinate columns of

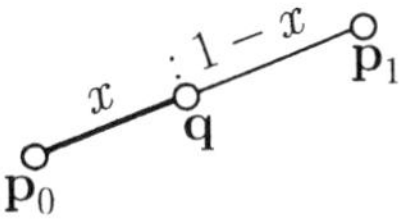

Figure 1.2: Linear interpolation and ratio.

any m points of $\mathcal{A}$. Then, the weighted sum

$$\mathbf{a} = \sum \mathbf{a}_i \alpha_i \quad \text{represents a} \quad \left\{ \begin{array}{l} \text{point} \\ \text{vector} \end{array} \right. \quad \text{if} \quad \sum \alpha_i = \left\{ \begin{array}{l} 1 \\ 0 \end{array} \right. .$$

If the weights sum to one, then $\mathbf{a} = \sum \mathbf{a}_i \alpha_i$ is called an **affine combination**. If, in addition, the weights are non-negative, then $\mathbf{a}$ is called a **convex combination**. It lies in the **convex hull** of the points $\mathbf{a}_i$, see Problem 4.

1.3 Affine maps

Let $\mathcal{A}$ and $\mathcal{B}$ be affine spaces, $\mathbf{U}$ and $\mathbf{V}$ be the underlying vector spaces and m and n be the corresponding dimensions, respectively. Then, a map $\Phi : \mathcal{A} \to \mathcal{B}$ is called **affine** if it can be represented by an $n \times m$ matrix A and a point $\mathbf{a}$ of $\mathcal{B}$ such that

$$\mathbf{y} = \Phi(\mathbf{x}) = \mathbf{a} + A\mathbf{x} \ ,$$

where $\mathbf{a}$ represents the image of the origin of $\mathcal{A}$.

The linear map $\varphi : \mathbf{U} \to \mathbf{V}$ given by

$$\mathbf{v} = \varphi(\mathbf{u}) = A\mathbf{u}$$

is called the **underlying linear map** of Φ. Using extended coordinates, both maps have the same matrix representation, which is given by

$$\begin{bmatrix} \mathbf{y} \\ 1 \end{bmatrix} = \begin{bmatrix} A & \mathbf{a} \\ \mathbf{o}^t & 1 \end{bmatrix} \begin{bmatrix} \mathbf{x} \\ 1 \end{bmatrix}, \quad \begin{bmatrix} \mathbf{v} \\ 0 \end{bmatrix} = \begin{bmatrix} A & \mathbf{a} \\ \mathbf{o}^t & 1 \end{bmatrix} \begin{bmatrix} \mathbf{u} \\ 0 \end{bmatrix},$$

and more concisely written as

$$\mathbb{y} = \mathbb{A}\mathbb{x} \ , \qquad \mathbb{v} = \mathbb{A}\mathbb{u} \ .$$

The following two properties are immediate consequences of the matrix representation.

An affine map Φ commutes with affine combinations, i.e.,

$$\Phi(\sum \mathbf{a}_i \alpha_i) = \sum \Phi(\mathbf{a}_i)\alpha_i \ .$$

And

an affine map is completely determined by a frame of $\dim 1 + \mathcal{A}$ independent points $\mathbf{p}_0 \ldots \mathbf{p}_m$ and its image $\mathbf{q}_0 \ldots \mathbf{q}_m$.

The first property even characterizes affine maps, see Problem 5. The second property is due to the fact that the matrix $\mathbb{A}$ can be written as

$$\underset{n+1}{\boxed{\mathbb{A}}}^{m+1} = \boxed{\mathbf{q}_0 \ \cdots \ \mathbf{q}_m} \boxed{\mathbf{p}_0 \ \cdots \ \mathbf{p}_m}^{-1} .$$

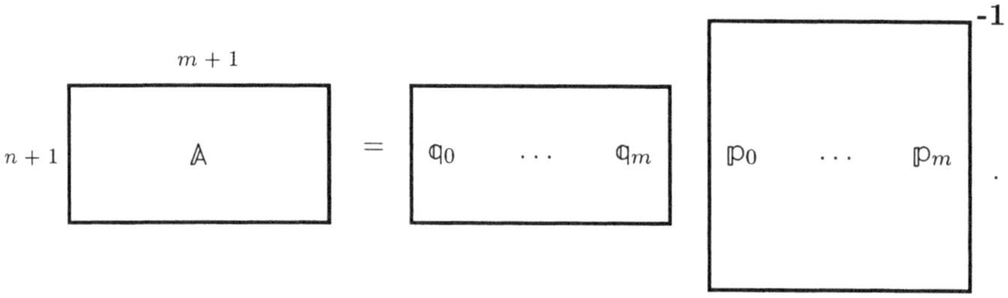

1.4 Parametric curves and surfaces

An element $\mathbf{x}$ of $\mathbb{R}^d$ whose coordinates depend on a parameter t traces out a **parametric curve**,

$$\mathbf{x}(t) = \begin{bmatrix} x_1(t) \\ \vdots \\ x_d(t) \end{bmatrix} .$$

Usually, we visualize $\mathbf{x}(t)$ as a point. In particular, if the coordinate functions $x_i(t)$ are polynomials of degree n or less than n, then $\mathbf{x}(t)$ is called a **polynomial curve** of degree n in t.

The **graph** of a function $x(t)$ is a special parametric curve in $\mathbb{R}^2$, which is given by

$$\mathbf{x}(t) = \begin{bmatrix} t \\ x(t) \end{bmatrix} .$$

Such a planar curve is often referred to as **functional curve**.

Figure 1.3 shows two examples of parametric curves: on the left Neil's parabola $\mathbf{x} = [t^2 \; t^3]^t$, and, on the right, the curve $\mathbf{x} = [t^2 \quad t^3 - t]^t$. These curves are non-functional.

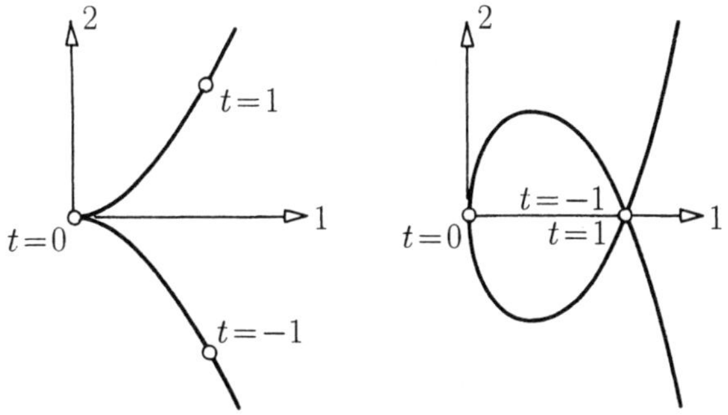

Figure 1.3: Parametric curves with a cusp and a loop.

Similarly, a column $\mathbf{x}$ depending on two parameters s and t traces out a (possibly degenerate) **parametric surface**

$$\mathbf{x}(s,t) = \begin{bmatrix} x_1(s,t) \\ \vdots \\ x_d(s,t) \end{bmatrix}.$$

It is called **polynomial** of total degree n if all x_i are polynomials of total degree n or less in s and t. Furthermore, the graph of a bivariate function $x(s,t)$ forms a special parametric surface in $\mathbb{R}^3$, which is given by

$$\mathbf{x}(s,t) = \begin{bmatrix} s \\ t \\ x(s,t) \end{bmatrix}.$$

Such a surface in $\mathbb{R}^3$ is often referred to as **functional surface**.

1.5 Problems

1 Show that a sequence of $m+1$ points $\mathbf{p}_0, \ldots, \mathbf{p}_m$ is affinely independent if and only if the sequence of their extended coordinate columns $\mathbb{p}_0, \ldots, \mathbb{p}_m$ is linearly independent.

2 The solutions of a homogeneous linear system form a vector space. Verify that the solutions of an inhomogeneous linear system form an affine space.

3 Consider the affine combination

$$\mathbf{p} = \mathbf{a}\alpha + \mathbf{b}\beta + \mathbf{c}\gamma \ , \qquad 1 = \quad \alpha + \beta + \gamma \ ,$$

of three independent points. Show that the barycentric coordinates α, β, γ determine the ratios shown in Figure 1.4.

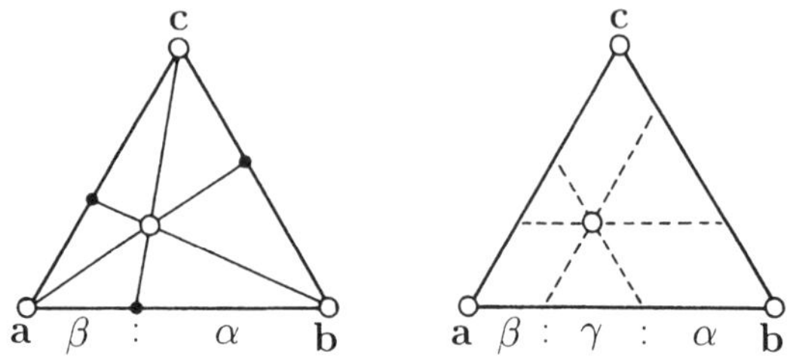

Figure 1.4: Ratios in a triangle.

4 The **convex hull** of r points $\mathbf{a}_1, \ldots, \mathbf{a}_r$ of some affine space $\mathcal{A}$ is the smallest subset of $\mathcal{A}$ containing $\mathbf{a}_1, \ldots, \mathbf{a}_r$, and for any two of its points, also the line segment in-between. Show that the convex hull of the $\mathbf{a}_i$ is formed by all their convex combinations.

5 Show that a map between two affine spaces which preserves affine combinations is an affine map.

6 Show that a map between two affine spaces which preserves affine combinations of any two points also preserves affine combinations of arbitrarily many points.

7 Show that the affine combinations of $r + 1$ independent points of some affine space $\mathcal{A}$ form an r-dimensional affine subspace of $\mathcal{A}$.

2 Bézier representation

Every polynomial curve segment can be represented by its so-called Bézier polygon. The curve and its Bézier polygon are closely related. They have common end points and end tangents, the curve segment lies in the convex hull of its Bézier polygon, etc. Furthermore, one of the fastest and numerically most stable algorithm used to render a polynomial curve is based on the Bézier representation.

2.1 Bernstein polynomials

Computing the binomial expansion

$$1 = (u + (1-u))^n = \sum_{i=0}^{n} \binom{n}{i} u^i (1-u)^{n-i}$$

leads to the **Bernstein polynomials** of degree n,

$$B_i^n(u) = \binom{n}{i} u^i (1-u)^{n-i} \ , \qquad i = 0, \ldots, n \ ,$$

which are illustrated in Figure 2.1 for $n = 4$.

The following properties of the Bernstein polynomials of degree n are important.

- *They are* **linearly independent**.

Namely, dividing $\sum_{i=0}^{n} b_i u^i (1-u)^{n-i} = 0$ by $(1-u)^n$, and setting $s =$

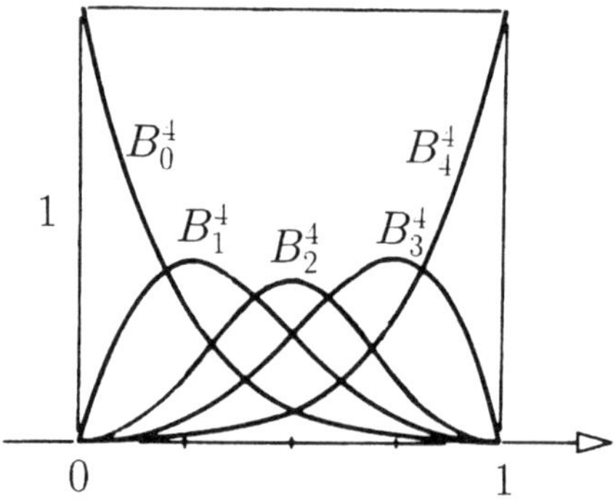

Figure 2.1: The Bernstein polynomials of degree 4 over $[0, 1]$.

$u/(1-u)$, gives $\sum_{i=0}^{n} b_i s^i = 0$, which implies $b_0 = \ldots = b_n = 0$.

- *They are* **symmetric**,

$$B_i^n(u) = B_{n-i}^n(1-u) \ .$$

- *They have* **roots** *at* 0 *and* 1 *only,*

$$B_i^n(0) = B_{n-i}^n(1) = \begin{cases} 1 & \text{for} \quad i = 0 \\ 0 & \phantom{\text{for}} \quad i > 0 \end{cases} \ .$$

- *They form a* **partition of unity**,

$$\sum_{i=0}^{n} B_i^n(u) = 1 \ , \quad \text{for all} \quad u \in \mathbb{R} \ .$$

- *They are* **positive** *in* $(0, 1)$,

$$B_i^n(u) > 0 \ , \quad \text{for} \quad u \in (0,1) \ .$$

- *They satisfy the* **recursion formula**

$$B_i^{n+1}(u) = uB_{i-1}^n(u) + (1-u)B_i^n(u) \ ,$$

  where $B_{-1}^n = B_{n+1}^n = 0$ *and* $B_0^0 = 1$.

This recursion formula follows directly from the identity

$$\binom{n+1}{i} = \binom{n}{i-1} + \binom{n}{i} \ .$$

Remark 1: The computation of the Bernstein polynomials up to degree n can be arranged in a triangular scheme, as shown below, where the recursion is represented by the "key" on the right:

$$\begin{array}{rccccc} 1 = & B_0^0 & B_0^1 & B_0^2 & \cdots & B_0^n \\ & & B_1^1 & B_1^2 & \cdots & B_1^n \\ & & & B_2^2 & \cdots & B_2^n \\ & & & & \ddots & \vdots \\ & & & & & B_n^n \end{array}$$

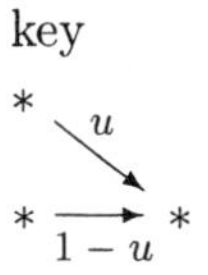

2.2 Bézier Representation

A dimension count shows that the $n+1$ (linearly independent) Bernstein polynomials B_i^n form a basis for all polynomials of degree $\leq n$. Hence, every polynomial curve $\mathbf{b}(u)$ of degree $\leq n$ has a unique nth degree **Bézier representation**

$$\mathbf{b}(u) = \sum_{i=0}^{n} \mathbf{c}_i B_i^n(u) \ .$$

Any affine parameter transformation

$$u = a(1-t) + bt \ , \qquad a \neq b \ ,$$

leaves the degree of the curve $\mathbf{b}$ unchanged. Consequently, $\mathbf{b}(u(t))$ has also an nth degree Bézier representation,

$$\mathbf{b}(u(t)) = \sum_{i=0}^{n} \mathbf{b}_i B_i^n(t) \ .$$

The coefficients $\mathbf{b}_i$ are elements of $\mathbb{R}^d$ and are called **Bézier points**. They are the vertices of the **Bézier polygon** of $\mathbf{b}(u)$ over the interval $[a, b]$. The parameter t is called the **local** and u the **global parameter** of $\mathbf{b}$, see Figure 2.2.

The properties of Bernstein polynomials summarized in 2.1 are passed on to the Bézier representation of a curve.

- *The* **symmetry** *of the Bernstein polynomials implies that*

$$\mathbf{b}(u) = \sum_{i=0}^{n} \mathbf{b}_i B_i^n(t) = \sum_{i=0}^{n} \mathbf{b}_{n-i} B_i^n(s) \ ,$$

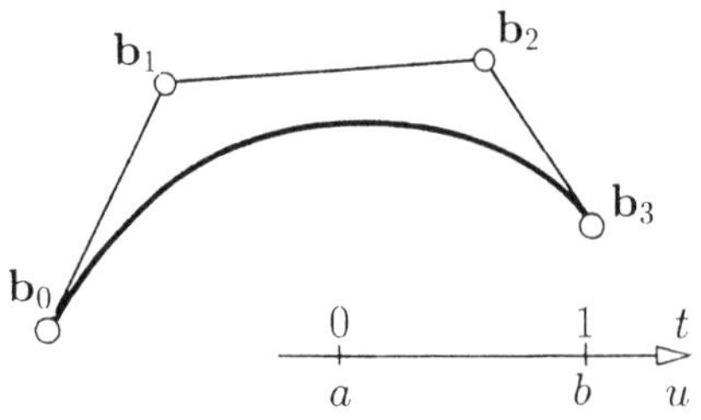

Figure 2.2: A cubic curve segment with its Bézier polygon over $[a, b]$.

where $u = a(1-t) + bt = b(1-s) + as$.

These two sums define the **Bézier representations of b** over $[a, b]$ and $[b, a]$, respectively. Thus, by using the oriented intervals $[a, b]$ and $[b, a]$, we can distinguish the two different parameter orientations of a polynomial curve.

- *For the* **end points** *of the curve segment* $\mathbf{b}[a, b]$*, one has*

$$\mathbf{b}(a) = \mathbf{b}_0 \qquad \text{and} \qquad \mathbf{b}(b) = \mathbf{b}_n \ .$$

Since the Bernstein polynomials sum to one,

- *any point* $\mathbf{b}(u)$ *is an* **affine combination** *of the Bézier points.*

As a consequence,

- *the Bézier representation is* **affinely invariant**, *i.e., given any affine map* Φ*, the image curve* $\Phi(\mathbf{b})$ *has the Bézier points* $\Phi(\mathbf{b}_i)$ *over* $[a, b]$.

Since the Bernstein polynomials are non-negative on $[0, 1]$,

- *one has for every* $u \in [a, b]$ *that* $\mathbf{b}(u)$ *is a* **convex combination** *of the* $\mathbf{b}_i$*. Hence, the curve segment* $\mathbf{b}[a, b]$ *lies in the convex hull of its Bézier points.*

This is illustrated in Figure 2.3.

Remark 2: Using the convex hull property separately for each coordinate function of the curve $\mathbf{b}(u)$, a **bounding box** is obtained for the curve segment $\mathbf{b}[a, b]$,

$$\mathbf{b}[a, b] \subset [\min_{i=0}^{n} \mathbf{b}_i, \ \max_{i=0}^{n} \mathbf{b}_i] \ , \qquad u \in [a, b] \ ,$$

as illustrated in Figure 2.4 for a planar curve.

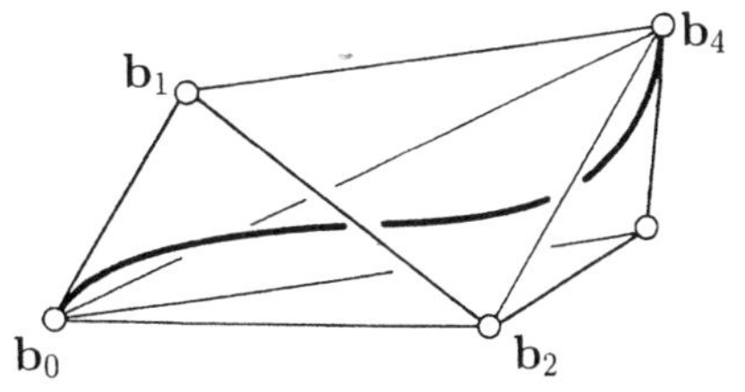

Figure 2.3: The convex hull of a Bézier polygon.

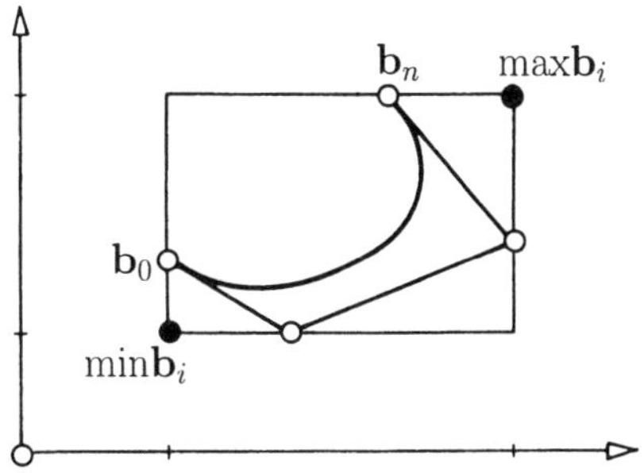

Figure 2.4: Bounding box.

2.3 The de Casteljau algorithm

A curve

$$\mathbf{b}(u) = \sum_{i=0}^{n} \mathbf{b}_i^0 B_i^n(t), \quad \text{with} \quad u = a(1-t) + bt,$$

can be evaluated easily by the **de Casteljau algorithm** [Casteljau '59]. Repeatedly using the recurrence relation of the Bernstein polynomials and collecting terms, one obtains

$$\mathbf{b}(u) = \sum_{i=0}^{n} \mathbf{b}_i^0 B_i^n(t) = \sum_{i=0}^{n-1} \mathbf{b}_i^1 B_i^{n-1}(t) = \cdots = \sum_{i=0}^{0} \mathbf{b}_i^n B_i^0(t) = \mathbf{b}_0^n \ ,$$

where

$$\mathbf{b}_i^{k+1} = (1-t)\mathbf{b}_i^k + t\,\mathbf{b}_{i+1}^k \ .$$

Two examples are shown in Figure 2.5, with $t = 0.4$ on the left and $t = 1.4$ on the right side.

The intermediate points $\mathbf{b}_i^k$ of the de Casteljau algorithm can be arranged in a triangular array, where each element is computed according to the "key"

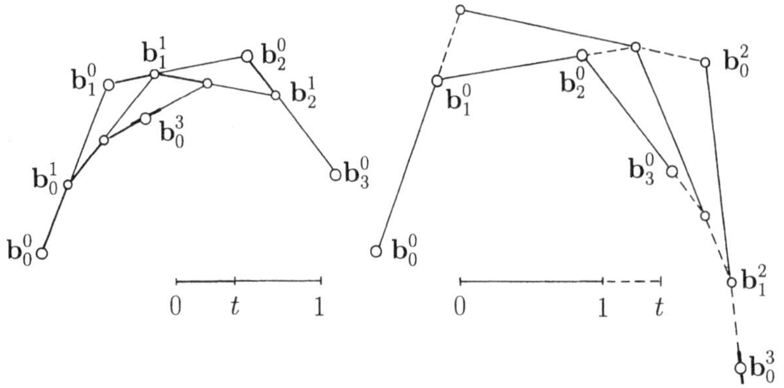

Figure 2.5: The de Casteljau construction.

on the right:

$$\begin{array}{llll}
\mathbf{b}_0^0 & & & \\
\mathbf{b}_1^0 & \mathbf{b}_0^1 & & \\
\mathbf{b}_2^0 & \mathbf{b}_1^1 & \mathbf{b}_0^2 & \\
\vdots & & \ddots & \\
\mathbf{b}_n^0 & \mathbf{b}_{n-1}^1 & \mathbf{b}_{n-2}^2 \cdots & \mathbf{b}_0^n
\end{array}$$

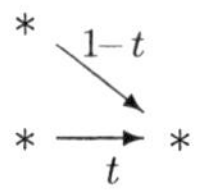

Remark 3: If t lies in $[0, 1]$, then the de Casteljau algorithm consists only of convex combinations, which accounts for the numerical stability of this algorithm.

Remark 4: Horner's scheme is a very effective method to evaluate a polynomial in monomial form. It can also be used for a curve $\mathbf{b}(t) = \sum \mathbf{b}_i B_i^n(t)$ in Bézier form. After writing $\mathbf{b}(t)$ as

$$\mathbf{b}(t) = (1-t)^n \left(\sum_{i=0}^{n} \mathbf{b}_i \binom{n}{i} \left(\frac{t}{1-t} \right)^i \right),$$

one first evaluates the sum in parentheses by Horner's scheme for the value $t/(1-t)$ and then multiplies the result by $(1-t)^n$.

This method fails, if t is close to 1. In this, case one can use the relationship

$$\mathbf{b}(t) = t^n \left(\sum_{i=0}^{n} \mathbf{b}_{n-i} \binom{n}{i} \left(\frac{1-t}{t} \right)^i \right).$$

2.4 Derivatives

The derivative of a Bernstein polynomial of degree n is simple to compute. From the definition of the Bernstein polynomials one gets

$$\frac{d}{dt}B_i^n(t) = n(B_{i-1}^{n-1}(t) - B_i^{n-1}(t)) \qquad \text{for} \quad i = 0, \dots, n \ ,$$

where $B_{-1}^{n-1} = B_n^{n-1} = 0$ as before. Thus, given a curve

$$\mathbf{b}(u) = \sum_{i=0}^{n} \mathbf{b}_i B_i^n(t) \ , \qquad t = \frac{u-a}{b-a} \ ,$$

one obtains for its derivative $\mathbf{b}'(u)$

$$\frac{d}{du}\mathbf{b}(u) = \frac{n}{b-a} \sum_{i=0}^{n-1} \Delta \mathbf{b}_i B_i^{n-1}(t) \ ,$$

where $\Delta \mathbf{b}_i = \mathbf{b}_{i+1} - \mathbf{b}_i$ This is illustrated in Figure 2.6.

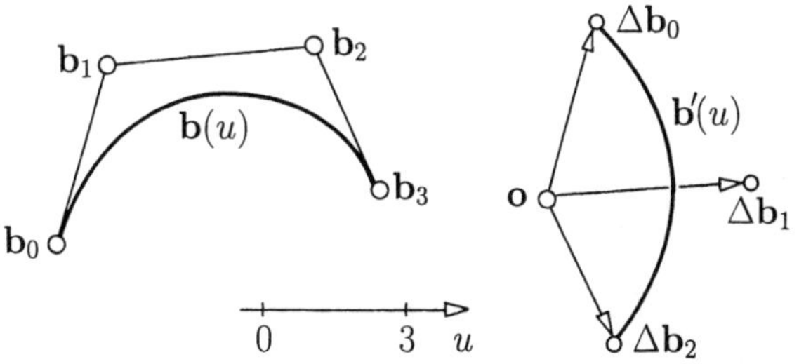

Figure 2.6: Bézier curve and its hodograph (1:3).

If the column $\mathbf{b}(u)$ is viewed as a point, then $\mathbf{b}'(u)$ is a vector. One obtains a point again if $\mathbf{b}'(u)$ is added to a point. In particular, $\mathbf{o} + \mathbf{b}'(u)$ is called the **(first) hodograph** of $\mathbf{b}$.

Applying the derivative formula above repeatedly, one obtains any rth derivative of $\mathbf{b}$,

$$\mathbf{b}^{(r)}(u) = \frac{n!}{(n-r)!(b-a)^r} \sum_{i=0}^{n-r} \Delta^r \mathbf{b}_i B_i^{n-r}(t) \ ,$$

where $\Delta^r \mathbf{b}_i = \Delta^{r-1}\mathbf{b}_{i+1} - \Delta^{r-1}\mathbf{b}_i$ denotes the **rth forward difference** of $\mathbf{b}_i$. As above one obtains the second and further hodographs.

Using the derivative formulas and the endpoint interpolation property of Bézier curves, we obtain a result that was fundamental for Bézier's first development.

> *The derivatives of* $\mathbf{b}$ *at* $t = 0$ *(or* $t = 1$*) up to order* r *depend only on the first (or last)* $r + 1$ *Bézier points, and vice versa.*

Geometrically, this means that, in general, the **tangents** of $\mathbf{b}$ at $t = 0$ and 1 are spanned by $\mathbf{b}_0, \mathbf{b}_1$ and $\mathbf{b}_{n-1}$, $\mathbf{b}_n$, respectively, and that the **osculating planes** of $\mathbf{b}$ at $t = 0$ and 1 are spanned by $\mathbf{b}_0, \mathbf{b}_1, \mathbf{b}_2$ and $\mathbf{b}_{n-2}, \mathbf{b}_{n-1}, \mathbf{b}_n$, respectively, and so forth. Figure 2.7 gives an illustration.

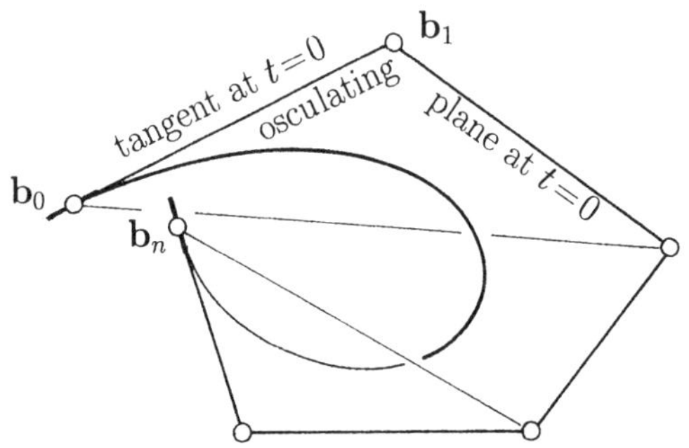

Figure 2.7: Tangents and osculating planes.

Remark 5: Viewing the Bézier polygon of a curve $\mathbf{b}(u) = \sum \mathbf{b}_i B_i^n(t)$, where $u = (1-t)a + t\,b$, as a piecewise linear function $\mathbf{p}(u)$ over $[a, b]$, one gets that

> *the derivative* $\mathbf{p}'(u)$ *of the Bézier polygon consists of the Bézier points of* $\mathbf{b}'(u)$.

This is illustrated in Figure 2.8 for a functional curve.

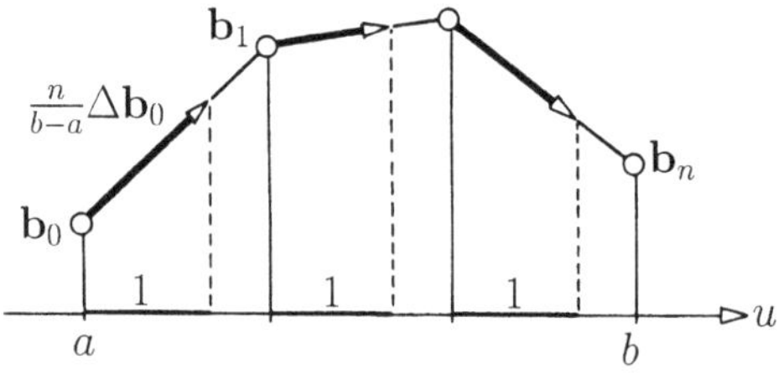

Figure 2.8: The derivative of a Bézier polygon.

2.5 Singular parametrization

Consider a polynomial curve

$$\mathbf{b}(t) = \sum_{i=0}^{n} \mathbf{b}_i B_i^n(t)$$

and its derivative

$$\dot{\mathbf{b}}(t) = n \sum_{i=0}^{n-1} \triangle \mathbf{b}_i B_i^{n-1}(t) \ ,$$

where the dot indicates differentiation with respect to the parameter t.

If $\triangle \mathbf{b}_0 = \mathbf{o}$, then $\dot{\mathbf{b}}(t)$ is zero at $t = 0$. However, with the singular reparametrization $t = \sqrt{s}$, one gets

$$\frac{d}{ds}\mathbf{b}(t(0)) = n \cdot \triangle \mathbf{b}_1 \ .$$

Thus, if $\triangle \mathbf{b}_0 = \mathbf{o}$ and $\triangle \mathbf{b}_1 \neq \mathbf{o}$, then the curve $\mathbf{b}(t)$ has a tangent at $t = 0$ that is directed towards $\mathbf{b}_2$, as illustrated in Figure 2.9.

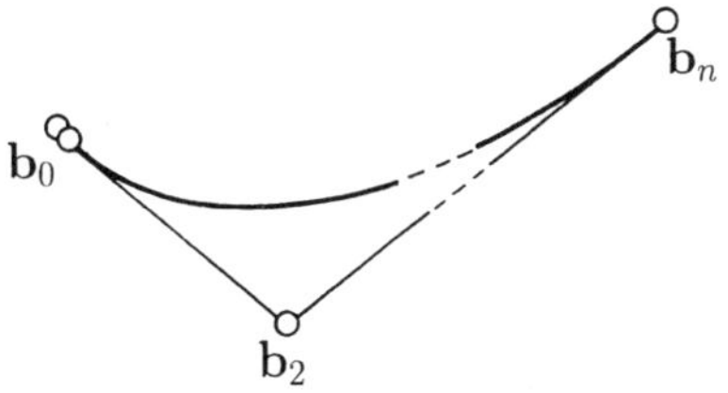

Figure 2.9: Singular parametrization.

Remark 6: If $\triangle \mathbf{b}_0 = \triangle \mathbf{b}_1 = \mathbf{o}$ and $\triangle \mathbf{b}_2 \neq \mathbf{o}$, then the tangent of $\mathbf{b}(t)$ at $t = 0$ is directed towards $\mathbf{b}_2$, etc.

2.6 A tetrahedral algorithm

Computing differences and the affine combinations of de Casteljau's algorithm can be combined. Namely, the rth derivative of a curve

$$\mathbf{b}(u) = \sum \mathbf{b}_i^0 B_i^n(t) \ , \quad t = \frac{u-a}{b-a} \ ,$$

at any u can be computed with de Casteljau's algorithm applied to multiples of the differences $\Delta^k \mathbf{b}_i$. Since the computation of affine combinations of affine combinations is commutative, i.e.,

$$\sum \alpha_i \sum \beta_j \mathbf{p}_{ij} = \sum \beta_j \sum \alpha_i \mathbf{p}_{ij} \ ,$$

the forward difference operator Δ commutes with the steps of de Casteljau's algorithm.

Hence, one can compute the rth derivative also by first computing $n - r$ steps of de Casteljau's algorithm, then r differencing steps and, finally, a multiplication by the factor $n \cdots (n - r + 1)/(b - a)^r$. Thus, it follows

$$\mathbf{b}^{(r)}(u) = \frac{n \cdots (n - r + 1)}{(b - a)^r} \quad \Delta^r \mathbf{b}_0^{n-r} \ ,$$

In particular, this formula says that the tangent and osculating plane of $\mathbf{b}$ at u are spanned by $\mathbf{b}_0^{n-1}, \mathbf{b}_1^{n-1}$ and $\mathbf{b}_0^{n-2}, \mathbf{b}_1^{n-2}, \mathbf{b}_2^{n-2}$, respectively, as illustrated in Figure 2.10 for a cubic.

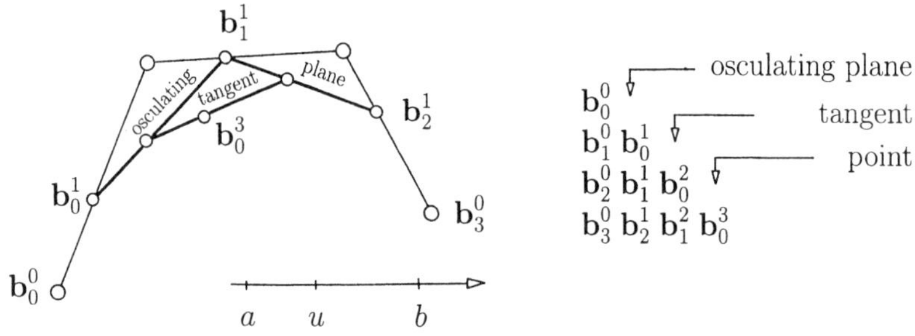

Figure 2.10: Osculating plane and tangent and the de Casteljau scheme.

Computing the points $\Delta^r \mathbf{b}_0^{n-k}$ for all k by successive de Casteljau and differencing steps, one also gets the intermediate points $\Delta^k \mathbf{b}_i^j$, $i + j + k \leq n$. All these points can be arranged conveniently in a tetrahedral array, as illustrated in Figure 2.11 for $n = 2$, where the key represents the recursion

$$\mathbf{c} = \mathbf{a}(1 - t) + \mathbf{b}t \quad \text{and} \quad \mathbf{d} = \mathbf{b} - \mathbf{a} \ .$$

This is not the only possible way to compute the tetrahedral array. Eliminating $\mathbf{a}$ or $\mathbf{b}$, one obtains

$$\mathbf{c} = \mathbf{b} + \mathbf{d}(t - 1) \quad \text{and} \quad \mathbf{c} = \mathbf{a} + \mathbf{d}t \ ,$$

respectively.

When we use one of these rules, instead of the differencing step, it suffices to compute only the points of the two triangular schemes given by the points

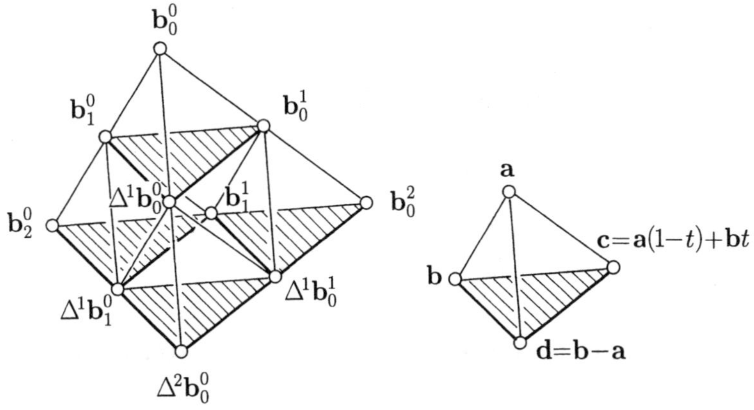

Figure 2.11: Tetrahedral algorithm.

on the left and bottom (or right) side of the tetrahedron.

Remark 7: It should be noted that differencing is not a numerically stable process, in general. Consequently, computing derivatives is not a numerically stable process either.

2.7 Integration

The integral of a polynomial curve in Bézier representation

$$\mathbf{b}(u) = \sum_{i=0}^{n} \mathbf{b}_i B_i^n(t), \quad t = \frac{u-a}{b-a} ,$$

has the Bézier representation

$$\mathbf{c}(u) = \int \mathbf{b}(u)du = \sum_{i=0}^{n+1} \mathbf{c}_i B_i^{n+1}(t) ,$$

where

$$\begin{aligned} \mathbf{c}_i &= \mathbf{c}_{i-1} + \frac{b-a}{n+1}\,\mathbf{b}_{i-1} \\ &= \mathbf{c}_0 + \frac{b-a}{n+1}\,(\mathbf{b}_0 + \cdots + \mathbf{b}_{i-1}) , \quad i = n+1,\ldots,1 , \end{aligned}$$

and $\mathbf{c}_0$ is an arbitrary integration constant. This is verified by differentiating $\mathbf{c}(u)$.

As a consequence of the integration formula and the endpoint interpolation property of the Bézier representation, one obtains

$$\int_a^b \mathbf{b}(u)du = \frac{b-a}{n+1}(\mathbf{b}_0 + \cdots + \mathbf{b}_n)$$

and, in particular, independently of i,

$$\int_0^1 B_i^n(t)dt = \frac{1}{n+1} .$$

2.8 Conversion to Bézier representation

Some older CAD data formats represent curves by monomials. Thus, data conversion between different CAD systems is an application where it is necessary to convert the monomial to the Bézier representation and vice versa. Let

$$\mathbf{b}(t) = \sum_{i=0}^{n} \mathbf{a}_i \binom{n}{i} t^i$$

be a curve in monomial form with binomial factors as in the Bézier representation. Since

$$\begin{aligned}\binom{n}{i} t^i (1-t+t)^{n-i} &= \sum_{k=0}^{n-i} \binom{n}{i}\binom{n-i}{n-i-k} t^{i+k}(1-t)^{n-i-k} \\ &= \sum_{k=0}^{n-i} \binom{i+k}{i} B_{i+k}^n \\ &= \sum_{j=0}^{n} \binom{j}{i} B_j^n ,\end{aligned}$$

one obtains the conversion formula

$$\mathbf{b}(t) = \sum_{j=0}^{n} \mathbf{b}_j B_j^n(t) ,$$

where

$$\mathbf{b}_j = \sum_{i=0}^{n} \binom{j}{i} \mathbf{a}_i$$

and $\binom{j}{i} = 0$ for $j < i$.

The formula for the conversion to monomial form can be derived similarly by multiplying out the Bernstein polynomials, see Problem 4. In Section 2.9, we present a different derivation of it.

Remark 8: If $\mathbf{a}_2 = \cdots = \mathbf{a}_n = \mathbf{o}$ and $\mathbf{a}_1 \neq \mathbf{o}$, then $\mathbf{b}(t)$ is a linear polynomial represented over $[0, 1]$ by the Bézier points

$$\mathbf{b}_j = \mathbf{a}_0 + j\mathbf{a}_1 \ ,$$

as illustrated in Figure 2.12.

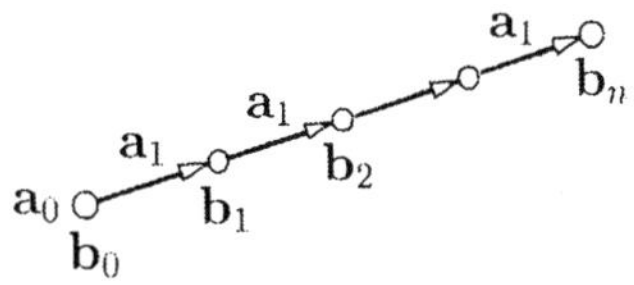

Figure 2.12: Equidistant Bézier points on a line.

Remark 9: Conversely, if the $n+1$ Bézier points $\mathbf{b}_i$ lie equidistantly on a line, then $\mathbf{b}(t)$ is a linear polynomial, which can be written as

$$\mathbf{b}(t) = (1-t)\mathbf{b}_0 + t\mathbf{b}_n \ .$$

This property is referred to as the **linear precision** of the Bézier representation.

Remark 10: As a consequence of Remark 9, the functional curve

$$\mathbf{b}(t) = \begin{bmatrix} t \\ b(t) \end{bmatrix}, \qquad b(t) = \sum b_i B_i^n(t) \ ,$$

has the Bézier points $[i/n \ \ b_i]^t$, as illustrated in Figure 2.13. The coefficients b_i are referred to as the **Bézier ordinates** of $b(t)$, and i/n as the **Bézier abscissae**.

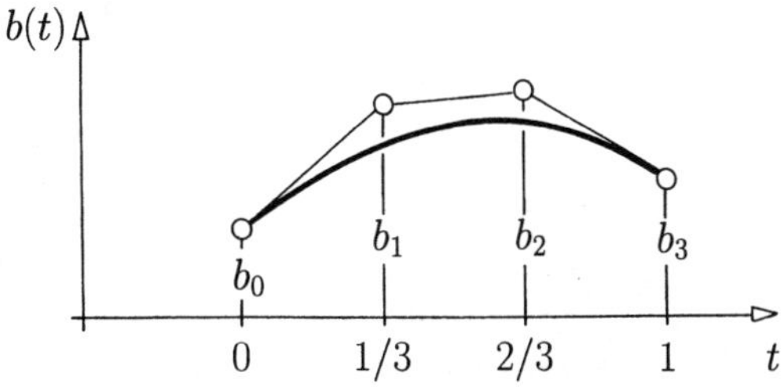

Figure 2.13: Bézier representation of a functional curve.

2.9 Conversion to monomial form

Given a polynomial curve in Bézier representation,

$$\mathbf{b}(u) = \sum_{i=0}^{n} \mathbf{b}_i B_i^n \left(\frac{u-a}{b-a} \right) ,$$

one can obtain its monomial form simply by a Taylor expansion,

$$\begin{aligned} \mathbf{b}(u) &= \sum_{i=0}^{n} \mathbf{b}^{(i)}(a) \frac{(u-a)^i}{i!} \\ &= \sum_{i=0}^{n} \binom{n}{i} \Delta^i \mathbf{b}_0 \frac{(u-a)^i}{(b-a)^i} . \end{aligned}$$

Since $\Delta^i \mathbf{b}_0 = \sum_{k=0}^{i} \binom{i}{k} (-1)^{i-k} \mathbf{b}_k$, see Problem 3, one can write this explicitly as

$$\mathbf{b}(u) = \sum_{i=0}^{n} \sum_{k=0}^{i} (-1)^{i-k} \binom{n}{i} \binom{i}{k} \mathbf{b}_k \, t^i .$$

Remark 11: Using the tetrahedral algorithm in 2.6, one can compute the Taylor expansion at u,

$$\mathbf{b}(u+h) = \sum_{i=0}^{n} \frac{1}{(b-a)^i} \Delta^i \mathbf{b}_0^{n-i} \binom{n}{i} h^i .$$

2.10 Problems

1 Show that the Bernstein polynomial $B_i^n(t)$ has only one maximum in $[0,1]$, namely at $t = i/n$.

2 The **Bernstein operator** $\mathcal{B}$ assigns to a function f on $[0,1]$ the polynomial

$$\mathcal{B}[f] = \sum_{i=0}^{n} f(i/n) B_i^n(t) .$$

If f is a polynomial of degree $m \le n$, then $\mathcal{B}[f]$ is also a polynomial of degree m, see also Problem 2 in 3.13. Show that this is true.

3 Show that

$$\Delta^i \mathbf{b}_0 = \sum_{k=0}^{i} \binom{i}{k} (-1)^{i-k} \mathbf{b}_k .$$

4 Derive the conversion formula in 2.9 to monomial form by elementary algebraic manipulations similar to what is done in 2.8.

5 Show that

$$n \ldots (n-k)\, t^{k+1} = \sum_{i=0}^{n} i \ldots (i-k)\, B_i^n(t) \ .$$

6 Show that a planar cubic $\mathbf{b}(t)$ has a **cusp** at $t = 0$, i.e., a point where it changes its direction, if $\dot{\mathbf{b}}(0) = \mathbf{o}$ and both coordinates of $\ddot{\mathbf{b}}(0)$ are non-zero. (The dots indicate differentiation with respect to t.)

7 Show that a planar cubic $\mathbf{b}(t) = \sum_{i=0}^{3} \mathbf{b}_i B_i^3(t)$ has a cusp if $\mathbf{b}_3$ lies on the parabola

$$\mathbf{p}(t) = (\mathbf{b}_0 + \mathbf{b}_1 - \mathbf{b}_2) B_0^2(t) + \mathbf{b}_1 B_1^2(t) + \mathbf{b}_2 B_2^2(t) \ ,$$

see [Pottmann & DeRose '91].

8 For which choices of $\mathbf{b}_3$ does the cubic $\mathbf{b}(t)$ have a loop?

9 Let $\sum_{i=0}^{n} \mathbf{a}_i \binom{n}{i} t^i = \sum_{i=0}^{n} \mathbf{b}_i B_i^n(t)$. Then, in matrix notation, one obtains

$$[\mathbf{a}_0 \ \ldots \ \mathbf{a}_n] = [\mathbf{b}_0 \ \ldots \ \mathbf{b}_n]\, \Delta \ ,$$

where $\Delta = \left[(-1)^{j-i}\binom{j}{i}\right]$ and $\Delta^{-1} = \left[\binom{i}{j}\right]$. The matrices Δ and Δ^{-1} are upper-triangular matrices.

3 Bézier techniques

3.1 Symmetric polynomials — 3.2 The main theorem — 3.3 Subdivision — 3.4 Convergence under subdivision — 3.5 Curve generation by subdivision — 3.6 Curve generation by forward differences — 3.7 Intersections — 3.8 The variation diminishing property — 3.9 The symmetric polynomial of the derivative — 3.10 Simple C^r joints — 3.11 Degree elevation — 3.12 Convergence under degree elevation — 3.13 Problems

Many algorithms for curves in Bézier representation can be understood and derived using symmetric polynomials. In this chapter, the relationship between univariate and symmetric multivariate polynomials is explained and the basic CAGD algorithms based on this relationship are presented. The most important algorithm is de Casteljau's. It has several applications and serves as an important theoretical tool.

3.1 Symmetric polynomials

Every polynomial curve $\mathbf{b}(u)$ of degree $\leq n$ can be associated with a unique n-variate symmetric polynomial $\mathbf{b}[u_1 \ldots u_n]$ having the following three properties.

- $\mathbf{b}[u_1 \ldots u_n]$ *agrees with* $\mathbf{b}(u)$ *on its* **diagonal**,

which means that $\mathbf{b}[u \ldots u] = \mathbf{b}(u)$.

- $\mathbf{b}[u_1 \ldots u_n]$ *is* **symmetric** *in its variables,*

which means that for any permutation $(v_1, \ldots, v_n)$ of $(u_1, \ldots, u_n)$

$$\mathbf{b}[v_1 \ldots v_n] = \mathbf{b}[u_1 \ldots u_n] \ .$$

- $\mathbf{b}[u_1 \ldots u_n]$ *is* **affine** *in each variable,* which means that

$$\mathbf{b}[(\alpha u + (1-\alpha)v)\, u_2 \,\ldots\, u_n] = \alpha \mathbf{b}[u\, u_2 \,\ldots\, u_n] + (1-\alpha)\mathbf{b}[v\, u_2 \,\ldots\, u_n] \ .$$

The symmetric polynomial $\mathbf{b}[u_1 \,\ldots\, u_n]$ is also referred to as the **polar form** [Casteljau '85] or **blossom** of $\mathbf{b}(u)$ [Ramshaw '87]. In order to show that such symmetric polynomials exist for all polynomials, it suffices to consider basis polynomials and to derive explicit representations of their symmetric forms.

Any linear combination

$$\mathbf{b}(u) = \sum_{i=0}^{n} \mathbf{c}_i\, C_i(u)$$

of nth degree polynomials $C_i(u)$ with blossoms $C_i[u_1 \,\ldots\, u_n]$ has the polar form

$$\mathbf{b}[u_1 \,\ldots\, u_n] = \sum_{i=0}^{n} \mathbf{c}_i\, C_i[u_1 \,\ldots\, u_n] \ ,$$

which clearly satisfies the three properties above.

Note that the diagonal $\mathbf{b}[u \,\ldots\, u]$ can be of lower degree than n, although $\mathbf{b}[u_1 \,\ldots\, u_n]$ depends on n variables.

In case the C_i are the weighted monomials $A_i^n = \binom{n}{i} u^i$, $i = 0, \ldots, n$, we obtain the **elementary symmetric polynomials**

$$A_i^n[u_1 \,\ldots\, u_n] = \sum_{1 \le j_1 < \cdots < j_i \le n} u_{j_1} \ldots u_{j_i} \ ,$$

which obviously satisfy the three properties above. This sum extends over $\binom{n}{i}$ products of i variables.

In case the C_i are the Bernstein polynomials

$$B_i^n(u) = \binom{n}{i} u^i (1-u)^{n-i} \ ,$$

we obtain

$$B_i^n[u_1 \,\ldots\, u_n] = \sum_{\substack{j_1 < \cdots < j_i \\ k_1 < \cdots < k_{n-i}}} u_{j_1} \ldots u_{j_i} \cdot (1 - u_{k_1}) \ldots (1 - u_{k_{n-i}}) \ ,$$

where $(j_1, \ldots, j_i, k_1, \ldots, k_{n-i})$ is a permutation of $(1, \ldots, n)$. Again, one can easily check the three properties above.

Remark 1: The symmetric polynomials $B_i^n[u_1 \,\ldots\, u_n]$ satisfy the recursion

$$B_i^{n+1}[u_0 \,\ldots\, u_n] = u_0 B_{i-1}^n[u_1 \,\ldots\, u_n] + (1 - u_0) B_i^n[u_1 \,\ldots\, u_n] \ .$$

3.2 The main theorem

The uniqueness of the symmetric polynomials and their relationship to the Bézier representation are given by the following **main theorem**, which is extended in several ways in other parts of the book.

For every polynomial curve $\mathbf{b}(u)$ *of degree* $\leq n$ *there exists a unique symmetric n-variate and n-affine polynomial* $\mathbf{b}[u_1 \ldots u_n]$ *with diagonal* $\mathbf{b}[u \ldots u] = \mathbf{b}(u)$. *Moreover, the points*

$$\mathbf{b}_i = \mathbf{b}[a \overset{n-i}{\ldots} a\, b \overset{i}{\ldots} b] \ , \qquad i = 0, \ldots, n \ ,$$

are the Bézier points of $\mathbf{b}(u)$ *over* $[a, b]$.

Proof: In 3.1 it is shown that a polar form $\mathbf{b}[u_1 \ldots u_n]$ exists for $\mathbf{b}(u)$. Hence, we may consider the points

$$\mathbf{b}_i^k = \mathbf{b}[a \overset{j}{\ldots} a\, u_1 \overset{k}{\ldots} u_k\, b \overset{i}{\ldots} b] \ , \qquad i + j + k = n \ .$$

Since $\mathbf{b}_0^n = \mathbf{b}[u_1 \ldots u_n]$ is symmetric and multi-affine, it can be computed from the points $\mathbf{b}_i^0$ by the following recursion formula

(1) $$\mathbf{b}_i^{k+1} = \mathbf{b}_i^k \cdot (1 - t_{k+1}) + \mathbf{b}_{i+1}^k \cdot t_{k+1} \ ,$$

where

$$t_k = \frac{u_k - a}{b - a} \ .$$

See Figure 3.1 for an illustration, where the points $\mathbf{b}[u_1 \ldots u_n]$ are labelled only by the arguments $u_1 \ldots u_n$ to simplify the notation.

Moreover, if all u_k equal u, then recursion formula(1) above reduces to de Casteljau's algorithm for the computation of $\mathbf{b}(u)$. Consequently, since the Bézier representation is unique, the points $\mathbf{b}_i$ are the Bézier points of $\mathbf{b}(u)$ over $[a, b]$. Hence, any two symmetric polynomials with the same diagonal $\mathbf{b}(u)$ agree for all arguments $[a \overset{n-i}{\ldots} a b \overset{i}{\ldots} b]$ and, because of recursion formula(1), they are also identical for all arguments $[u_1 \ldots u_n]$. Hence, $\mathbf{b}(u)$ has a unique n-affine polar form. ◇

3.3 Subdivision

Recursion formula (1), which is illustrated in Figure 3.1, reveals a very important additional property of de Casteljau's algorithm.

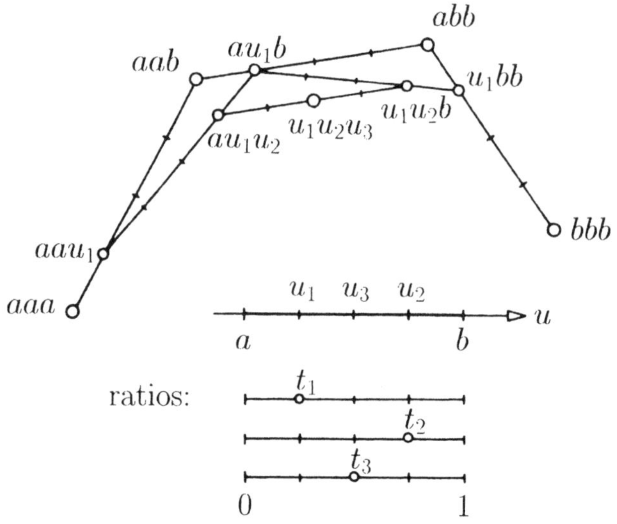

Figure 3.1: Main theorem, illustration.

In de Casteljau's array,

$$\begin{array}{llll} \mathbf{b}_0^0 & & & \\ \mathbf{b}_1^0 & \mathbf{b}_0^1 & & \\ \vdots & & \ddots & \\ \mathbf{b}_n^0 & \mathbf{b}_{n-1}^1 \cdots & \mathbf{b}_0^n = \mathbf{b}(c) & \end{array}$$

used to compute a point $\mathbf{b}(c)$, the Bézier points

$$\mathbf{b}_0^i = \mathbf{b}[a \overset{n-i}{\cdots} a\, c \overset{i}{\cdots} c] \quad \text{and} \quad \mathbf{b}_j^{n-i} = \mathbf{b}[c \overset{n-i}{\cdots} c\, b \overset{i}{\cdots} b]$$

of the curve segments over $[a, c]$ and $[c, b]$ are found in the upper diagonal and bottom row, respectively.

The computation of the Bézier points over the two intervals $[a, c]$ and $[c, b]$ is called **subdivision**. The corresponding construction is shown in Figure 3.2. Again, the points

$$\mathbf{b}_i^k = \mathbf{b}[a \overset{n-i-k}{\cdots} a\, c \overset{k}{\cdots} c\, b \overset{i}{\cdots} b]$$

are labelled by their arguments. Note that the figure is still correct if we interchange b and c.

On subdividing $\mathbf{b}(u)$ repeatedly, one can obtain the Bézier polygons of $\mathbf{b}(u)$ over any number of abutting intervals $[a_0, a_1], [a_1, a_2], \ldots, [a_{k-1}, a_k]$.

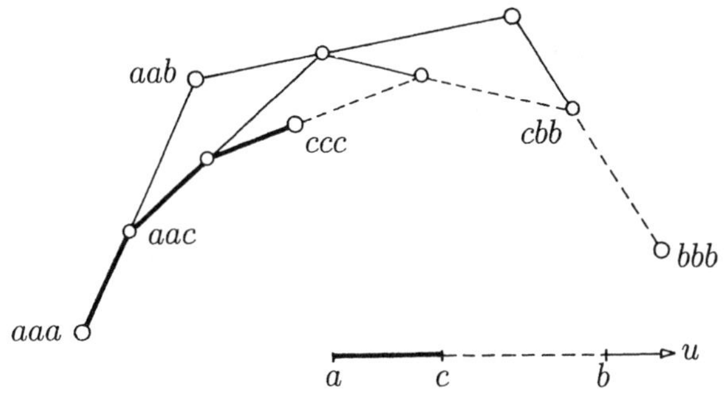

Figure 3.2: Subdivision by de Casteljau's algorithm.

Together, these polygons form the **composite Bézier polygon** of $\mathbf{b}$ over $[a_0, a_1, \ldots, a_k]$. In general, this composite polygon has $kn+1$ distinct vertices.

3.4 Convergence under subdivision

The Bézier polygon of a small curve segment is a fairly good approximation of this segment. To make this precise, let $\mathbf{b}_0, \ldots, \mathbf{b}_n$ be the Bézier points of some curve $\mathbf{b}(u)$ over any subinterval $[c, c+nh]$ of some fixed interval $[a, b]$. Furthermore, let $c_i = c + ih$, $i = 0, \ldots, n$. Then,

there is a constant M not depending on c such that

$$\max_i \|\mathbf{b}(c_i) - \mathbf{b}_i\| \leq Mh^2 \ .$$

Proof: We expand the symmetric polynomial $\mathbf{b}[u_1 \ldots u_n]$ at $[c_i \ldots c_i]$ and obtain

$$\begin{aligned} \mathbf{b}_i &= \mathbf{b}[c \overset{n-i}{\ldots} c \; c{+}nh \overset{i}{\ldots} c{+}nh] \\ &= \mathbf{b}[c_i \ldots c_i] - \sum_{j=1}^{n-i} ih \frac{\partial}{\partial u_j} \mathbf{b}[c_i \ldots c_i] + \sum_{j=n-i+1}^{n} (n-i)h \frac{\partial}{\partial u_j} \mathbf{b}[c_i \ldots c_i] \\ &\quad + O(h^2) \ , \end{aligned}$$

which proves the claim since all partial derivatives are equal. ◇

Applications of this approximation property are discussed in the following sections. A more general version of this proof can be found in 6.3 for splines.

Remark 2: With the maximum norm $\|\cdot\|_\infty$, the lowest possible constant M for which the above estimate holds for all curves is derived in [Nairn et al. '99] and [Reif '00] and found to be

$$\max_{i=0,\ldots,n-2} \|\Delta^2 \mathbf{b}_i\|_\infty \cdot \lfloor n/2 \rfloor \cdot \lceil n/2 \rceil / 2n \ .$$

Remark 3: Quadratic convergence is best possible as one sees with the parabola $p(u) = u^2$ whose middle Bézier point over $[0, 2h]$ is zero, while $p(h) = h^2$.

3.5 Curve generation by subdivision

Subdivision provides a very fast method for generating an approximant of a Bézier curve. From 3.4, it follows that the Bézier polygons over

$$[0, \frac{1}{2^k}, \frac{2}{2^k}, \ldots, 1]$$

of some curve

$$\mathbf{b}(t) = \sum \mathbf{b}_i B_i^n(t)$$

converge to the curve segment $\mathbf{b}[0,1]$ with order $1/4^k$. This suggests the following plotting routine, see [Lane & Riesenfeld '80].

PLOT BÉZIER $(\mathbf{b}_0, \ldots, \mathbf{b}_n; k)$

if	$k = 0$
then	plot the polygon $\mathbf{b}_0, \ldots, \mathbf{b}_n$
else	compute the composite Bézier polygon $\mathbf{a}_0, \ldots, \mathbf{a}_{2n}$
	of $\sum \mathbf{b}_i B_i^n(t)$ over $[0, 0.5, 1]$
	PLOT BÉZIER $(\mathbf{a}_0, \ldots, \mathbf{a}_n, k-1)$
	PLOT BÉZIER $(\mathbf{a}_n, \ldots, \mathbf{a}_{2n}, k-1)$

Figure 3.3 was produced by applying this routine to the Bézier polygon of a functional cubic curve for $k = 3$.

In addition to the number of iterations, one may use further stopping criteria. For example, one may stop if the input Bézier polygon is close to a line segment. A simple measure for straightness is based on the second forward differences. Thus, one may change the first line of the algorithm above to

$$\textbf{if} \quad k = 0 \quad \textbf{or} \quad \max\{\|\Delta^2 \mathbf{b}_i\| \mid i = 0, \ldots, n-2\} < \varepsilon \ .$$

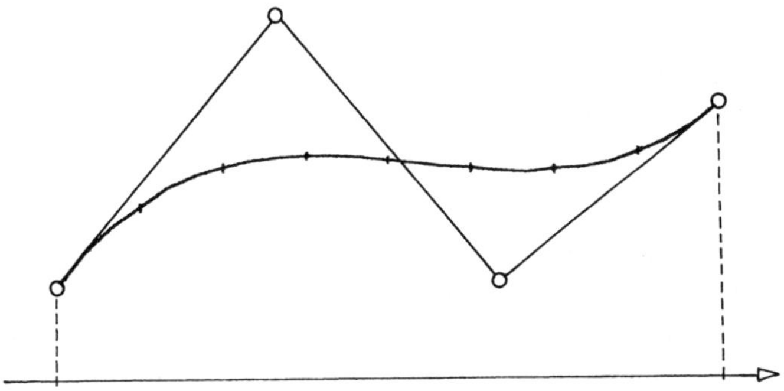

Figure 3.3: Subdividing a Bézier polygon three-times.

Instead of drawing the Bézier polygon, one can simply draw the line segment $\mathbf{b}_0\mathbf{b}_n$ when this criterion is satisfied. A bound of its deviation from the curve is given by the following theorem:

Let $\mathbf{l}(t) = \mathbf{b}_0(1-t) + \mathbf{b}_n t$ *be the linear interpolant of* $\mathbf{b}(t)$. *Then,*

$$\begin{aligned} \sup_{0\le t\le 1} \|\mathbf{b}(t) - \mathbf{l}(t)\| &\le \frac{1}{8} \sup_{0\le t\le 1} \|\ddot{\mathbf{b}}(t)\| \\ &\le \frac{1}{8} n(n-1) \max_{i=0,\dots,n-2} \|\Delta^2 \mathbf{b}_i\| \ , \end{aligned}$$

where $\|\cdot\|$ *denotes the supremum, sum, or Euclidean norm.*

For a proof, we refer to [de Boor '78, p. 39] and [Filip et al. '86].

Remark 4: If $\mathbf{b}(u)$ has the Bézier points $\mathbf{b}_i$ over $[a, b]$ and the Bézier points $\mathbf{c}_i$ over some subinterval $[c, c+h]$, then the differences $\|\Delta^2 \mathbf{c}_i\|$, is bounded by $h^2 \max \|\Delta^2 \mathbf{b}_i\|$, see Problem 3. Hence, the approximation order of a linear interpolant is quadratic. Moreover, the approximation order is only quadratic, in general. Thus, due to Remark 3, the composite Bézier polygon over $[0, \frac{1}{2^m}, \dots, 1]$ is, asymptotically, as good an approximation as the secant polygon with vertices

$$\mathbf{b}(\frac{i}{(n2^m)}), \quad i = 0, 1, \dots, n2^m \ .$$

Remark 5: Essentially, the above plotting routine only evaluates convex combinations of the form $(\mathbf{a}+\mathbf{b})/2$. Thus, one can accelerate the procedure if the divisions by 2 are realized by bit shifts. Then there are roughly $(n+1)/2$ vector additions and 1 division per vertex of the polygon in this plotting routine.

3.6 Curve generation by forward differences

Another quick plotting algorithm for polynomial curves $\mathbf{b}(u)$ is based on forward differences. Let

$$\mathbf{p}_i = \mathbf{b}(a + ih) \ , \quad i = 0, \ldots, m \ ,$$

be points on the curve $\mathbf{b}(u)$ with equidistant parameter values. If $\mathbf{b}(u)$ is of degree n, then $\Delta^{n+1}\mathbf{p}_i = 0$ and $\Delta^n \mathbf{p}_i$ = constant for all i, see Problem 1.

This can be used to compute the points $\mathbf{p}_i$, $i = n+1, \ldots, m$, from the points $\mathbf{p}_0, \ldots, \mathbf{p}_n$. First, one computes the constant $\Delta^n \mathbf{p}_0$ by repeated differencing and then the points $\mathbf{p}_i$, $i > n$, "backwards" by repeated summation. The computation is conveniently arranged in the following array:

$$\begin{array}{llll}
\mathbf{p}_0 & & & \\
\mathbf{p}_1 & \Delta^1\mathbf{p}_0 & & \\
\vdots & & \ddots & \\
\mathbf{p}_n & \Delta^1\mathbf{p}_{n-1} & \cdots & \Delta^n\mathbf{p}_0 \\
\mathbf{p}_{n+1} & \Delta^1\mathbf{p}_n & \cdots & \Delta^n\mathbf{p}_1 \\
\vdots & & & \vdots \\
\mathbf{p}_m & \Delta^1\mathbf{p}_{m-1} & \cdots & \Delta^n\mathbf{p}_{m-n}
\end{array}$$

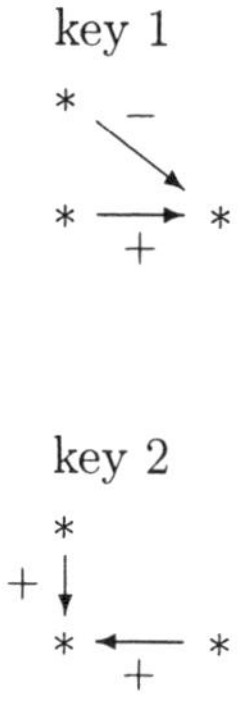

Remark 6: An operation count shows that, except for the computation of $\mathbf{p}_0, \ldots, \mathbf{p}_n$ and the triangular scheme, one needs to perform n vector additions for each point on the curve. Thus, curve generation by subdivision is nearly twice as fast as by forward differencing, and it is numerically more stable, see Remark 5.

3.7 Intersections

Subdivision is also useful in computing the intersection of two planar Bézier curves, say

$$\mathbf{b}(s) = \sum \mathbf{b}_i B_i^m(s) \ , \qquad s \in [0, 1] \ ,$$

and

$$\mathbf{c}(t) = \sum \mathbf{c}_i B_i^n(t) \ , \qquad t \in [0, 1] \ .$$

The idea is to compare the convex hulls of both Bézier polygons. If they are

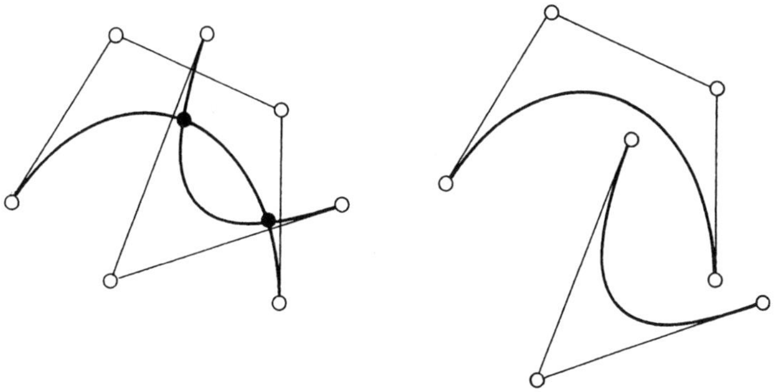

Figure 3.4: Intersecting and non-intersecting planar curves.

disjoint, there are no intersections. If they overlap, the curves may or may not intersect. In the latter case, one subdivides at $t = 1/2$ and compares the convex hulls of $\mathbf{b}[0, 1/2]$ and $\mathbf{b}[1/2, 1]$ with $\mathbf{c}[0, 1/2]$ and $\mathbf{c}[1/2, 1]$. This process is repeated for every pair of curve segments where the convex hulls of the Bézier polygons overlap. If, eventually, the convex hulls are small and flat, the curves are approximated by straight line segments, whose intersections can easily be determined.

Instead of using convex hulls, it is much easier to use their axis-parallel **bounding boxes** $[\min \mathbf{b}_i, \max \mathbf{b}_i]$ and $[\min \mathbf{c}_i, \max \mathbf{c}_i]$, see Remark 1 in 2.2. This idea leads to the following simpler intersection routine.

INTERSECT($\mathbf{b}_0, \ldots, \mathbf{b}_m; \mathbf{c}_0, \ldots, \mathbf{c}_n; \varepsilon$)

if $[\min \mathbf{b}_i, \max \mathbf{b}_i] \cap [\min \mathbf{c}_j, \max \mathbf{c}_j] \neq \emptyset$
then **if** $m(m-1) \max \|\Delta^2 \mathbf{b}_i\| > \varepsilon$
  then compute the composite Bézier polygon $\mathbf{b}'_0, \ldots, \mathbf{b}'_{2m}$
   of $\sum \mathbf{b}_i B_i^m(s)$ over $[0, 0.5, 1]$,
   INTERSECT ($\mathbf{b}'_0, \ldots, \mathbf{b}'_m; \mathbf{c}_0, \ldots, \mathbf{c}_n; \varepsilon$)
   INTERSECT($\mathbf{b}'_m, \ldots, \mathbf{b}'_{2m}; \mathbf{c}_0, \ldots, \mathbf{c}_n; \varepsilon$)
  else **if** $n(n-1) \max \|\Delta^2 \mathbf{c}_i\| > \varepsilon$
   then compute the composite Bézier polygon $\mathbf{c}'_0, \ldots, \mathbf{c}'_{2n}$
    of $\sum \mathbf{c}_i B_i^n(t)$ over $[0, 0.5, 1]$,
    INTERSECT ($\mathbf{b}_0, \ldots, \mathbf{b}_m; \mathbf{c}'_0, \ldots, \mathbf{c}'_n; \varepsilon$)
    INTERSECT ($\mathbf{b}_0, \ldots, \mathbf{b}_m; \mathbf{c}'_n, \ldots, \mathbf{c}'_{2n}; \varepsilon$)
   else intersect the line segments $\mathbf{b}_0\mathbf{b}_m$ and $\mathbf{c}_0\mathbf{c}_n$

3.8 The variation diminishing property

Subdivision is not only a practical but also a useful theoretical tool. In the following, we use it to derive the **geometric variation diminishing property**, which is illustrated in Figure 3.5.

> *A curve* $\mathbf{b}(t)$, $t \in [0,1]$, *is cut by an arbitrary hyperplane* $\mathcal{H}$ *no more often than its Bézier polygon is.*

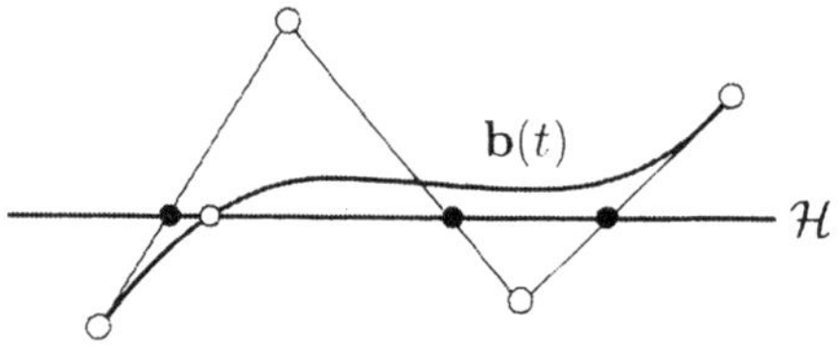

Figure 3.5: Intersections with a hyperplane.

For a proof, we first observe that de Casteljau's algorithm is a repeated **corner cutting** process for any $t \in [0,1]$, as illustrated in Figure 3.6 for a single corner.

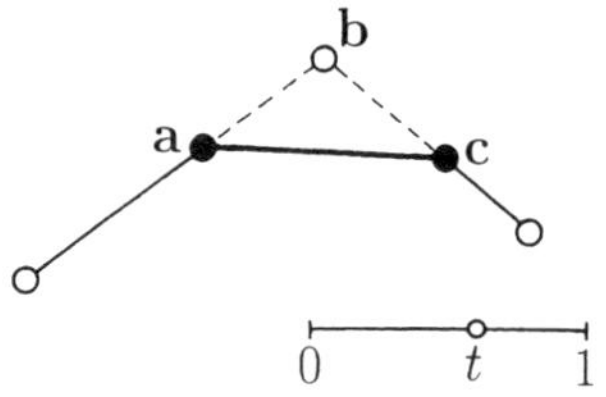

Figure 3.6: Cutting a corner.

If the line segment $\mathbf{ac}$ crosses $\mathcal{H}$, then the polygon $\mathbf{abc}$ also crosses $\mathcal{H}$. The converse, however, is not true in general. Consequently, the Bézier polygon over any subdivision $[0, t_1, \dots, t_k, 1]$ of $[0,1]$ has no more intersections with $\mathcal{H}$ than the Bézier polygon has.

In particular, if the t_i are chosen such that $\mathbf{b}(t_1), \dots, \mathbf{b}(t_k)$ are the intersections of $\mathbf{b}$ with $\mathcal{H}$, then it follows that the Bézier polygon over $[0,1]$ also has at least k intersections with $\mathcal{H}$. ◇

Remark 7: If a curve or polygon in $\mathbb{R}^d$ meets every hyperplane in at most two points or lies completely in this plane, it is called **convex**. Convex curves are planar, i.e., they lie in two-dimensional planes. As a consequence of the variation diminishing property, any curve with a convex Bézier polygon is convex itself. However, the converse is not true in general, as the example in Figure 3.7 shows, see also Problem 11.

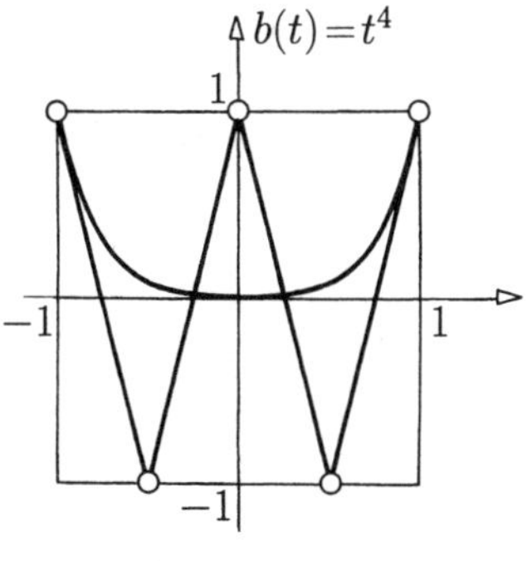

Figure 3.7: Convex quartic curve with non-convex Bézier polygon.

Remark 8: The graph of a polynomial

$$b(t) = \sum b_i B_i^n(t), \quad t \in [0,1] ,$$

is convex if and only if $\ddot{b}(t) \geq 0$ or $\ddot{b}(t) \leq 0$. Its Bézier polygon is convex if and only if all $\Delta^2 b_i \geq 0$ or all $\Delta^2 b_i \leq 0$.

3.9 The symmetric polynomial of the derivative

The derivatives of a polynomial curve $\mathbf{b}(u)$ can be written in terms of its polar form $\mathbf{b}[u_1 \ldots u_n]$. From 2.6 or simply by differentiating the polar form, it follows that

$$\mathbf{b}'(u) = \frac{n}{b-a} (\mathbf{b}[b\, u \ldots u] - \mathbf{b}[a\, u \ldots u]) ,$$

and, more specifically,

$$\mathbf{b}'(u) = n(\mathbf{b}[1u \ldots u] - \mathbf{b}[0u \ldots u]) .$$

Checking the three characterizing properties of a polar form, we find that the multi-affine symmetric polynomial of $\mathbf{b}'(u)$ is given by

$$\mathbf{b}'[u_2 \ldots u_n] = n(\mathbf{b}[1\, u_2 \ldots u_n] - \mathbf{b}[0\, u_2 \ldots u_n]) .$$

The symmetric polynomial $\mathbf{b}[u_1 u_2 \ldots u_n]$ of the initial curve $\mathbf{b}(u)$ represents an affine map in u_1 if $u_2, \dots, u_n$ are fixed. Consequently,

$$\mathbf{b}[\delta\, u_2 \,\dots\, u_n] = \mathbf{b}[b\, u_2 \,\dots\, u_n] - \mathbf{b}[a\, u_2 \,\dots\, u_n]$$

represents the underlying linear map, where $\delta = b - a$. To avoid misinterpretations, we use Greek letters in the arguments of a polar form to denote differences of affine parameters, i.e., vectors. In particular, we use the notation $\varepsilon = 1 - 0$. Thus, the derivative can be written as

$$\mathbf{b}'(u) = n\,\mathbf{b}[\varepsilon\, u \,\dots\, u] \ .$$

Differentiating further, we get the polar form of the rth derivative of $\mathbf{b}(u)$

$$\mathbf{b}^{(r)}[u_{r+1} \,\dots\, u_n] = \frac{n!}{(n-r)!}\mathbf{b}[\varepsilon \overset{r}{\dots} \varepsilon\, u_{r+1} \,\dots\, u_n] \ ,$$

where

$$\mathbf{b}[\varepsilon \overset{r}{\dots} \varepsilon\, u_{r+1} \,\dots\, u_n] = \mathbf{b}[\varepsilon \overset{r-1}{\dots} \varepsilon\, 1\, u_{r+1} \,\dots\, u_n] - \mathbf{b}[\varepsilon \overset{r-1}{\dots} \varepsilon\, 0\, u_{r+1} \,\dots\, u_n] \ .$$

Remark 9: Since $\mathbf{b}[u_1 \,\dots\, u_n]$ is affine in each variable, the first partial derivative, for example, is given by

$$\begin{aligned}
\frac{\partial}{\partial u_1}\mathbf{b}[u_1 \,\dots\, u_n] &= \mathbf{b}[1\, u_2 \,\dots\, u_n] - \mathbf{b}[0\, u_2 \,\dots\, u_n] \\
&= \mathbf{b}[\varepsilon\, u_2 \,\dots\, u_n] \\
&= \frac{1}{n}\mathbf{b}'[u_2 \,\dots\, u_n] \ .
\end{aligned}$$

Hence, it follows that

$$\frac{\partial^r}{\partial u_1 \,\dots\, \partial u_r}\,\mathbf{b}[u_1 \,\dots\, u_n] = \frac{(n-r)!}{n!}\,\mathbf{b}^{(r)}[u_{r+1} \,\dots\, u_n] \ .$$

3.10 Simple C^r joints

Subdivision also provides a convenient tool to describe certain differentiability conditions of two polynomial curves $\mathbf{b}(u)$ and $\mathbf{c}(u)$ given by their Bézier polygons $\mathbf{b}_0, \dots, \mathbf{b}_n$ over $[a, b]$ and $\mathbf{c}_0, \dots, \mathbf{c}_n$ over $[b, c]$, respectively.

From 2.4, it follows that the derivatives up to order r at $u = b$ determine, and are determined by, the Bézier points $\mathbf{b}_{n-r}, \dots, \mathbf{b}_n$ and $\mathbf{c}_0, \dots, \mathbf{c}_r$. This leads us to **Stärk's theorem** [Stärk '76]:

The derivatives of $\mathbf{b}$ *and* $\mathbf{c}$ *up to order* r *agree at* $u = b$ *if and only if* $\mathbf{c}_0, \ldots, \mathbf{c}_r$ *are the first* $r+1$ *Bézier points of* $\mathbf{b}$ *over* $[b, c]$, *which means that*

$$\mathbf{b}[b \overset{n-i}{\cdots} b\, c \overset{i}{\cdots} c] = \mathbf{c}_i \quad \textit{for} \quad i = 0, \ldots, r \ .$$

Using the main theorem 3.2, Stärk's theorem can be rephrased in the following way.

The derivatives of $\mathbf{b}$ *and* $\mathbf{c}$ *up to order* r *agree at* $u = b$ *if and only if both polynomials* $\mathbf{b}[b \overset{n-r}{\cdots} b\; u \overset{r}{\cdots} u]$ *and* $\mathbf{c}[b \overset{n-r}{\cdots} b\; u \overset{r}{\cdots} u]$ *are equal.*

Over $[a, b, c]$, the polynomial $\mathbf{b}[b \overset{n-r}{\cdots} b\; u \overset{r}{\cdots} u]$ has the composite Bézier polygon $\mathbf{b}_{n-r}, \ldots, \mathbf{b}_n, \mathbf{c}_1, \ldots, \mathbf{c}_r$. The $\mathbf{c}_i$, $i \le r$, can be computed from the $\mathbf{b}_{n-i}$ by applying the algorithm of de Casteljau, see Figures 3.8 and 3.9.

Figures 3.8 and 3.9 show simple C^r joints obtained by **Stärk's construction**. Figure 3.9 left represents the so called **A-frame**.

Remark 10: Since two polynomials are equal if and only if their polar forms are equal, we see that $\mathbf{b}(u)$ and $\mathbf{c}(u)$ have identical derivatives up to order r at $u = b$ if and only if their polar forms satisfy

$$\mathbf{b}[b \overset{n-r}{\cdots} b\; u_1 \ldots u_r] = \mathbf{c}[b \overset{n-r}{\cdots} b\; u_1 \ldots u_r]$$

for arbitrary values of the variables $u_1, \ldots, u_r$.

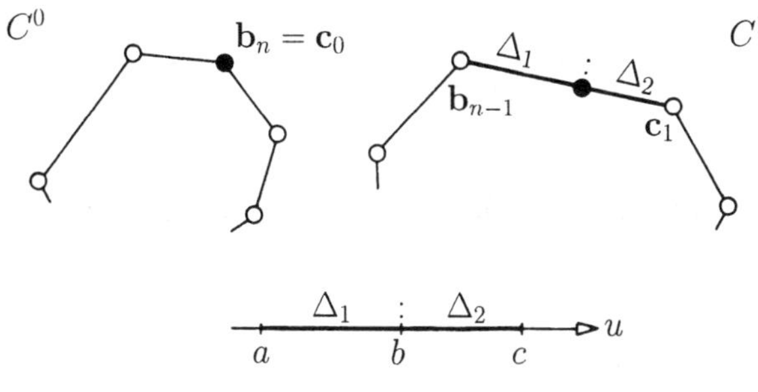

Figure 3.8: Simple C^0 and C^1 joints.

3.11 Degree elevation

For every curve of degree n and every $m \ge n$, there is an mth degree Bézier representation.

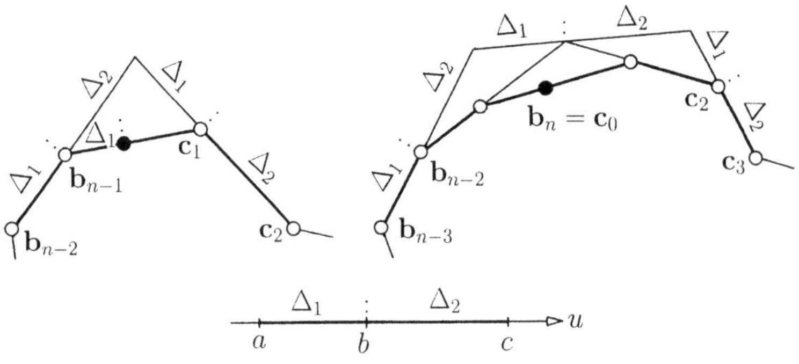

Figure 3.9: Simple C^2 and C^3 joints.

The conversion to a higher degree representation is used in certain surface constructions and is sometimes necessary to exchange data between different CAD systems. Such a conversion is called **degree elevation**.

Given an nth degree Bézier representation

$$\mathbf{b}(u) = \sum \mathbf{b}_i B_i^n(t)$$

of some curve $\mathbf{b}(u)$, we show how to raise its degree by one, i.e., we write $\mathbf{b}(u)$ in terms of the Bernstein polynomials $B_i^{n+1}(t)$. Again, the symmetric polynomial $\mathbf{b}[u_1 \ldots u_n]$ of $\mathbf{b}(u)$ is used.

We denote the absence of a term in a sequence by an asterisk and define

$$\text{(2)} \qquad \mathbf{c}[u_0 \ldots u_n] = \frac{1}{n+1} \sum_{i=0}^{n} \mathbf{b}[u_0 \ldots u_i^* \ldots u_n] \; .$$

It is easy to check that this $(n+1)$-variate polynomial is symmetric and multi-affine and has the "diagonal" $\mathbf{b}(u)$. Consequently, $\mathbf{c}[u_0 \ldots u_n]$ is the $(n+1)$-variate polar form of $\mathbf{b}(u)$. This equation is due to [Ramshaw '87, p. 58]. Hence, by the main theorem in 3.2 and some counting it follows that

$$\begin{aligned}
\mathbf{c}_i &= \mathbf{c}[a \overset{n+1-i}{\ldots} a \; b \overset{i}{\ldots} b] \\
&= \tfrac{i}{n+1} \mathbf{b}[a \overset{n+1-i}{\ldots} a \; b \overset{i-1}{\ldots} b] + \tfrac{n+1-i}{n+1} \mathbf{b}[a \overset{n-i}{\ldots} a \; b \overset{i}{\ldots} b] \\
&= \tfrac{i}{n+1} \mathbf{b}_{i-1} + \tfrac{n+1-i}{n+1} \mathbf{b}_i
\end{aligned}$$

are the Bézier points of $\mathbf{b}(u)$ over $[a, b]$ in its representation of degree $n + 1$. Figure 3.10 illustrates the corresponding geometric construction for $n = 3$.

Remark 11: Approximating a polynomial of exact degree m by a polynomial

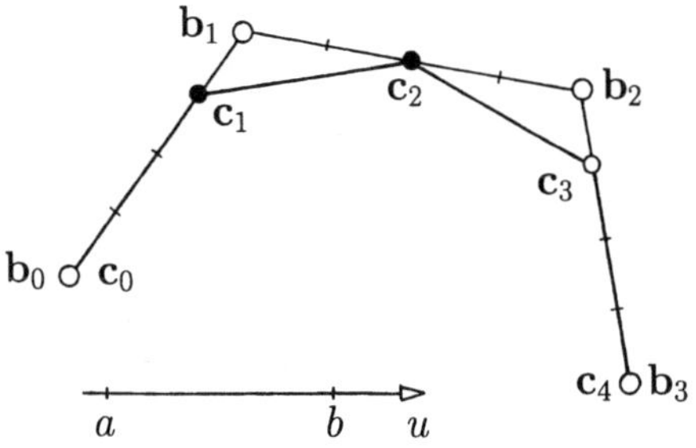

Figure 3.10: Degree elevation.

of lower degree $n < m$ is called **degree reduction**, see for example [Eck '93, Eck '95, Lutterkort et al '99].

3.12 Convergence under degree elevation

Repeating the degree elevation process, one can obtain any higher degree representation

$$\mathbf{b}(t) = \sum_{k=0}^{m} \mathbf{d}_k B_k^m(t) \ , \quad m > n \ ,$$

with Zhou's simple expression

$$\mathbf{d}_k = \sum_{i=0}^{n} \mathbf{b}_i \beta_{ik} \ ,$$

where

$$\beta_{ik} = \binom{n}{i}\binom{m-n}{k-i} / \binom{m}{k}$$

is the so-called **polyhypergeometric distribution** known from probability theory. A derivation is straightforward:

$$\begin{aligned} \mathbf{b}(t) &= \sum_{i=0}^{n} \mathbf{b}_i B_i^n(t)(1-t+t)^{m-n} \\ &= \sum_{i=0}^{n} \sum_{j=0}^{m-n} \mathbf{b}_i \binom{n}{i}\binom{m-n}{j} t^{i+j}(1-t)^{m-i-j} \\ &= \sum_{k=0}^{m} \left(\sum_{i=0}^{n} \mathbf{b}_i \beta_{ik} \right) B_k^m(t) \ , \quad \text{where} \quad k = i+j \ , \end{aligned}$$

see [Farin '86, de Boor '87].

Similarly to the convergence resulting from repeated subdivision, the Bézier polygon of the mth degree representation of $\mathbf{b}(t)$ converges to $\mathbf{b}[0,1]$ as m goes to ∞, see [Farin '79, Trump & Prautzsch '96]: On rewriting the β_{ik}, one obtains

$$\begin{aligned} \beta_{ik} &= \binom{n}{i} \prod_{\alpha=0}^{i-1} \frac{k-\alpha}{m-\alpha} \prod_{\alpha=i}^{n-1} \frac{m-k+i-\alpha}{m-\alpha} \\ &= \binom{n}{i} (k/m)^i (1-k/m)^{n-i} + O(1/m) \\ &= B_i^n(k/m) + O(1/m) \ . \end{aligned}$$

Substituting this into the equation for the $\mathbf{d}_k$, gives

$$\max_{k=0}^{m} \|\mathbf{d}_k - \mathbf{b}(k/m)\| = O(1/m) \ ,$$

see also 11.8.

A different, more general, proof of this fact is given in 6.6 for splines. See Problem 6 in 6.9 for an efficient construction of the $\mathbf{d}_k$.

3.13 Problems

1 Consider the equidistant shifts $\mathbf{p}_i(x) = \mathbf{p}(x - ih)$ of some polynomial curve $\mathbf{p}$ of genuine degree n. Show that the curve $\Delta^k \mathbf{p}_i(x)$ is of genuine degree $n - k$.

2 Show that the **Bernstein operator**

$$\mathcal{B}[f](u) = \sum_{i=0}^{n} f(ih) B_i^n(t) \ , \quad u = nh \cdot t \ ,$$

has approximation order 2, i.e., if f is twice differentiable, then

$$\max_{u \in [0,nh]} \|\mathcal{B}[f](u) - f(u)\| = O(h^2) \ .$$

3 Let $\mathbf{b}_0, \ldots, \mathbf{b}_n$ be the Bézier points of some curve over an interval $[a, b]$, and let $\mathbf{c}_0, \ldots, \mathbf{c}_n$ be the Bézier points of the same curve over any subinterval $[c, c + h]$ of $[a, b]$. Show that

$$\max_{i=0,\ldots,n-k} \|\Delta^k \mathbf{c}_i\| \le \left(\frac{h}{b-a}\right)^k \max_{i=0,\ldots,n-k} \|\Delta^k \mathbf{b}_i\| \ ,$$

where $\|\cdot\|$ denotes the maximum, sum, or Euclidean norm.

4 Let $\mathbf{b}$ be a polynomial curve. Show that the arc lengths of its Bézier polygons over $[0, \frac{1}{m}, \frac{2}{m}, \ldots, 1]$ converge quadratically in $1/m$ to the arc length $\int_0^1 \|\dot{\mathbf{b}}(t)\|_2 \, dt$ of $\mathbf{b}[0,1]$, see also [Kobbelt & Prautzsch '95, Gravesen '97].

5 Show that the arc length of the Bézier polygons of the mth degree representation of $\mathbf{b}$ converges linearly in $1/m$ to the arc length of $\mathbf{b}[0,1]$.

6 Come up with an algorithm that finds the self-intersections of a planar Bézier curve. If a curve intersects itself, what can be said about the hodograph?

7 There exist pairs of cubics such that both curves over $[0,1]$ and also their Bézier polygons have 9 intersections each. Find an example.

8 There exist pairs of cubics, over $[0,1]$, which have common end points and which intersect each other more often than their Bézier polygons do. Come up with an example.

9 Describe an algorithm which checks whether two axes parallel boxes intersect.

10 Consider a curve $\mathbf{b}$ in $\mathbb{R}^n$ which intersects or touches any hyperplane in at most m points or segments. Show that $\mathbf{b}$ lies in an m-dimensional subspace.

11 Show that the mth degree Bézier polygon of $b(x) = x^4$ over $[-1,1]$ is non-convex for all $m \geq 4$.

12 Consider the mth degree Bézier representation of an nth degree polynomial

$$\mathbf{b}(t) = \sum_{i=0}^{m} \mathbf{b}_i B_i^m(t) \ .$$

Show that there is an nth degree polynomial $\mathbf{p}$ such that $\mathbf{p}(i/m) = \mathbf{b}_i$.

13 Show that $\mathbf{b}(t) = [t^2 \ t]^t$ and $\mathbf{b}(t^2)$ parametrize the same curve over $[0,1]$ but have different Bézier polygons of degree 4.

14 Use symmetric polynomials to prove the degree elevation formula given in 3.11,

$$\mathbf{d}_k = \sum_{i=0}^{n} \mathbf{b}_i \beta_{ik} \ .$$

4 Interpolation and approximation

4.1 Interpolation — 4.2 Lagrange interpolation — 4.3 Newton interpolation — 4.4 Hermite interpolation — 4.5 Piecewise cubic Hermite interpolation — 4.6 Approximation — 4.7 Least squares fitting — 4.8 Improving the parameter — 4.9 Problems

In geometric modeling and other applications, one often needs to find an analytic, usually polynomial, representation of curves for which no mathematical or only complicated descriptions are known. To construct such a representation, one usually measures or evaluates a given curve at a number of points and determines an interpolant or approximant. This chapter reviews some basic techniques.

4.1 Interpolation

A sequence of n functions $C_1(u), \ldots, C_n(u)$ is said to be linearly independent over the sequence $u_1, \ldots, u_n$ if the matrix

$$C = \begin{bmatrix} C_1(u_1) & \cdots & C_n(u_1) \\ \vdots & & \vdots \\ C_1(u_n) & \cdots & C_n(u_n) \end{bmatrix}$$

is non-singular. In this case, to any n points $\mathbf{p}_1, \ldots, \mathbf{p}_n \in \mathbb{R}^d$ there is a unique curve

$$\mathbf{p}(u) = \sum_{i=1}^{n} \mathbf{x}_i C_i(u)$$

interpolating the points $\mathbf{p}_i$ at the nodes u_i, i.e.,

$$\mathbf{p}(u_i) = \mathbf{p}_i \ , \quad i = 1, \ldots, n \ .$$

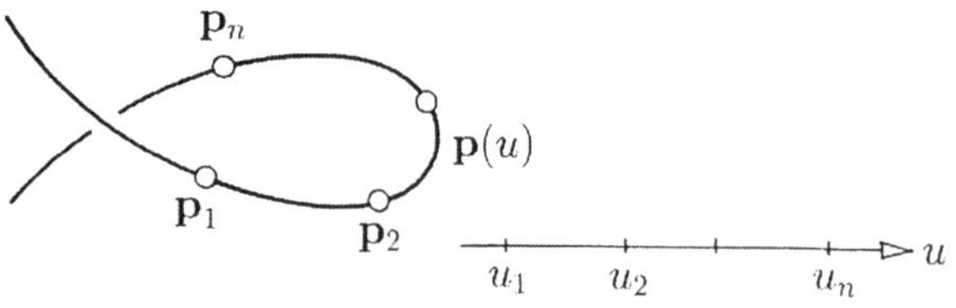

Figure 4.1: Interpolating curve

For a proof, write the **interpolation conditions** in matrix notation,

$$\begin{bmatrix} C_1(u_1) & \dots & C_n(u_1) \\ \vdots & & \vdots \\ C_1(u_n) & \dots & C_n(u_n) \end{bmatrix} \begin{bmatrix} \mathbf{x}_1^t \\ \vdots \\ \mathbf{x}_n^t \end{bmatrix} = \begin{bmatrix} \mathbf{p}_1^t \\ \vdots \\ \mathbf{p}_n^t \end{bmatrix} ,$$

or more concisely,

$$CX = P ,$$

which represents d simultaneous linear systems for the d columns of X. The existence of the solution follows from the linear independence of $C_1, \dots, C_n$ over $u_1, \dots, u_n$.

Remark 1: If the C_i are linearly independent polynomials of degree $n-1$, then the matrix C is invertible for any n distinct values $u_1, \dots, u_n$. Namely, the homogeneous system $C\mathbf{x} = \mathbf{o}$ (for a single column $\mathbf{x}$) has only the trivial solution $\mathbf{x} = \mathbf{o}$ since the zero polynomial is the only polynomial of degree $n-1$ with n roots.

Remark 2: Two points can be interpolated by a line, three points by a parabola, four points by a cubic, and so on.

4.2 Lagrange form

A fundamental and simple method to obtain a polynomial interpolant is due to Lagrange. Given $n+1$ points $\mathbf{p}_i$ with corresponding parameter values u_i, $i = 0, \dots, n$, the interpolating polynomial curve of degree n is

$$\mathbf{p}(u) = \sum_{i=0}^{n} \mathbf{p}_i L_i^n(u) ,$$

where the **Lagrange polynomials** $L_i^n(u), i = 0, \ldots, n$, are defined by the interpolation conditions

$$L_i^n(u_k) = \begin{cases} 1 & \text{if} \quad k = i \\ 0 & \phantom{\text{if}} \quad k \neq i \end{cases} .$$

Figure 4.2 shows an example.

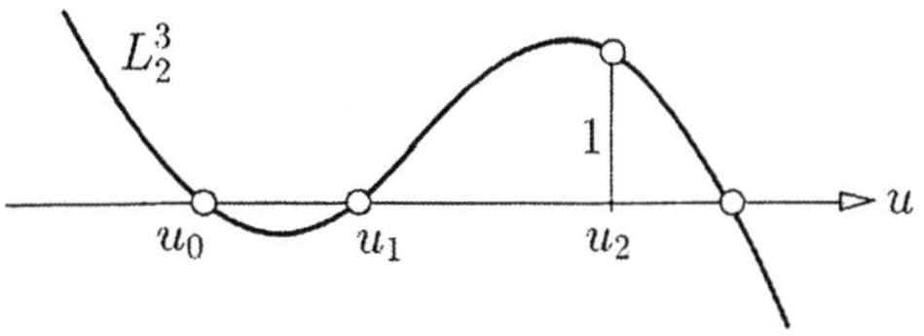

Figure 4.2: A cubic Lagrange polynomial.

Checking the definition, we easily verify that

$$L_i^n(u) = \frac{\prod_{j=0,\, j\neq i}^n (u - u_j)}{\prod_{j=0,\, j\neq i}^n (u_i - u_j)} .$$

There are different ways to evaluate Lagrange polynomials. One possibility is to use the following recursion. First observe that the Lagrange polynomials sum to one,

$$\sum_{i=0}^n L_i^n(u) \equiv 1 .$$

Then it follows from the definition of the Lagrange polynomials that

$$(1) \quad L_i^k = \begin{cases} L_i^{k-1}\alpha_{ik} & \text{for} \quad i = 0, \ldots, k-1 \\ 1 - \sum_{j=0}^{k-1} L_j^k = \sum_{j=0}^{k-1} L_j^{k-1}(1 - \alpha_{jk}) & \text{for} \quad i = k \ , \end{cases}$$

where α_{ik} represents the local parameter over $[u_k, u_i]$, i.e.,

$$\alpha_{ik} = \frac{u - u_k}{u_i - u_k} .$$

In analogy to the derivation of de Casteljau's algorithm in 2.3, one obtains from recurrence relation (1) an iterative method to compute $\mathbf{p}(u)$ from the points

$$\mathbf{p}_i^0 = \mathbf{p}_i$$

by affine combinations:

$$\begin{aligned}\mathbf{p}(u) &= \sum_{i=0}^{n} \mathbf{p}_i^0 L_i^n(u) \\ &= \sum_{i=0}^{n-1} \mathbf{p}_i^1 L_i^{n-1}(u) \\ &\vdots \\ &= \sum_{i=0}^{0} \mathbf{p}_i^n L_i^0(u) = \mathbf{p}_0^n \ ,\end{aligned}$$

where

$$\mathbf{p}_i^{k+1} = \mathbf{p}_i^k \alpha_{i,n-k} + \mathbf{p}_{n-k}^k (1 - \alpha_{i,n-k}) \ .$$

This computation of $\mathbf{p}(u)$ is known as **Aitken's algorithm**. The polynomials $\mathbf{p}_i^k = \mathbf{p}_i^k(u)$ have degree $\leq k$ and interpolate the points $\mathbf{p}_i$ and $\mathbf{p}_n, \ldots, \mathbf{p}_{n-k+1}$.

Remark 3: The computation of the Lagrange polynomials by recursion formula(1) can be conveniently arranged in a triangular array,

$$\begin{array}{lllllll} 1 = & L_0^0 & L_0^1 & L_0^2 & \cdots & L_0^n \\ & & L_1^1 & L_1^2 & \cdots & L_1^n \\ & & & L_2^2 & \cdots & L_2^n \\ & & & & \ddots & \vdots \\ & & & & & L_n^n \ . \end{array}$$

Remark 4: Similarly, one can arrange the computation of the $\mathbf{p}_i^k$ in a triangular array,

$$\begin{array}{lllll} \mathbf{p}_0^0 & & & & \\ \mathbf{p}_1^0 & \mathbf{p}_0^1 & & & \\ \mathbf{p}_2^0 & \mathbf{p}_1^1 & \mathbf{p}_0^2 & & \\ \vdots & & & \ddots & \\ \mathbf{p}_n^0 & \mathbf{p}_{n-1}^1 & \mathbf{p}_{n-2}^2 \cdots & & \mathbf{p}_0^n \ . \end{array}$$

4.3 Newton form

Another polynomial basis, also depending on the interpolation abscissae, has been introduced by Newton, namely the $n+1$ monic polynomials $P_i(u)$ of degree i with zeroes at the first i interpolation abscissae $u_0, \ldots, u_{i-1}$,

$$P_0 = 1 \quad \text{and, for} \quad i \geq 1, \quad P_i(u) = (u - u_0) \cdots (u - u_{i-1}) \ .$$

Let $[u_0 \ \ldots \ u_i]\mathbf{q}$ denote the leading coefficient of the ith degree polynomial that interpolates some given curve $\mathbf{q} = \mathbf{q}(u)$ at $u_0, \ldots, u_i$. Then, it follows that the polynomial curve $\mathbf{p}(u)$ of degree $\leq n$ interpolating $\mathbf{q}(u)$ at $u_0, \ldots, u_n$ can be written as

$$\mathbf{p}(u) = \sum_{i=0}^{n} [u_0 \ \ldots \ u_i]\mathbf{q} \cdot P_i(u) \ .$$

From the Lagrange representation, it follows that the leading coefficient is of the form

$$[u_0 \ \ldots \ u_n]\mathbf{q} = \sum_{k=0}^{n} \frac{\mathbf{q}(u_k)}{(u_k - u_0) \cdots (u_k - u_k)^* \cdots (u_k - u_n)} \ ,$$

where the asterisk * means the absence of this term. Using this explicit representation, it is easy to verify the recursion formula

$$\begin{aligned} [u_0 \ \ldots \ u_n]\mathbf{q} &= \frac{[u_1 \ \ldots \ u_n]\mathbf{q} - [u_0 \ \ldots \ u_{n-1}]\mathbf{q}}{u_n - u_0} \\ &\vdots \\ [u_i \, u_j]\mathbf{q} &= \frac{[u_j]\mathbf{q} - [u_i]\mathbf{q}}{u_j - u_i} \\ [u_j]\mathbf{q} &= \mathbf{q}(u_j) \ . \end{aligned}$$

Because of this recursion, $[u_0 \ \ldots \ u_i]\mathbf{q}$ is called the ith **divided difference** of $\mathbf{q}(u)$ at $u_0, \ldots, u_i$. Note that the divided difference $[u_0 \ \ldots \ u_n]\mathbf{q}$ is symmetric with respect to its knots $u_0, \ldots, u_n$.

Remark 5: If $\mathbf{q}(u)$ is sufficiently smooth, then the divided difference $[u_0 \ \ldots \ u_n]\mathbf{q}$ depends continuously on the abscissae u_i, see Problems 1 and 2. Consequently, $[u_0 \ \ldots \ u_n]\mathbf{q}$ is also defined if some or all u_i coalesce. In particular, one has

$$[u_i \overset{n+1}{\ldots} u_i]\mathbf{q} = \mathbf{q}^{(n)}(u_i)/n! \ .$$

Remark 6: For $u_0 = \cdots = u_k$ it follows from Remark 5 that

$$\mathbf{p}(u) = \sum_{j=0}^{n} [u_0 \ \ldots \ u_j]\mathbf{q} \cdot P_j(u)$$

interpolates all derivatives up to order k of $\mathbf{q}(u)$ at $u = u_0$. Since $\mathbf{p}(u)$ does not depend on the order of the abscissae u_i, it also interpolates any derivative $\mathbf{q}^{(r)}(u_i)$, where $u_i = \cdots = u_{i+r}$.

Remark 7: If the values u_i are equidistant, $u_i = u_0 + ih$, then it follows that

$$[u_0 \ldots u_n]\mathbf{q} = \frac{1}{n!\,h^n}\Delta^n \mathbf{q}(u_0) \ .$$

4.4 Hermite interpolation

From Remark 6 in Section 4.3, it follows that the curve

$$\mathbf{p}(u) = \sum_{i=0}^{n} [u_0 \ldots u_i]\mathbf{q} \cdot P_i$$

interpolates points and, if interpolation abscissae coalesce, also derivatives of $\mathbf{q}$. This type of interpolation is called **Hermite interpolation**.

In particular, it is very common to interpolate all derivatives up to some order k at the end points of some interval $[a, b]$. The corresponding interpolant is of degree $n = 2k + 1$ and can be written as

$$\mathbf{p}(u) = \sum_{i=0}^{k} \mathbf{q}_a^{(i)} H_i^n + \sum_{i=0}^{k} \mathbf{q}_b^{(i)} H_{n-i}^n \ ,$$

where $\mathbf{q}_u^{(i)}$ denotes the ith derivative of $\mathbf{q}$ at u and where the so-called **Hermite polynomials** H_i^n of degree n are defined by

$$\frac{d^r}{du^r} H_i^n(a) = \frac{d^r}{du^r} H_{n-i}^n(b) = \begin{cases} 1 & \text{if} \quad i = r \\ 0 & \text{otherwise} \end{cases} \ , \qquad r = 0, \ldots, k \ .$$

Figure 4.3 shows a cubic Hermite interpolant and the corresponding four cubic Hermite polynomials.

More conveniently, one can describe the interpolant by its Bézier points $\mathbf{b}_j$ over $[a, b]$. From Section 2.8, it follows that

$$\mathbf{b}_j = \sum_{i=0}^{j} \binom{j}{i} \frac{(b-a)^i}{n \cdots (n-i+1)} \mathbf{q}_a^{(i)}$$

and

$$\mathbf{b}_{n-j} = \sum_{i=0}^{j} \binom{j}{i} \frac{(a-b)^i}{n \cdots (n-i+1)} \mathbf{q}_b^{(i)} \ ,$$

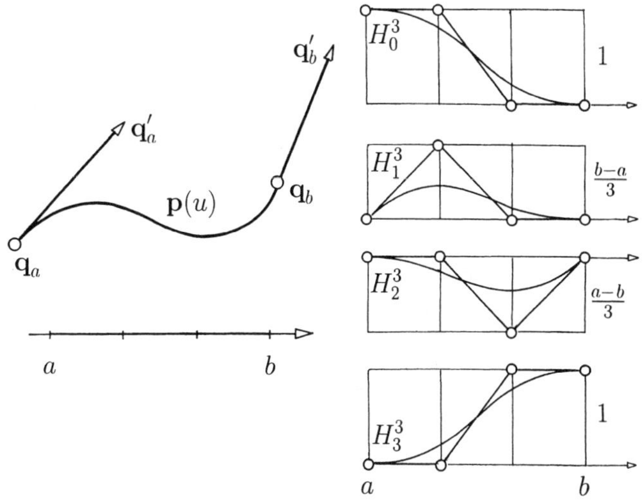

Figure 4.3: Cubic Hermite interpolation.

i.e., for $j = 0, \ldots, (n-1)/2$, one has

$$\begin{aligned}
\mathbf{b}_0 &= \mathbf{q}_a \\
\mathbf{b}_1 &= \mathbf{q}_a + \frac{b-a}{n}\mathbf{q}'_a \\
\mathbf{b}_2 &= \mathbf{q}_a + \frac{b-a}{n}(2\mathbf{q}'_a + \frac{b-a}{n-1}\mathbf{q}''_a) \\
&\vdots \\
\mathbf{b}_n &= \mathbf{q}_b \,.
\end{aligned}$$

Remark 8: Instead of interpolating the second derivatives $\mathbf{q}''_a$ and $\mathbf{q}''_b$, one can interpolate the curvatures at a and b. The **curvature** at a is given by

$$x_a = \frac{n-1}{n} \frac{\| \Delta\mathbf{b}_0 \times \Delta^2\mathbf{b}_0 \|}{\| \Delta\mathbf{b}_0 \|^3} ,$$

where $\times$ denotes the cross product.

4.5 Piecewise cubic Hermite interpolation

A function $f(u)$ can be approximated by a polynomial p of degree n interpolating f at some given abscissae $u_0 < \cdots < u_n$. A standard result from numerical analysis states that the deviation of p from f over $[u_0, u_n]$ can be written as

$$p(u) - f(u) = \frac{f^{(n+1)}(v)}{(n+1)!}(u - u_0)\cdots(u - u_n) \ ,$$

where $v = v(u)$ lies in $[u_0, u_n]$. Thus, the error becomes smaller if the abscissae move closer. On the other hand, a higher degree polynomial interpolating f at more points is, in general, not a better approximant. Therefore, one resorts to piecewise polynomials of low degree.

For example, cubics are very popular since they are of low degree and, for many applications, of sufficient flexibility. In the sequel, we describe the piecewise cubic Hermite interpolation.

Given $m+1$ points $\mathbf{p}_0, \ldots, \mathbf{p}_m$ and derivatives $\mathbf{d}_0, \ldots, \mathbf{d}_m$ with corresponding parameter values $u_0 < \cdots < u_m$, there is a unique piecewise cubic C^1 curve $\mathbf{s}(u)$ over $[u_1, u_m]$ such that

$$\mathbf{s}(u_i) = \mathbf{p}_i \ , \quad \text{and} \quad \mathbf{s}'(u_i) = \mathbf{d}_i \ .$$

Its Bézier representation is readily obtained from Section 4.4, and it is

$$\mathbf{s}(u) = \sum_{j=0}^{3} \mathbf{b}_{3i+j} B_j^3(t_i) \ , \quad u \in [u_i, u_{i+1}] \ ,$$

where $t_i = (u - u_i)/(u_{i+1} - u_i)$ represents the local parameter over $[u_i, u_{i+1}]$, $i = 0, \ldots, m-1$, and

$$\begin{aligned}
\mathbf{b}_{3i} &= \mathbf{p}_i \\
\mathbf{b}_{3i+1} &= \mathbf{p}_i + \mathbf{d}_i \Delta u_i/3 \\
\mathbf{b}_{3i+2} &= \mathbf{p}_{i+1} - \mathbf{d}_{i+1} \Delta u_i/3 \ ,
\end{aligned}$$

see Figure 4.4 for an illustration.

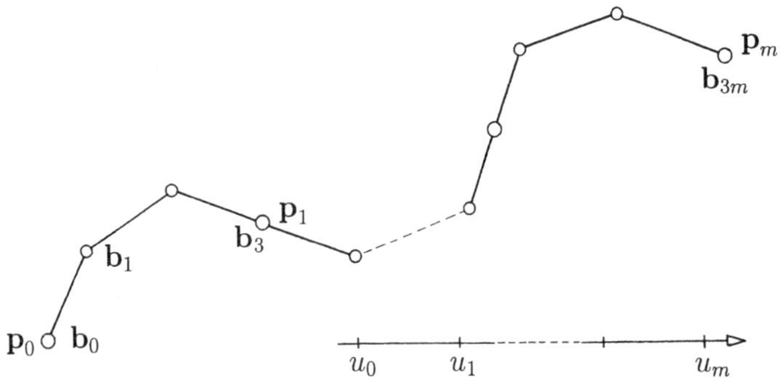

Figure 4.4: Composite Bézier polygon.

Often, the derivatives $\mathbf{d}_i$ are not given, but have to be determined from the data points. A simple method to estimate the $\mathbf{d}_i$ is to compute the derivatives of the quadratic interpolant to the three consecutive points $\mathbf{p}_{i-1}, \mathbf{p}_i, \mathbf{p}_{i+1}$, as illustrated in Figure 4.5.

This means to set

$$\mathbf{d}_i = (1-\alpha_i)\,\frac{\Delta\mathbf{p}_{i-1}}{\Delta u_{i-1}} + \alpha_i \frac{\Delta\mathbf{p}_i}{\Delta u_i}, \quad \text{for} \quad i = 1, \ldots, m-1 \ ,$$

where

$$\alpha_i = \frac{\Delta u_{i-1}}{\Delta u_{i-1} + \Delta u_i} \ .$$

At the endpoints, one adjusts these estimations to

$$\mathbf{d}_0 = 2\,\frac{\Delta\mathbf{p}_0}{\Delta u_0} - \mathbf{d}_1 \ , \quad \text{and} \quad \mathbf{d}_m = 2\,\frac{\Delta\mathbf{p}_{m-1}}{\Delta u_{m-1}} - \mathbf{d}_{m-1} \ .$$

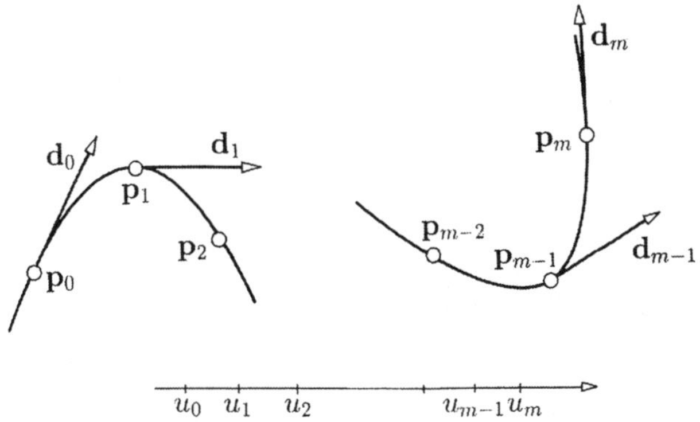

Figure 4.5: Obtaining derivative estimates from parabolas.

One often also must determine the interpolation abscissae u_i. Some simple choices are

the **equidistant parametrization**, for which

$$\Delta u_i = \text{constant} \ ,$$

the **chord length parametrization**, for which

$$\Delta u_i = \|\Delta\mathbf{p}_i\| \ ,$$

and the **centripetal parametrization** [Lee '89], for which

$$\Delta u_i = \sqrt{\|\Delta \mathbf{p}_i\|} \ .$$

These and further methods are discussed at length in [Foley & Nielson '89, Farin '02, Hoschek & Lasser '92].

Remark 9: An interpolation scheme with fixed interpolation abscissae u_i is called **linear** if the Bézier points $\mathbf{b}_0, \mathbf{b}_1, \ldots, \mathbf{b}_{3m}$ depend linearly on the interpolation points $\mathbf{p}_0, \ldots, \mathbf{p}_m$. Thus, a **linear interpolation scheme** is given by an $m{+}1 \times 3m{+}1$ matrix S that establishes the relationship between the $\mathbf{b}_i$ and $\mathbf{p}_k$

$$[\mathbf{b}_0 \ \ldots \ \mathbf{b}_{3m}] = [\mathbf{p}_0 \ \cdots \ \mathbf{p}_m]S \ .$$

Note that the interpolation scheme above, with the derivatives $\mathbf{d}_i$ obtained from quadratic interpolants, is linear.

4.6 Approximation

In general, given any sequence of n basis functions $C_1, \ldots, C_n$, there is no curve

$$\mathbf{p}(u) = \sum_{i=1}^{n} \mathbf{x}_i C_i(u)$$

that interpolates more than n points $\mathbf{p}_1, \ldots, \mathbf{p}_m$, $m > n$, at prescribed parameter values $u_1, \ldots, u_m$. In this case, one determines an **approximant** $\mathbf{p}(u)$, usually under the condition that the differences

$$\mathbf{p}(u_i) - \mathbf{p}_i = \mathbf{r}_i$$

become minimal in some sense.

Using matrix notation, this means to minimize

$$\begin{bmatrix} C_1(u_1) & \ldots & C_n(u_1) \\ \vdots & & \vdots \\ C_1(u_m) & \ldots & C_n(u_m) \end{bmatrix} \begin{bmatrix} \mathbf{x}_1^t \\ \vdots \\ \mathbf{x}_n^t \end{bmatrix} - \begin{bmatrix} \mathbf{p}_1^t \\ \vdots \\ \mathbf{p}_m^t \end{bmatrix} = \begin{bmatrix} \mathbf{r}_1^t \\ \vdots \\ \mathbf{r}_m^t \end{bmatrix}$$

or, more concisely,

$$CX - P = R = [r_{ij}] \ .$$

There are several choices that can be used to measure the "size" of R. Most commonly one minimizes the square sum

$$\sum_{i=1}^{m} \mathbf{r}_i^2 = \sum_{i,j} r_{ij}^2 = \text{trace}\, R^t R = \text{trace}\, RR^t \ .$$

This method is due to Gauss and it is called **least squares fitting**.

A slight modification of this approach is to minimize the weighted square sum

$$\sum w_i^2 \mathbf{r}_i^2 = \text{trace}\,(WR)^t(WR) \ ,$$

with some weight matrix

$$W = \begin{bmatrix} w_1 & & \\ & \ddots & \\ & & w_m \end{bmatrix} .$$

Both approaches result in linear systems as discussed in the next section. Other approaches are more complicated. For example, one could minimize the maximum distance

$$\max \|\mathbf{r}_i\|_\infty = \max \|r_{ij}\| \ .$$

This leads to a system of linear inequalities, which can be solved by the simplex algorithm, see Problem 8.

Remark 10: It is important to note that the jth column of R only depends on the jth column of X for all j. Therefore, minimizing R using any of the three distances above means to minimize each column of R separately.

4.7 Least squares fitting

In this section, we will show how to find a least squares fit. However, more generally than in the previous section, we also require the solution X to satisfy further linear constraints $DX = Q$, which would typically force the approximant $\mathbf{p}(u)$ to interpolate some of the points $\mathbf{p}_i$. As explained in Remark 10 of 4.6, it suffices to consider a column $\mathbf{x}$ of X and the corresponding columns $\mathbf{p}, \mathbf{q}$, and $\mathbf{r}$ of P, Q, and R, respectively.

The following theorem shows how one can find the minimal **residual**

$$\mathbf{r} = C\mathbf{x} - \mathbf{p}$$

under the constraints $D\mathbf{x} = \mathbf{q}$.

The sum of squares $\mathbf{r}^t\mathbf{r} = \sum r_i^2$ *is minimal for the solution* $\mathbf{x}, \mathbf{y}$

of the linear system

$$\left[\begin{array}{c|c} C^tC & D^t \\ \hline D & O \end{array}\right] \left[\begin{array}{c} \mathbf{x} \\ \hline \mathbf{y} \end{array}\right] = \left[\begin{array}{c} C^t\mathbf{p} \\ \hline \mathbf{q} \end{array}\right] . \tag{2}$$

Proof: Let $\mathbf{x}, \mathbf{y}$ be a solution of (2) such that $D\mathbf{x} = \mathbf{q}$. Further, let $\mathbf{x} + \mathbf{h}$ be any other point satisfying the constraints

$$D[\mathbf{x} + \mathbf{h}] = \mathbf{q} .$$

This implies $D\mathbf{h} = \mathbf{o}$. Further, let

$$\begin{aligned} \bar{\mathbf{r}} &= C[\mathbf{x} + \mathbf{h}] - \mathbf{p} \\ &= \mathbf{r} + C\mathbf{h} . \end{aligned}$$

Then it follows that

$$\bar{\mathbf{r}}^t\bar{\mathbf{r}} = \mathbf{r}^t\mathbf{r} + 2\mathbf{r}^tC\mathbf{h} + \mathbf{h}^tC^tC\mathbf{h} .$$

The last term is non-negative, and the second term is zero since, under the assumption (2),

$$\mathbf{h}^tC^t\mathbf{r} = \mathbf{h}^tC^t[C\mathbf{x} - \mathbf{p}] = -\mathbf{h}^tD^t\mathbf{y} = 0 .$$

Hence, $\mathbf{r}^t\mathbf{r}$ is minimal. ◇

Remark 11: If there are no constraints, i.e., $D = O$ and $\mathbf{q} = \mathbf{o}$, then the linear system (2) reduces to the so-called **Gaussian normal equations**

$$C^tC\mathbf{x} = C^t\mathbf{p} .$$

Remark 12: If C a is the identity matrix and if $D\mathbf{x} = \mathbf{q}$ is an underdetermined system, then (2) consists of the **correlate equations**

$$\mathbf{x} = \mathbf{p} - D^t\mathbf{y}$$

and the **normal equations**

$$D\mathbf{p} - DD^t\mathbf{y} = \mathbf{q} ,$$

which one obtains by substituting the correlate equations into the constraints $D\mathbf{x} = \mathbf{q}$.

Remark 13: Let W be an $m \times m$ diagonal weight matrix. Then, as a consequence of the theorem above, the weighted residual

$$W\mathbf{r} = WC\mathbf{x} - W\mathbf{p}$$

becomes minimal under the condition $D\mathbf{x} = \mathbf{q}$ for the solution $\mathbf{x}, \mathbf{y}$ of the weighted equation

$$\left[\begin{array}{c|c} C^tW^2C & D^t \\ \hline D & O \end{array}\right] \left[\begin{array}{c} \mathbf{x} \\ \hline \mathbf{y} \end{array}\right] = \left[\begin{array}{c} C^tW^2\mathbf{p} \\ \hline \mathbf{q} \end{array}\right] .$$

4.8 Improving the parameter

The quality of an approximation can often be improved by a different choice of the knots u_i. Let $\mathbf{p}(u)$ be a curve approximating some points $\mathbf{p}_i$ at certain parameter values u_i. Further, let v_i be the parameter values of the points on $\mathbf{p}(u)$ closest to the $\mathbf{p}_i$. Then, in general, the u_i and v_i are different. Hence, a least squares fit based on the v_i would lead to a curve that lies closer to the $\mathbf{p}_i$ since the new set of curves over which one minimizes also contains $\mathbf{p}(u)$.

A simple method to determine the v_i from the u_i is to linearize $\mathbf{p}(u)$ at u_i and compute the closest point to $\mathbf{p}_i$ on the tangent of $\mathbf{p}$ at u_i, as illustrated in Figure 4.6 left. This means solving the equation

$$[\mathbf{p}(u_i) + \Delta_i \mathbf{p}'(u_i) - \mathbf{p}_i]^t \mathbf{p}'(u_i) = 0$$

for Δ_i, which gives

$$\Delta_i = [\mathbf{p}_i - \mathbf{p}(u_i)]^t \frac{\mathbf{p}'(u_i)}{\|\mathbf{p}'(u_i)\|_2^2} .$$

Then, $u_i + \Delta_i$ is an approximation for v_i.

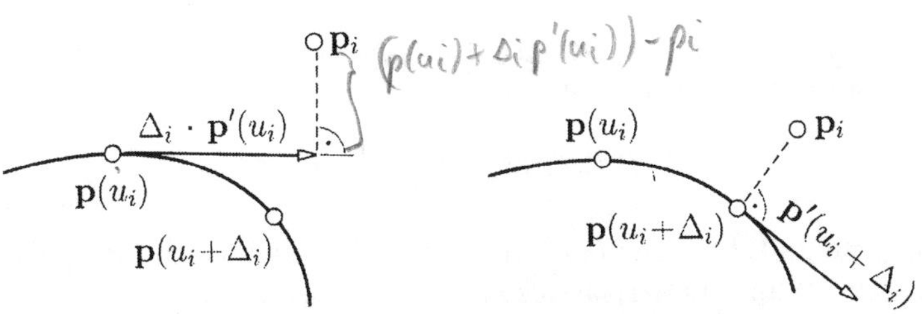

Figure 4.6: Improving a parameter.

Another method is to determine Δ_i such that the line from $\mathbf{p}_i$ to $\mathbf{p}(u_i + \Delta_i)$ is perpendicular to the tangent at $\mathbf{p}(u_i + \Delta_i)$, as illustrated in Figure 4.6 right. This is expressed by the condition

$$f(\Delta_i) = [\mathbf{p}(u_i + \Delta_i) - \mathbf{p}_i]^t \mathbf{p}'(u_i + \Delta_i) = 0 \ .$$

Using Newton's method, one obtains

$$\Delta_i \approx -\frac{f(0)}{f'(0)} = -\frac{[\mathbf{p}(u_i) - \mathbf{p}_i]^t \, \mathbf{p}'(u_i)}{\mathbf{p}'(u_i) \cdot \mathbf{p}'(u_i) + [\mathbf{p}(u_i) - \mathbf{p}_i]^t \, \mathbf{p}''(u_i)} \ .$$

One could iterate both processes to determine better approximations for the v_i, but usually one first computes a new approximating curve using the parameters $u_i + \Delta_i$ each time before doing so.

4.9 Problems

1 Use induction over n to show the **Hermite-Genocchi formula** for divided differences [de Boor '84, p. 21]

$$[u_0 \ \dots \ u_n]\mathbf{q} = \int_{0 \le x_n \le \cdots \le x_1 \le 1} \mathbf{q}^{(n)}(u_0 + (u_1 - u_0)x_1 + \cdots + (u_n - u_{n-1})x_n) d\mathbf{x} \ .$$

2 Observe from the Hermite-Genocchi formula that

$$\lim_{u_0, \dots, u_n \to u} [u_0 \ \dots \ u_n]\mathbf{q} = \mathbf{q}^{(n)}(u)/n! \ .$$

3 Sketch the quintic Hermite polynomials $H_0^5, \dots, H_5^5$ over some interval $[a, b]$ with their Bézier points.

4 For any three points $\mathbf{b}_0, \mathbf{b}_1, \mathbf{c}$ compute a point $\mathbf{b}_2$ on the line $\mathbf{b}_1\mathbf{c}$ such that the Bézier points $\mathbf{b}_0, \mathbf{b}_1, \mathbf{b}_2$ define a quadratic curve $\mathbf{b}(t)$ with prescribed curvature $\kappa_0 = \| \dot{\mathbf{b}} \times \ddot{\mathbf{b}} \| \ / \ \| \dot{\mathbf{b}} \|^3$ at $\mathbf{b}_0$.

5 Compute and draw the interpolating polynomial which interpolates the **Gaussian error function**

$$\exp(-t^2/2) \quad \text{at} \quad t = -7, -5, -3, \dots, 7 \ .$$

6 Compute and draw the piecewise cubic C^1 curve which interpolates $\exp(-t^2/2)$ and its derivatives at $t = -7, -1, +1, +7$.

7 Evaluate the interpolant of Problem 6 by de Casteljau's algorithm at $\pm 6, \pm 4, \pm 2, 0$.

8 Let $\mathbf{r} = C\mathbf{x} - \mathbf{p}$. Show that $\|\mathbf{r}\|_\infty = \max |r_i|$ is minimal for the solution $\mathbf{x}, \rho$ of the linear program

$$\begin{aligned} C\mathbf{x} - \mathbf{p} - \mathbf{e}\rho &\leq \mathbf{o} \\ -C\mathbf{x} + \mathbf{p} - \mathbf{e}\rho &\leq \mathbf{o} \\ \rho &= \min! \ , \end{aligned}$$

where $\mathbf{e} = [1 \ \ldots \ 1]^t$, see, for example, [Boehm & Prautzsch '93, 10.4].

5 B-spline representation

5.1 Splines — 5.2 B-splines — 5.3 A recursive definition of B-splines — 5.4 The de Boor algorithm — 5.5 The main theorem — 5.6 Derivatives — 5.7 B-spline properties — 5.8 Conversion to B-spline form — 5.9 The complete de Boor algorithm — 5.10 Conversions between Bézier and B-spline representations — 5.11 B-splines as divided differences — 5.12 Problems

Splines are piecewise polynomial curves that are differentiable up to a prescribed order. The simplest example is a piecewise linear C^0 spline, i.e., a polygonal curve. Other examples are the piecewise cubic C^1 splines, as constructed in 4.5.

The name spline is derived from elastic beams, so-called splines, used by draftsmen to lay out broad sweeping curves in ship design. Held in place by a number of heavy weights, these physical splines assume a shape that minimizes the strain energy. This property is approximately shared by the mathematical cubic C^2 splines.

5.1 Splines

A curve $\mathbf{s}(u)$ is called a **spline of degree** n with the **knots** $a_0, \ldots, a_m$, where $a_i \le a_{i+1}$ and $a_i < a_{i+n+1}$ for all possible i, if

> $\mathbf{s}(u)$ *is* $n - r$ *times differentiable at any* r*-fold knot*[1], *and* $\mathbf{s}(u)$ *is a polynomial of degree* $\le n$ *over each knot interval* $[a_i, a_{i+1}]$, *for* $i = 0, \ldots, m-1$.

It is also common to refer to a spline of degree n as a **spline of order** $n+1$. Figures 5.1 and 5.2 show examples of splines with simple knots obtained by

[1] *A knot* a_{i+1} *is called* r*-fold if* $a_i < a_{i+1} = \cdots = a_{i+r} < a_{i+r+1}$.

Stärk's construction, see Figures 3.8 and 3.9. The inner and end Bézier points are marked by hollow and solid dots, respectively.

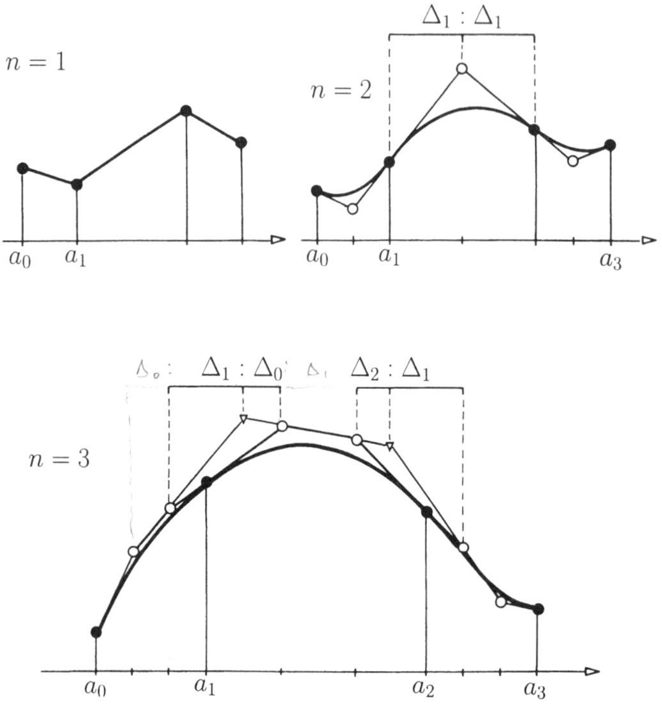

Figure 5.1: Spline functions of degree 1, 2 and 3.

5.2 B-splines

As with the Bézier representation of polynomial curves, it is desirable to write a spline $\mathbf{s}(u)$ as an affine combination of some control points $\mathbf{c}_i$, namely

$$\mathbf{s}(u) = \sum \mathbf{c}_i N_i^n(u) ,$$

where the $N_i^n(u)$ are basis spline functions with minimal support and certain continuity properties. Schoenberg introduced the name B-splines for these functions [Schoenberg '67]. Their Bézier polygons can be constructed by Stärk's theorem.

Figure 5.3 shows a piecewise cubic C^2 B-spline. Stärk's theorem is only needed for the Bézier ordinates, while the abscissae are given by Remark 8 in 2.3.

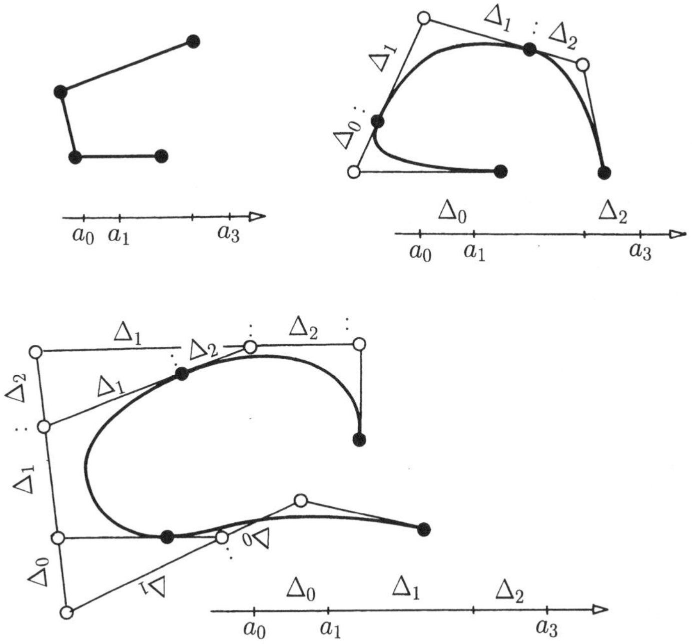

Figure 5.2: Parametric splines of degree 1, 2, and 3.

For higher degree this construction, albeit possible, becomes much less obvious and more complicated, see [Prautzsch '89]. Therefore, we use a recurrence relation, which was found independently by de Boor and Mansfield [de Boor '72] in 1970 and Cox [Cox '72] in 1971. We define B-splines from that relation and derive all important properties from that relation.

5.3 A recursive definition of B-splines

To define B-splines, let (a_i) be a, for simplicity, biinfinite and strictly increasing sequence of knots, which means $a_i < a_{i+1}$, for all i. We define the **B-splines** N_i^n with these knots by the recursion formula

$$N_i^0(u) = \begin{cases} 1 & \text{if} \quad u \in [a_i, a_{i+1}) \\ 0 & \text{otherwise} \end{cases}$$

and

$$N_i^n(u) = \alpha_i^{n-1} N_i^{n-1}(u) + (1 - \alpha_{i+1}^{n-1}) N_{i+1}^{n-1}(u) \ ,$$

where

$$\alpha_i^{n-1} = (u - a_i)/(a_{i+n} - a_i)$$

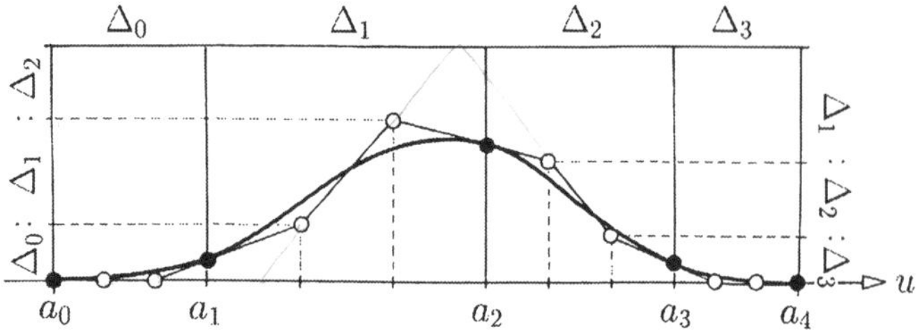

Figure 5.3: Bézier points of the B-spline $N_0^3(u)$.

is the local parameter with respect to the support of N_i^{n-1}. Figure 5.4 shows B-splines of degree 0, 1 and 2.

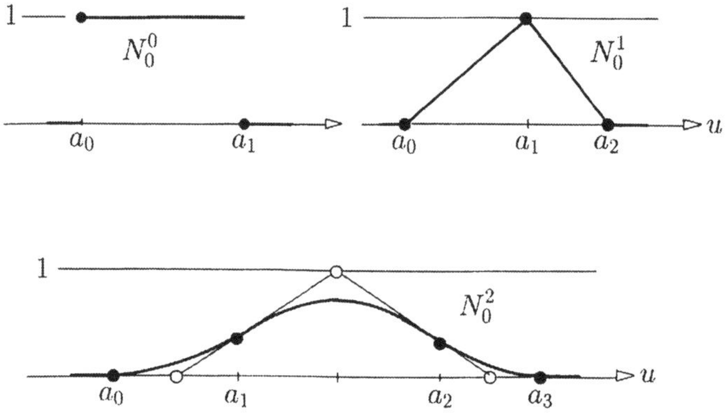

Figure 5.4: B-splines of degree 0, 1 and 2.

In case of multiple knots, the B-splines $N_i^n(u)$ are defined by the same recursion formula and the convention

$$N_i^{r-1} = N_i^{r-1}/(a_{i+r} - a_i) = 0 \quad \text{if} \quad a_i = a_{i+r} \ .$$

Figure 5.5 shows B-splines with multiple knots.

From the definition above, the following properties of B-splines are evident.

- *$N_i^n(u)$ is piecewise polynomial of degree n,*
- *$N_i^n(u)$ is positive in (a_i, a_{i+n+1}),*
- *$N_i^n(u)$ is zero outside of $[a_i, a_{i+n+1}]$,*
- *$N_i^n(u)$ is right side continuous.*

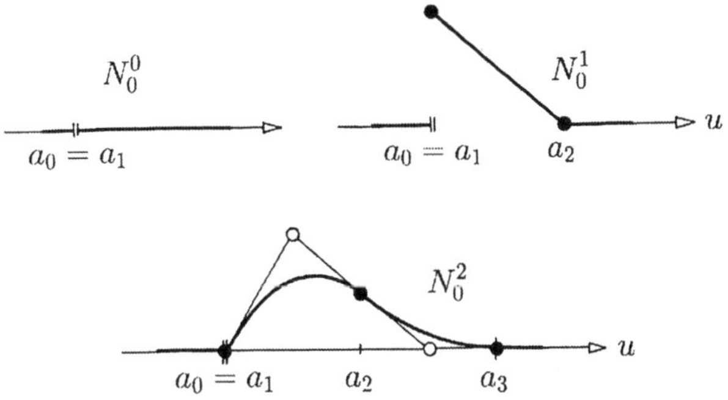

Figure 5.5: Some B-splines with multiple knots.

In Sections 5.5 and 5.6, we will see that the B-splines are $n - r$ times differentiable at r-fold knots, and that every spline is a unique combination of B-splines.

Remark 1: If, in particular, $a_1 = \ldots = a_n = 0$ and $a_{n+1} = \ldots = a_{2n} = 1$, then the above recursion formula for $N_0^n, \ldots, N_n^n$ and $u \in [0, 1)$ coincides with the recursion formula of the Bernstein polynomials. Hence, we have

$$N_i^n(u) = B_i^n(u) \quad \text{for} \quad i = 0, \ldots, n \quad \text{and} \quad u \in [0, 1) \ .$$

5.4 The de Boor algorithm

Consider a linear combination

$$\mathbf{s}(u) = \sum_i \mathbf{c}_i^0 N_i^n(u)$$

of the nth degree B-splines over some knot sequence (a_i). Since any finite sum can be converted to a formally biinfinite sum by adjunction of zero terms, we assume, without loss of generality, that the knot sequence and, hence, the sum above are biinfinite. Since the $N_i^n(u)$ have local supports, this sum is actually finite for any given u. In particular, let $u \in [a_n, a_{n+1})$, then $\mathbf{s}(u)$ can be written as

$$\mathbf{s}(u) = \sum_{i=0}^{n} \mathbf{c}_i^0 N_i^n(u) \ .$$

Using the B-spline recursion repeatedly and collecting terms, one obtains

$$\begin{aligned} \mathbf{s}(u) &= \sum_{i=1}^{n} \mathbf{c}_i^1 N_i^{n-1}(u) \\ &\;\vdots \\ &= \sum_{i=n}^{n} \mathbf{c}_i^n N_i^0(u) = \mathbf{c}_n^n \ , \end{aligned}$$

where the $\mathbf{c}_i^r$ are given by the affine combinations

$$\mathbf{c}_i^r = (1-\alpha)\mathbf{c}_{i-1}^{r-1} + \alpha\, \mathbf{c}_i^{r-1} \ , \quad \alpha = \alpha_i^{n-r} = \frac{u - a_i}{a_{i+n+1-r} - a_i} \ .$$

Note that $\alpha \in [0,1]$ since $u \in [a_n, a_{n+1})$, i.e., the affine combinations are actually convex.

This algorithm was developed by de Boor in 1972 [de Boor '72]. The points $\mathbf{c}_i^r$ are conveniently arranged in a triangular array, as illustrated below, where the recursion formula is represented by the key on the right,

$$\begin{array}{lllll} \mathbf{c}_0^0 & & & & \\ \mathbf{c}_1^0 & \mathbf{c}_1^1 & & & \\ \mathbf{c}_2^0 & \mathbf{c}_2^1 & \mathbf{c}_2^2 & & \\ \vdots & & & \ddots & \\ \mathbf{c}_n^0 & \mathbf{c}_n^1 & \mathbf{c}_n^2 & \cdots & \mathbf{c}_n^n \end{array}$$

key

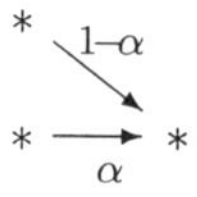

(α depending on key position)

An important consequence of de Boor's algorithm is that the spline $\mathbf{s}(u)$ over each knot interval is an affine and actually convex combination of $n+1$ consecutive coefficients $\mathbf{c}_i$. Hence, if the $\mathbf{c}_i$ represent points of some affine space, then $\mathbf{s}(u)$ is also a point. For this reason, we refer to the $\mathbf{c}_i$ as the **control points** of $\mathbf{s}(u)$.

Further, the spline lies in the affine hull of its control points, which implies that

$$\sum_{i=0}^{n} 1 \cdot N_i^n(u) = 1 \quad \text{for} \quad u \in [a_n, a_{n+1}) \ ,$$

i.e., the B-splines form a partition of unity. Figure 5.6 illustrates the geometric interpretation of de Boor's algorithm, which was first given in [Gordon & Riesenfeld '74].

Remark 2: For arbitrary $u \in \mathbb{R}$, de Boor's algorithm applied, as described above, to $\mathbf{c}_0^0, \ldots, \mathbf{c}_n^0$ does not compute $\mathbf{s}(u)$ in general, but the polynomial $\mathbf{s}_n(u)$ that agrees with $\mathbf{s}(u)$ over $[a_n, a_{n+1})$.

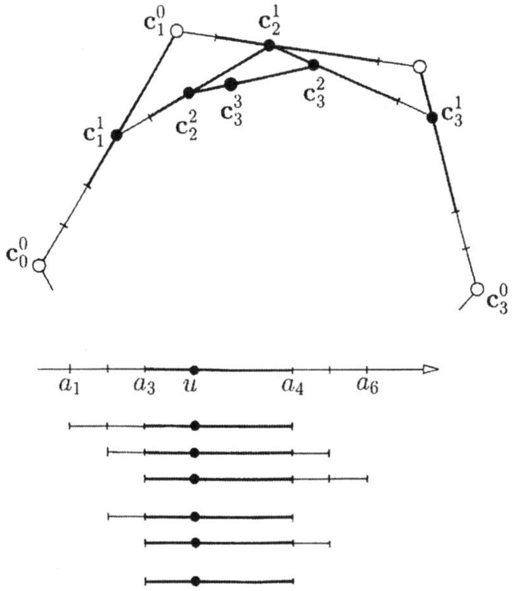

Figure 5.6: The convex combinations of de Boor's algorithm for $n = 3$.

5.5 The main theorem in its general form

Symmetric polynomials will help to see de Boor's algorithm in a wider context. As before, let

$$\mathbf{s}(u) = \sum_i \mathbf{c}_i N_i^n(u)$$

be an nth degree spline with knots a_i, and let $\mathbf{s}_i[u_1 \ldots u_n]$ be the polar form that agrees on its diagonal with $\mathbf{s}(u)$ over $[a_i, a_{i+1})$. Then we have the following general form of the **main theorem** 3.2.

As illustrated in Figure 5.7, the control points of $\mathbf{s}$ *are given by*

$$\mathbf{c}_i = \mathbf{s}_j[a_{i+1} \ldots a_{i+n}] \ , \qquad i = j - n, \ldots, j \ .$$

For a proof, let

$$\mathbf{p}_i^r = \mathbf{s}_j[a_{i+1} \ldots a_{i+n-r}\, u \overset{r}{\ldots} u]$$

and

$$u = (1 - \alpha)a_i + \alpha a_{i+n-r+1} \ .$$

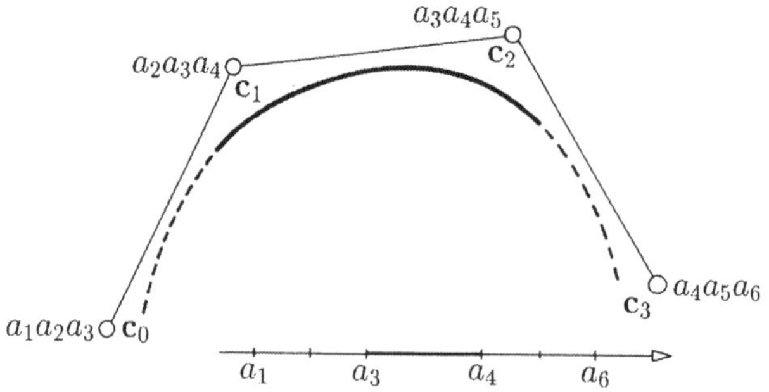

Figure 5.7: The main theorem for a cubic spline.

Then, since $\mathbf{s}_j$ is multiaffine and symmetric, it follows that

$$\mathbf{p}_i^r = (1-\alpha)\mathbf{p}_{i-1}^{r-1} + \alpha\,\mathbf{p}_i^{r-1} \quad , \quad \alpha = \alpha_i^{n-r} = \frac{u - a_i}{a_{i+n-r+1} - a_i} \quad ,$$

and, in particular,

$$\mathbf{p}_i^0 = \mathbf{s}_j[a_{i+1} \ldots a_{i+n}] \quad \text{and} \quad \mathbf{p}_j^n = \mathbf{s}_j(u) \quad .$$

For $u \in [a_j, a_{j+1})$, this construction agrees with de Boor's algorithm and can be used to compute any polynomial $\mathbf{s}_j[u \ldots u]$ of degree n. Hence, every polynomial of degree n can be written over $[a_j, a_{j+1})$ as a linear combination of the B-splines $N_{j-n}^n(u), \ldots, N_j^n(u)$. A dimension count shows that this linear combination is unique whence the assertion follows. ◇

Remark 3: In the proof, we showed that, over $[a_n, a_{n+1}]$, the B-splines $N_0^n(u), \ldots, N_n^n(u)$ form a basis for the space of all polynomial up to degree n. This result is due to [Curry & Schoenberg '66].

Remark 4: The spline "segment" $\mathbf{s}_i$ determines the control points $\mathbf{c}_{i-n}, \ldots, \mathbf{c}_i$. Conversely, every point $\mathbf{c}_j$ is determined by any "segment" $\mathbf{s}_j, \ldots, \mathbf{s}_{j+n}$, i.e.,

$$\mathbf{c}_i = \mathbf{s}_i[a_{i+1} \ldots a_{i+n}] = \cdots = \mathbf{s}_{i+n}[a_{i+1} \ldots a_{i+n}] \quad .$$

Remark 5: The proof above shows that the symmetric polynomial $\mathbf{s}_n[u_1 \ldots u_n]$ can be computed by a generalization of de Boor's algorithm. All one has to do, is to replace $\alpha = \alpha(u)$ in the recursion formula by

$$\alpha(u_r) = \frac{u_r - a_i}{a_{i+n-r+1} - a_i} \quad .$$

If m of the n variables $u_1, \ldots, u_n$ are knots, then only $n-m$ recursion steps are needed to compute $\mathbf{s}_j[u_1 \ldots u_n]$. The computation can be arranged in a triangular array consisting of $1+2+\cdots+(n-m+1)$ points.

5.6 Derivatives and smoothness

Because of the basis property of B-splines, see Remark 3, the derivative of the polynomial spline segment $\mathbf{s}_n$ can be written as

$$\mathbf{s}'_n(u) = \sum_{i=1}^{n} \mathbf{d}_i N_i^{n-1}(u) \ , \quad u \in [a_n, a_{n+1}) \ ,$$

where the unknown vectors $\mathbf{d}_i$ can easily be expressed in terms of the $\mathbf{c}_i$. Let $\mathbf{s}'_n[u_2 \ \ldots \ u_n]$ be the symmetric polynomial of $\mathbf{s}'_n(u)$, and let the direction $\Delta = a_{i+n} - a_i$ be given by the support of the B-spline $N_i^{n-1}(u)$. Then, it follows from the main theorem and 3.9 that

$$\begin{aligned} \mathbf{d}_i &= \mathbf{s}'_n[a_{i+1} \ \ldots \ a_{i+n-1}] \\ &= \frac{n}{\Delta} \mathbf{s}_n[\Delta \ a_{i+1} \ \ldots \ a_{i+n-1}] \\ &= \frac{n}{a_{i+n} - a_i} (\mathbf{c}_i - \mathbf{c}_{i-1}) \ . \end{aligned}$$

Since the $\mathbf{d}_i$ do not depend on the knot interval $[a_n, a_{n+1})$, the derivative of the spline $\mathbf{s}$ can be written for all $u \in \mathbb{R}$ as

(1) $$\mathbf{s}'(u) = \sum_i \frac{n}{a_{i+n} - a_i} \nabla \mathbf{c}_i N_i^{n-1}(u) \ ,$$

where $\nabla \mathbf{c}_i = \mathbf{c}_i - \mathbf{c}_{i-1}$ denotes the **first backward difference**.

One can differentiate further so as to obtain the B-spline representation of higher derivatives. This is also useful in showing that the B-splines have the desired smoothness properties:

An nth degree spline $\mathbf{s}$ is continuous at any n-fold knot. Namely if $a_0 < a_1 = \cdots = a_n < a_{n+1}$, then it follows from Remark 4 in 5.5 that

$$\begin{aligned} \mathbf{s}_0(a_1) &= \mathbf{s}_0[a_1 \ \ldots \ a_n] = \mathbf{c}_0 \\ &= \mathbf{s}_n[a_1 \ \ldots \ a_n] \\ &= \mathbf{s}_n(a_n) \ . \end{aligned}$$

Thus, if a_i is an r-fold knot, then the $(n-r)$th derivative of $\mathbf{s}$ is continuous at a_i. In other words,

a B-spline satisfies the smoothness criteria of a spline given in 5.1.

5.7 B-spline properties

We summarize the basic properties of B-splines.

- The B-splines of degree n with a given knot sequence that do not vanish over some knot interval are **linearly independent** over this interval.
- A dimension count shows that the B-splines $N_0^n, \dots, N_m^n$ with the knots $a_0, \dots, a_{m+n+1}$ form a **basis** for all splines of degree n with support $[a_0, a_{m+n+1}]$ and the same knots.
- Similarly, the B-splines $N_0^n, \dots, N_m^n$ over the knots $a_0, \dots, a_{m+n+1}$ restricted to the interval $[a_n, a_{m+1})$ form a **basis** for all splines of degree n restricted to the same interval.
- The B-splines of degree n form a **partition of unity**, i.e.,

$$\sum_{i=0}^{m} N_i^n(u) = 1, \quad \text{for} \quad u \in [a_n, a_{m+1}) \ .$$

- A spline $\mathbf{s}[a_n, a_{m+1}]$ of degree n with **n-fold end knots,**

$$(a_0 =)a_1 = \ \dots \ = a_n \quad \text{and} \quad a_{m+1} = \ \dots \ = a_{m+n}(= a_{m+n+1})$$

  has the same end points and end tangents as its control polygon.
- The **end knots** a_0 and a_{m+n+1} **have no influence** on N_0^n and N_m^n over the interval $[a_n, a_{m+1}]$.
- The B-splines are **positive** over the interior of their support,

$$N_i^n(u) > 0 \quad \text{for} \quad u \in (a_i, a_{i+n+1}) \ .$$

- The B-splines have **compact support**,

$$\text{supp} N_i^n = [a_i, a_{i+n+1}] \ .$$

- The B-splines satisfy the **de Boor, Mansfield, Cox recursion formula**

$$N_i^n(u) = \alpha_i^{n-1} N_i^{n-1}(u) + (1 - \alpha_{i+1}^{n-1}) N_{i+1}^{n-1}(u) \ ,$$

  where $\alpha_i^{n-1} = (u - a_i)/(a_{i+n} - a_i)$ represents the local parameter over the support of N_i^{n-1}.
- The **derivative** of a single B-spline is given by

$$\frac{d}{du} N_i^n(u) = \frac{n}{u_{i+n} - u_i} N_i^{n-1}(u) - \frac{n}{u_{i+n+1} - u_{i+1}} N_{i+1}^{n-1}(u) \ .$$

- The B-spline representation of a spline curve is **invariant under affine maps**.
- Any segment $\mathbf{s}_j[a_j, a_{j+1})$ of an nth degree spline lies in the **convex hull** of its $n+1$ control points $\mathbf{c}_{j-n}, \dots, \mathbf{c}_j$.
- A **degree elevation** formula is given in 6.5.

5.8 Conversion to B-spline form

Since any polynomial of degree n can be viewed as a spline of degree n or higher with an arbitrary sequence of knots, one can express the monomials as linear combinations of B-splines over any knot sequence (a_i). To do so, recall from 3.1 that the monomials $A_j^n(u) = \binom{n}{j} u^j$ have the polar forms

$$A_j^n[u_1 \dots u_n] = \sum_{i< \dots <k} u_i \overset{j}{\dots} u_k \ .$$

Thus, it follows from the main theorem 5.5 that

$$A_j^n(u) = \sum_i \alpha_{ji} N_i^n(u) \ ,$$

where $\alpha_{ji} = A_j^n[a_{i+1} \dots a_{i+n}]$, and, consequently,

$$\begin{aligned} \mathbf{a}(u) &= \mathbf{a}_0 A_0^n(u) + \cdots + \mathbf{a}_n A_n^n(u) \\ &= \sum_i (\mathbf{a}_0 \alpha_{0i} + \cdots \mathbf{a}_n \alpha_{ni}) N_i^n(u) \ , \end{aligned}$$

which generalizes Marsden's identity given in Problem 4 below. In particular, one obtains

$$\begin{aligned} u &= \frac{1}{n} A_1^n(u) \\ &= \sum_i \gamma_i N_i^n(u) \ , \end{aligned}$$

Farin: $\gamma_i = (a_i + \dots + a_{i+n-1})/n$

Indexfehler

where $\gamma_i = \alpha_{1i}/n = (a_{i+1} + \cdots + a_{i+n})/n$. The γ_i are the so-called **Greville abscissae** [Greville '67].

Remark 6: The Greville abscissae show up naturally in the control points of the graph of a spline function

$$s(u) = \sum_i c_i N_i^n \ .$$

Namely, the graph $\mathbf{s}(u) = [u \ s(u)]^t$ has the control points $\mathbf{c}_i = [\gamma_i \ c_i]^t$. Figure 5.8 shows the example $s(u) = N_2^3(u)$. Other examples are shown in Figure 5.1.

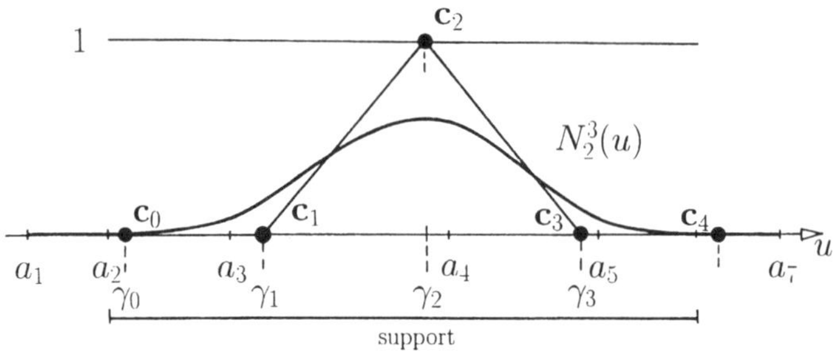

Figure 5.8: Control points of the cubic B-spline $N_2^3(u)$.

5.9 The complete de Boor algorithm

The Taylor expansion of a polynomial spline segment

$$\mathbf{s}_n(u) = \sum_{i=0}^{n} \mathbf{c}_i N_i^n(u) \ , \qquad u \in [a_n, a_{n+1}) \ ,$$

can be computed at any $u \in \mathbb{R}$ following the ideas presented in 2.6 for Bézier curves.

Let $\mathbf{s}_n[u_1 \ \dots \ u_n]$ be the polar form of $\mathbf{s}_n$ and consider the points and vectors

$$\mathbf{c}_{r,i,k} = \mathbf{s}_n[\varepsilon \overset{r}{\dots} \varepsilon\, a_i \ \dots \ a_{i+n-r-k}\, u \overset{k}{\dots} u],$$

where ε denotes the direction $1 - 0$, for $i = r + k, \dots, n$. It follows that

$$\frac{d^r}{du^r} \mathbf{s}_n(u) \quad = \quad \frac{n!}{(n-r)!} \mathbf{c}_{r,n,n-r} \ ,$$

and the Taylor expansion is given by

$$\mathbf{s}_n(u + h) = \sum_{r=0}^{n} \mathbf{c}_{r,n,n-r} \binom{n}{r} h^r \ .$$

The points and vectors $\mathbf{c}_{rik}$ can again be arranged conveniently in a tetrahedral array, see Figure 5.9, where $n = 2$ and $\varepsilon 4$ stands for $\mathbf{s}_n[\varepsilon\, a_4]$, etc.

This array, first considered by Sablonniere in 1978 [Sablonniere '78], is com-

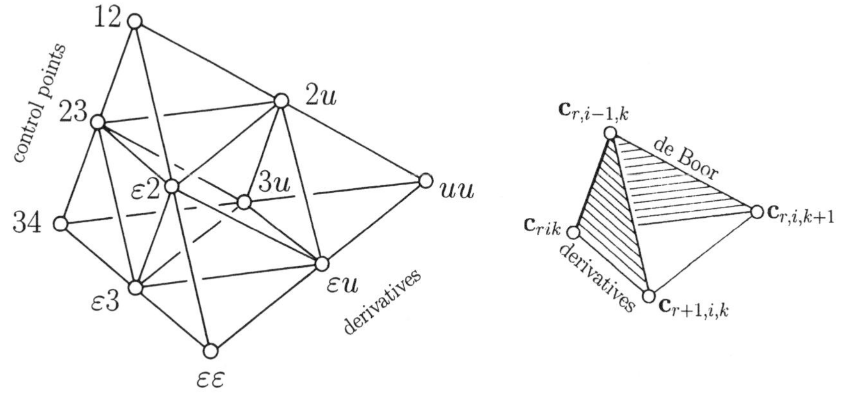

Figure 5.9: The complete de Boor algorithm.

posed of $\binom{n+2}{3}$ subtetrahedra and contains the given control points $\mathbf{c}_i = \mathbf{c}_{0,i,0}$ on the "left" edge and the multiples

$$\frac{(n-r)!}{n!}\,\mathbf{s}_n^{(r)}(u)$$

of the derivatives on the opposite edge.

Any two of the four points of a subtetrahedron can be computed from the two other points. The computation rules follow directly from the properties of multiaffine symmetric polynomials. For example, one has

$$\mathbf{c}_{r+1,i,k} = \frac{1}{a_{i+n-r-k} - a_i}(\mathbf{c}_{r,i,k} - \mathbf{c}_{r,i-1,k})$$

on the "left" face,

$$\mathbf{c}_{r,i,k+1} = (1-\alpha)\mathbf{c}_{r,i-1,k} + \alpha\,\mathbf{c}_{r,i,k}\ , \qquad \alpha = \frac{u - a_i}{a_{i+n-r-k} - a_i}\ ,$$

on the "rear" face,

(2) $$\mathbf{c}_{r,i,k} = \mathbf{c}_{r,i,k+1} + (a_{i+n-r-k} - u)\mathbf{c}_{r+1,i,k}\ ,$$

on the "bottom" face, and

(3) $$\mathbf{c}_{r,i-1,k} = \mathbf{c}_{r,i,k+1} + (a_i - u)\mathbf{c}_{r+1,i,k}$$

on the "top" face.

Remark 7: One can use the above formulae to convert a B-spline representation to monomial representation and vice versa.

Remark 8: In order to obtain the derivatives from the control points or vice versa, it suffices to compute, for example, only the left and top faces of the tetrahedral array, see [Lee '82, Boehm '84b].

Remark 9: If one first computes the rear face and then the bottom (or top) face of the tetrahedral array, one needs to solve formula (2) above (or (3)) for $\mathbf{c}_{r+1,i,k}$ (or $c_{r,i-1,k}$). This is impossible if $u = a_{i+n-r-k}$ (or $u = a_i$). Hence, the derivatives of the polynomial $\mathbf{s}_n$ cannot be computed in this fashion for $u = a_{n+1}, \ldots, a_{2n}$ (or $u = a_0, \ldots, a_{n-1}$).

5.10 Conversions between Bézier and B-spline representations

There is also a tetrahedral algorithm to convert a B-spline representation into a Bézier representation and vice versa [Boehm '77, Sablonniere '78]. It can be derived similarly as the algorithm in 5.9. Let the notations be as in 5.9 and let

$$\mathbf{q}_{rik} = \mathbf{s}_n[a \overset{r}{\ldots} a\, a_{i+1} \ldots a_{i+n-r-k}\, b \overset{k}{\ldots} b]$$

for $i = r + k, \ldots, n$. Thus, the control points of the spline are given by

$$\mathbf{c}_i = \mathbf{q}_{0i0} \;,$$

and the Bézier points of the polynomial $\mathbf{s}_n$ over $[a, b]$ are given by

$$\mathbf{b}_j = \mathbf{q}_{n-j,n,j} \;.$$

Again, the points $\mathbf{q}_{rik}$ are conveniently arranged in a tetrahedral array, as illustrated below in Figure 5.10 for $n = 2$, where $a3, ab$, etc. stand for $\mathbf{q}_{120}$, $\mathbf{q}_{101}$, etc.

The left face is computed according to the rule

$$\mathbf{q}_{r+1,i,k} = (1-\alpha)\mathbf{q}_{r,i-1,k} + \alpha\, \mathbf{q}_{r,i,k} \;, \qquad \alpha = \frac{a - a_i}{a_{i+n-r-k} - a_i} \;,$$

and the bottom face according to the rule

$$\mathbf{q}_{r,i,k+1} = (1-\gamma)\mathbf{q}_{r+1,i,k} + \gamma\, \mathbf{q}_{r,i,k} \;, \qquad \gamma = \frac{b - a}{a_{i+n-r-k} - a} \;.$$

Conversely, one can compute the B-spline control points from the Bézier points. First, one solves the two formulae above for $\mathbf{q}_{r,i-1,k}$ and $\mathbf{q}_{rik}$. Second, one applies the formulae to compute the bottom and then the left face.

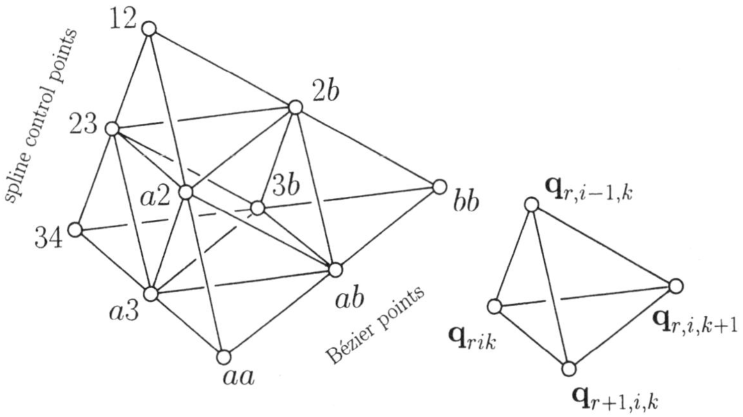

Figure 5.10: Conversion between Bézier and B-spline representations.

5.11 B-splines as divided differences

The standard definition of B-splines uses divided differences, and the calculus of divided differences has been heavily used in developing the univariate spline theory [de Boor '78]. In particular, divided differences were used by de Boor, Cox, and Mansfield to derive the recurrence relation.

Using the derivative formula (1) in 5.6, we show that B-splines are divided differences of the **truncated power function**

$$f(a) = (a-u)_+^n := \begin{cases} (a-u)^n & \text{if} \quad a > u \\ 0 & \text{otherwise} \end{cases} ,$$

shown in Figure 5.11. Note that f is a function of a while u is some fixed parameter.

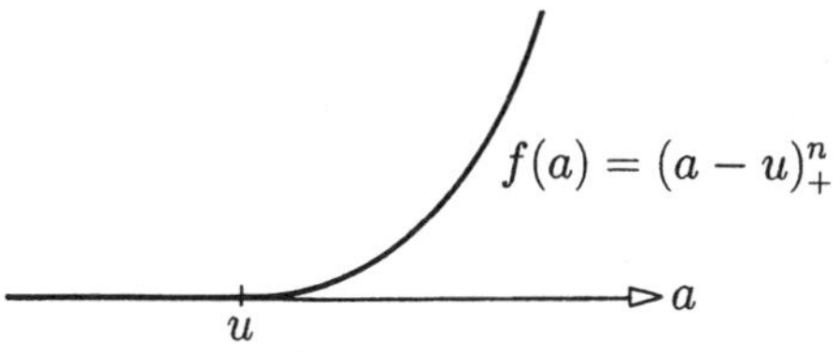

Figure 5.11: A truncated power function.

With the divided differences given in 4.3, one obtains that

the B-spline N_0^n with the knots $a_0, \ldots, a_{n+1}$ can be written as

$$N_0^n(u) = (a_{n+1} - a_0)[a_0 \ \ldots \ a_{n+1}](a-u)_+^n \ .$$

One can prove this fact by induction over n. For $n = 0$, the identity is easily checked. For the induction step from $n-1$ to n, we recall from 4.3 that $[a_0 \ \ldots \ a_{n+1}]f(a)$ is the leading coefficient of the $(n+1)$th degree polynomial interpolating $f(a)$ at $a_0, \ldots, a_{n+1}$. Hence, one can substitute $f(a)$ by the monomial $(a-u)^n$ of degree n in a if $u < a_0$ and by the zero function if $u \geq a_{n+1}$. This shows that the identity above holds for $u < a_0$ and $u \geq a_{n+1}$.

Thus, it suffices to show that the derivative of the claimed identity holds. Note that the divided difference is a linear combination of possibly differentiated power functions. Hence, the divided difference is differentiable in u, except at an $(n+1)$-fold knot. This causes no problem since there is at most one $(n+1)$-fold knot.

Using the recursive definition of divided differences, the induction hypothesis and the derivative formula for B-splines, we obtain

$$\begin{aligned}
&\frac{d}{du}(a_{n+1} - a_0)[a_0 \ \ldots \ a_{n+1}](a-u)_+^n \\
&\quad = -n(a_{n+1} - a_0)[a_0 \ \ldots \ a_{n+1}](a-u)_+^{n-1} \\
&\quad = n([a_0 \ \ldots \ a_n](a-u)_+^{n-1} - [a_1 \ \ldots \ a_{n+1}](a-u)_+^{n-1}) \\
&\quad = \frac{n}{a_n - a_0} N_0^{n-1} - \frac{n}{a_{n+1} - a_1} N_1^{n-1} \\
&\quad = \frac{d}{du} N_0^n(u) \ ,
\end{aligned}$$

which proves the assertion. ◇

5.12 Problems

1 Consider a cubic C^2 spline $s(u)$ with the single knots $a_0, \ldots, a_m$. Show that every C^2 function $f(u) \neq s(u)$ which interpolates s at all knots and also the derivative of $s(u)$ at $u = a_0$ and a_m has greater strain energy than s, i.e.,

$$\int_{a_0}^{a_m} |f''(u)|^2 du > \int_{a_0}^{a_m} |s''(u)|^2 du \ .$$

For a solution, we refer to literature in numerical analysis, for example, [Boehm & Prautzsch '93, pp. 125 f].

2 Given a spline $s(u) = \sum_{i=0}^{m} c_i N_i^n(u)$ with the knots $a_0, \ldots, a_{m+n+1}$, show that

$$\int_{a_0}^{a_{m+n+1}} s(u)du = \sum_{i=0}^{m} \frac{a_{i+n+1} - a_i}{n+1} c_i \ .$$

3 Sketch the cubic B-splines with the knots $0,0,0,0,1$; $0,0,0,1,2$; $0,0,1,2,3$ and $0,1,2,3,4$ with their Bézier polygons. Compute the values of their Bézier ordinates.

4 Use symmetric polynomials to prove **Marsden's identity**

$$(u-a)^n = \sum_i (a_{i+1} - a) \dots (a_{i+n} - a) N_i^n(u) \ .$$

5 Use the derivative formula of B-splines to derive the recursion formula of de Boor, Mansfield and Cox by induction.

6 Use Leibniz's identity for the product $f = gh$ of two functions, i.e.,

$$[a_i \ \dots \ a_{i+k}]f = \sum_{r=i}^{i+k} ([a_i \ \dots \ a_r]g)([a_r \ \dots \ a_{i+k}]h) \ ,$$

to derive[de Boor '72] the recursion formula of de Boor, Mansfield and Cox, see also [de Boor '72].

7 Let $s(u) = \sum_{i=0}^{3} c_i N_i^3$ with the knots $0,1,2,\dots,7$ be given by $c_0,\dots,c_3 = 4,7,-2,1$.

a) Sketch $s[3,4]$ with its control polygon.

b) Compute s, s', s'' and s''' at $u = 3$.

c) Compute the monomial representation of $s(u)$ over $[3,4]$.

d) Compute the Bézier representation of $s(u)$ over $[3,4]$.

8 Show that if a multiaffine symmetric polynomial can be computed from $n+1$ points $\mathbf{p}[a_{i,1} \ \dots \ a_{i,n}], i = 0,\dots,n$, by affine combinations as in de Boor's algorithm, then there are real numbers $a_1,\dots,a_{2n}$ such that $a_{i,j} = a_{i+j}$.

6 B-spline techniques

6.1 Knot insertion — 6.2 The Oslo algorithm — 6.3 Convergence under knot insertion — 6.4 A degree elevation algorithm — 6.5 A degree elevation formula — 6.6 Convergence under degree elevation — 6.7 Interpolation — 6.8 Cubic spline interpolation — 6.9 Problems

Most algorithms for curves in Bézier representation have a generalized form for splines. One of the most important spline algorithms is knot insertion. It can be used for degree elevation, the de Boor algorithm and subdivision. In particular, de Casteljau's algorithm can be understood as a special multiple knot insertion.

6.1 Knot insertion

Because of the basis property of B-splines, see 5.7, a spline

$$\mathbf{s}(u) = \sum_i \mathbf{c}_i N_i^n(u)$$

with a knot sequence (a_i) is also a spline with any finer knot sequence $(\hat{a}_j)$, which contains (a_i) as a subsequence, see Figure 6.1 for an illustration.

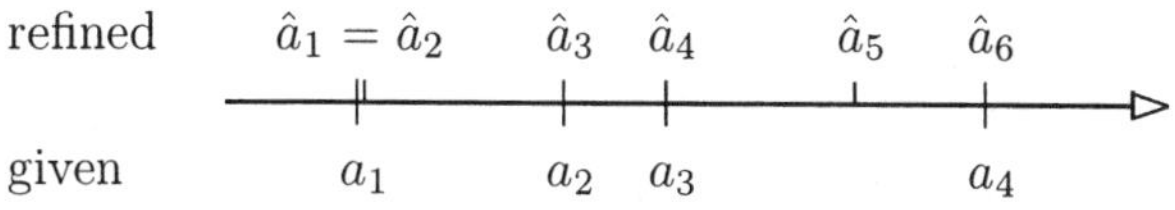

Figure 6.1: Refinement of a knot sequence.

Let $\hat{N}_j^n(u)$ denote the B-spline of degree n with the knots $\hat{a}_j, \ldots, \hat{a}_{j+n+1}$, then $\mathbf{s}$ can be written as

$$\mathbf{s}(u) = \sum_j \hat{\mathbf{c}}_j \hat{N}_j^n(u) \ .$$

The new control points $\hat{\mathbf{c}}_j$ are best computed by repeatedly inserting single knots [Boehm '80]. Without loss of generality, we can assume that $(\hat{a}_j)$ contains just one more knot, say $\hat{a}$, than (a_i). After an appropriate index shift, we can assume $\hat{a} = \hat{a}_{n+1}$ and $a_n \le \hat{a} < a_{n+1}$.

As an immediate consequence of the main theorem in 5.5, one has

$$\hat{\mathbf{c}}_j = \begin{cases} \mathbf{c}_j & \text{for} \quad j \le 0 \\ \mathbf{c}_{j-1} \cdot (1-\alpha_j) + \mathbf{c}_j \cdot \alpha_j & \text{for} \quad j = 1, 2, \ldots, n \ , \\ \mathbf{c}_{j-1} & \text{for} \quad j \ge n+1 \end{cases}$$

where

$$\alpha_j = \frac{\hat{a} - a_j}{a_{j+n} - a_j}$$

is the local coordinate of $\hat{a}$ with respect to the support $[a_j, a_{j+n}]$ of $\hat{N}_j^n$. Figure 6.2 gives an illustration for $n = 3$.

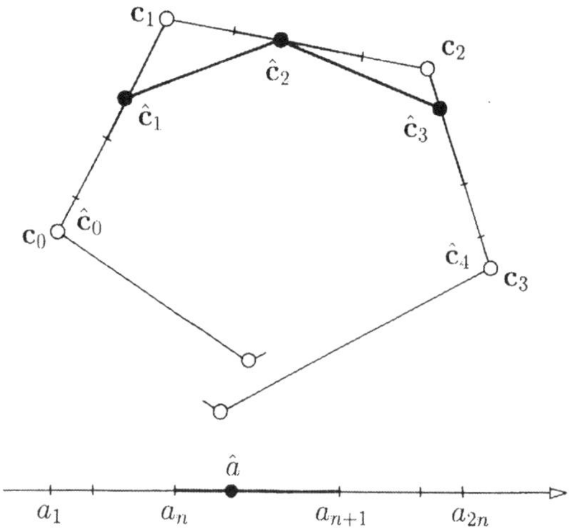

Figure 6.2: Inserting a new knot.

Comparing knot insertion with de Boor's algorithm in 5.4, one sees that the new control points $\hat{\mathbf{c}}_1, \ldots, \hat{\mathbf{c}}_n$ are the points $\mathbf{c}_1^1, \ldots, \mathbf{c}_n^1$ that appear in the second column of de Boor's triangular array [Boehm '80]. Moreover, the points

$$\mathbf{c}_1^1, \ldots, \mathbf{c}_r^r, \ldots, \mathbf{c}_n^r, \ldots, \mathbf{c}_n^1$$

of de Boor's array are identical with the new control points of $\mathbf{s}$ if $\hat{a}$ is inserted r times. These points replace the old control points $\mathbf{c}_1, \ldots, \mathbf{c}_{n-1}$, while the other control points remain.

In particular, for $r = n$, it follows that de Boor's algorithm can be viewed as the n-fold repeated insertion of the knot $\hat{a}$.

Remark 1: If all knots in the refined sequence $(\hat{a}_i)$ have multiplicity n, then the control polygon $\hat{\mathbf{c}}_i$ represents the Bézier polygon of $\mathbf{s}$. This follows from Remark 1 in 5.3. Hence, raising the multiplicities of all knots to n generates the Bézier representation of a spline as first observed by Cohen et al [Cohen et al. '80].

Remark 2: A single B-spline can be expressed as a linear combination of B-splines with any finer knot sequence $(\hat{a}_j)$. The corresponding formula,

$$N_i^n(u) = \sum_j \alpha_{ij} \hat{N}_j^n(u) \;,$$

is due to de Boor [de Boor '76b]. The coefficients α_{ij} are **discrete B-splines** with knots $\hat{a}_j$. This name is introduced in [Schumaker '73], for the case that the $\hat{a}_j$ are equidistant knots .

Remark 3: In particular, if $s(u) = N_j^n(u)$ is a single B-spline, then the knot insertion construction above implies the identity

$$N_j^n(u) = \begin{cases} \hat{N}_j^n(u) & \text{for} \quad j \le 0 \\ \alpha_j \hat{N}_j^n(u) + (1 - \alpha_{j+1}) \hat{N}_{j+1}^n(u) & \text{for} \quad j = 1, \ldots, n \\ \hat{N}_{j+1}^n(u) & \text{for} \quad j \ge n+1 \end{cases} .$$

This identity represents a **knot deletion** for B-splines [Boehm '80].

6.2 The Oslo algorithm

While repeated knot insertion of single knots provides, in general, the better method to compute the control points $\hat{\mathbf{c}}_i$ with respect to any refinement $(\hat{a}_j)$ of (a_i), one may also compute each new control point $\hat{\mathbf{c}}_j$ by the generalization of de Boor's algorithm given in 5.5, Remark 5.

To compute the $\hat{\mathbf{c}}_j$, one needs some knot interval $[a_k, a_{k+1}]$ which intersects the support $[\hat{a}_j, \hat{a}_{j+n+1}]$ of $\hat{N}_j^n$, as illustrated in Figure 6.3. The control points

$$\hat{\mathbf{c}}_j = \mathbf{s}_k[\hat{a}_{j+1} \, \ldots \, \hat{a}_{j+n}]$$

can be computed by the generalized de Boor algorithm. This recursion formula for the $\hat{\mathbf{c}}_j$ was found by Cohen, Lyche and Riesenfeld in Oslo 1980 by successive evaluations of discrete B-splines and has been named the **Oslo algorithm** [Cohen et al. '80]. Note that the affine combinations of the Oslo-algorithm are, in general, not convex. Further improvements are necessary to avoid non-convex combinations.

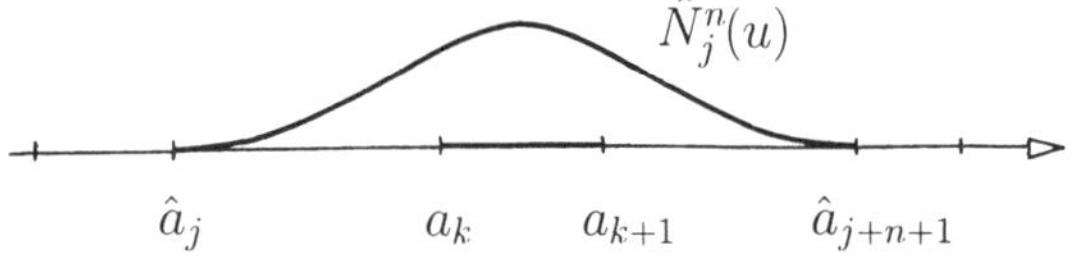

Figure 6.3: Choosing a_k for the construction of $\hat{\mathbf{c}}_j$.

6.3 Convergence under knot insertion

In this section, we generalize the ideas used in 3.3. Consider the spline $\mathbf{s}(u) = \sum_i \mathbf{c}_i N_i^n(u)$ with some knot sequence (a_i). On inserting more and more knots such that the knot sequence becomes dense eventually, the sequence of the corresponding control polygons converges to the spline $\mathbf{s}$ with the rate of convergence being quadratic in the maximum knot distance.

More precisely, let $[a, b]$ be some interval, let $h = \max\{\Delta a_i | [a_i, a_{i+1}] \subset [a, b]\}$, and let $\gamma_i = (a_{i+1} + \cdots + a_{i+n})/n$ be the Greville abscissae.

Then one has

$$\max \|\mathbf{s}(\gamma_i) - \mathbf{c}_i\| = O(h^2) \ ,$$

where the maximum is taken over all i such that $[a_{i+1}, a_{i+n}] \subset [a, b]$.

For the proof, which is due to [Schaback '93], consider a control point $\mathbf{c}_i = \mathbf{s}_r[a_{i+1} \ldots$
$a_{i+n}]$, where $\mathbf{s}_r$ is the symmetric polynomial of $\mathbf{s}$ restricted to the knot interval $[a_r, a_{r+1})$ containing γ_i. Since

$$\frac{\partial}{\partial u_1}\mathbf{s}_r[u \ldots u] = \cdots = \frac{\partial}{\partial u_n}\mathbf{s}_r[u \ldots u] \ ,$$

the Taylor expansion of $\mathbf{s}_r$ around $[\gamma_i \ldots \gamma_i]$ is of the form

$$\begin{aligned} \mathbf{c}_i &= \mathbf{s}_r[\gamma_i \ldots \gamma_i] + \sum_{j=i+1}^{i+n} (a_j - \gamma_i)\frac{\partial}{\partial u_1}\mathbf{s}_r[\gamma_i \ldots \gamma_i] + O(h^2) \\ &= \mathbf{s}(\gamma_i) + O(h^2) \ , \end{aligned}$$

which proves the assertion. ◇

6.4 A degree elevation algorithm

Due to the basis properties of B-splines, we can write a spline

$$\mathbf{s}(u) = \sum_i \mathbf{c}_i N_i^n(u)$$

with some knot sequence (a_i) also as a linear combination of B-splines of degree $n+1$,

$$\mathbf{s}(u) = \sum_j \mathbf{d}_j \hat{N}_j^{n+1}(u) \ ,$$

with the knot sequence $(\hat{a}_j)$ obtained from (a_i) by raising the multiplicity of each knot a_i by one, as illustrated in Figure 6.4.

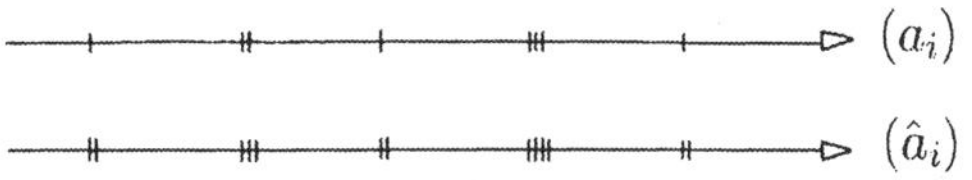

Figure 6.4: Knot sequences for degree elevation.

The main theorem from 5.5 and formula (2) from Section 3.11 tell us that

$$\begin{aligned}\mathbf{d}_j &= \mathbf{s}_r[\hat{a}_{j+1} \quad \dots \quad \hat{a}_{j+n+1}] \\ &= \frac{1}{n+1} \sum_{k=1}^{n+1} \mathbf{s}_r[\hat{a}_{j+1} \dots \hat{a}_k^* \dots \hat{a}_{j+n}] \ ,\end{aligned}$$

where $\mathbf{s}_r[u_1 \dots u_n]$ and $\mathbf{s}_r[u_1 \dots u_{n+1}]$, respectively, denote the n- and $(n+1)$-variate polar forms of a polynomial segment of the spline $\mathbf{s}(u)$ that depends on $\mathbf{d}_j$. Each point

$$\mathbf{s}_r[\hat{a}_{j+1} \dots \hat{a}_k^* \dots \hat{a}_{j+n}]$$

can be computed by the generalized de Boor algorithm given in Remark 5 from Section 5.5, where one needs to insert at most $\lfloor (n-1)/2 \rfloor$ knots. For a uniform cubic spline with two segments, this algorithm is first described in [Ramshaw '87, pp. 109f] and in [Seidel '89] for five segments.

Remark 4: The number of operations for the algorithm above is of order $O(n^2)$ per new control point $\hat{\mathbf{c}}_i$. It is possible to organize the computations more efficiently such that only $O(n)$ operations are needed, see [Prautzsch & Piper '91, Liu '97, Trump '01].

Remark 5: Different and in the citing order increasingly slower $O(n^2)$ algorithms are derived in [Prautzsch '84a, Cohen et al. '85, Piegl & Tiller '94].

6.5 A degree elevation formula

For a single B-spline there is a simple expression in terms of higher degree B-splines. Let $N^n(u|a_i \ldots a_{i+n+1})$ denote the B-spline of degree n over the knots $a_i, \ldots, a_{i+n+1}$. Then, it follows that

$$(1) \qquad N^n(u|a_0 \ldots a_{n+1}) = \frac{1}{n+1} \sum_{i=0}^{n+1} N^{n+1}(u|a_0 \ldots a_i\, a_i \ldots a_{n+1}) \ ,$$

which was found by Micchelli [Micchelli '79]. Figure 6.5 illustrates the example

$$N^1(u|abc) = \frac{1}{2}(N^2(u|aabc) + N^2(u|abbc) + N^2(u|abcc)) \ .$$

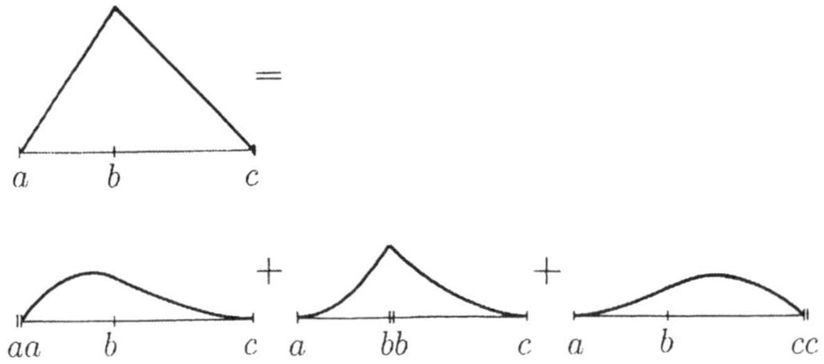

Figure 6.5: Degree elevation formula, example.

For the proof, we follow [Lee '94] and use divided differences. Setting $f(a) = (a-u)_+^{n+1}$, the degree elevation formula (1) takes the form

$$[a_0 \ldots a_{n+1}]f' = \sum_{i=0}^{n+1} [a_0 \ldots a_i\, a_i \ldots a_{n+1}]f \ .$$

More generally, this formula holds for any differentiable function f, as we can verify by induction over n. For $n = -1$, the formula follows from the definition of divided differences,

$$[a_0]f' = f'(a_0) = [a_0\, a_0]f \ .$$

For $n \geq 0$, we use the recursion formula of divided differences, see 4.3, the induction hypothesis, and again the recursion,

$$\begin{aligned}
&[a_0 \dots a_{n+1}]f' \\
&= \frac{1}{a_{n+1} - a_0}([a_1 \dots a_{n+1}]f' - [a_0 \dots a_n]f') \\
&= \frac{1}{a_{n+1} - a_0}\Big(\sum_{i=1}^{n}([a_1 \dots a_i\, a_i \dots a_{n+1}]f - [a_0 \dots a_i\, a_i \dots a_n]f) \\
&\qquad + [a_1 \quad \dots \quad a_{n+1}\, a_{n+1}]f - [a_0 \quad \dots \quad a_{n+1}]f \\
&\qquad + [a_0 \quad \dots \quad a_{n+1}]f - [a_0\, a_0 \quad \dots \quad a_n]f\Big) \\
&= \sum_{i=0}^{n+1}[a_0 \dots a_i\, a_i \dots a_{n+1}]f \ . \quad \diamond
\end{aligned}$$

6.6 Convergence under degree elevation

Repeated degree elevation leads to a sequence of control polygons that converges to the spline. More precisely, consider a spline $\mathbf{s}(u)$ of degree n with its mth degree representation,

$$\mathbf{s}(u) = \sum \mathbf{c}_i N_i^m(u) \ , \quad m > n \ ,$$

and associated knots a_i generated by successive degree elevation. Hence, the supports of the N_i^m are not longer than $h := \sup |a_{i+n+1} - a_i|$. Let $\gamma_i := \frac{1}{m}(a_{i+1} + \cdots + a_{i+m})$ denote the Greville abscissae. If h and all derivatives of $\mathbf{s}$ are bounded, one obtains

$$\sup \|\mathbf{c}_i - \mathbf{s}(\gamma_i)\| = O(1/m) \ .$$

Proof: We proceed as in 6.3 and consider, for instance, the control point $\mathbf{c}_0 = \mathbf{s}[a_1 \dots a_m]$, where $\mathbf{s}[u_1 \dots u_m]$ denotes the polar form of $\mathbf{s}$ over some suitable knot interval. Since, due to Remark 9 in 3.9,

$$\frac{\partial^k}{\partial u_{i_1} \dots \partial u_{i_k}} \mathbf{s}[\gamma_0 \dots \gamma_0] = \frac{\varepsilon_{\mathbf{i}}}{m \dots (m-k+1)} \mathbf{s}^{(k)}(\gamma_0) \ ,$$

where $\varepsilon_{\mathbf{i}} = 1$ if all coordinates of $\mathbf{i} = (i_1, \dots, i_k)$ are different and $\varepsilon_{\mathbf{i}} = 0$ otherwise, the Taylor expansion of $\mathbf{s}[a_1 \dots a_m]$ around $[\gamma_0 \dots \gamma_0]$ is given by

$$\mathbf{c}_0 = \mathbf{s}(\gamma_0) + \sum_{k=1}^{n} \sum_{i_1=1}^{m} \cdots \sum_{i_k=1}^{m} \frac{\varepsilon_{\mathbf{i}}}{k!} \frac{(a_{i_1} - \gamma_0) \dots (a_{i_k} - \gamma_0)}{m \dots (m-k+1)} \mathbf{s}^{(k)}(\gamma_0) \ .$$

Exploiting that $\varepsilon_{\mathbf{i}}$ is zero for $m^k - (m \dots (m-k+1))$ different $\mathbf{i}$ and using the fact that $\sum_{i_1=1}^{m} \cdots \sum_{i_k=1}^{m}(a_{i_1} - \gamma_0) \ldots (a_{i_k} - \gamma_0) = 0$, we obtain

$$\|\mathbf{c}_0 - \mathbf{s}(\gamma_0)\| \le \sum_{k=2}^{n} \frac{h^k}{k!} \frac{m^k - (m \ldots (m-k+1))}{m \ldots (m-k+1)} \sup \|\mathbf{s}^k(u_0)\| \ ,$$

which proves the claim above. ◇

6.7 Interpolation

Splines are often used to solve interpolation problems. Of particular interest is the uniqueness of a spline interpolant. Let $N_0^n, \ldots, N_m^n$ be the B-splines of degree n with the knots $a_0, \ldots, a_{m+n+1}$ and let $\mathbf{p}_0, \ldots, \mathbf{p}_m$ be given points with associated interpolation abscissae $u_0 < \ldots < u_m$. We wish to find a spline $\mathbf{s} = \sum_{i=0}^{m} \mathbf{c}_i N_i^n$ solving the interpolation problem

$$\mathbf{s}(u_j) = \sum_{i=0}^{m} \mathbf{c}_i N_i^n(u_j) = \mathbf{p}_j \ .$$

This means solving the following linear system

$$(2) \qquad \begin{bmatrix} N_0^n(u_0) & \cdots & N_m^n(u_0) \\ \vdots & & \vdots \\ N_0^n(u_m) & \cdots & N_m^n(u_m) \end{bmatrix} \begin{bmatrix} \mathbf{c}_0^t \\ \vdots \\ \mathbf{c}_m^t \end{bmatrix} = \begin{bmatrix} \mathbf{p}_0^t \\ \vdots \\ \mathbf{p}_m^t \end{bmatrix} ,$$

which we abbreviate by $NC = P$. Note that this linear system consists of several systems, one for each column of C. The matrix N is called **collocation matrix**.

The **Schoenberg-Whitney theorem** from 1953 establishes when the interpolation problem has a unique solution [Schoenberg & Whitney '53].

> *The matrix N is invertible if and only if N has a positive diagonal, which means that $N_i^n(u_i) \neq 0$ for all i.*

Note that if the N_i^n are continuous, then the condition $N_i^n(u_i) \neq 0$ is equivalent to the requirement $u_i \in (a_i, a_{i+n+1})$.

For a proof of the theorem, we follow [Powell '81]. Let $N_i^n(u_i) = 0$ for some i and assume $a_{i+n+1} \le u_i$. Then, it follows that

$$N_1^n(u_j) = \cdots = N_i^n(u_j) = 0 \quad \text{for all} \quad j \ge i \ ,$$

i.e.,

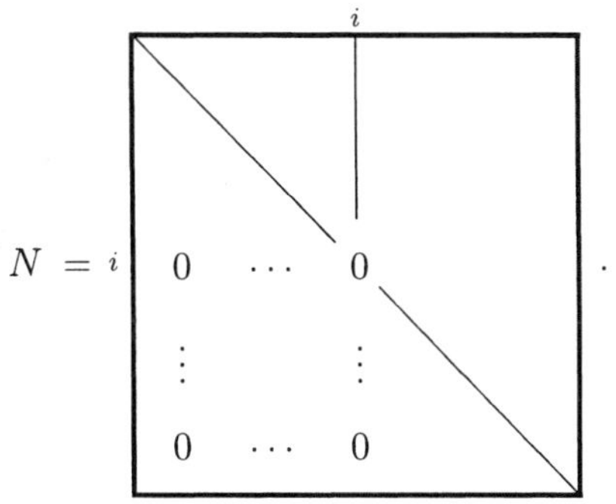

Hence, standard linear algebra shows that N is singular. Similarly, N is also singular if $u_i \leq a_i$.

For the converse, let N be singular and $\mathbf{c}$ be a non-trivial column solving the (single) homogenous system

$$N\mathbf{c} = \mathbf{o} \ .$$

For the moment, we assume that no $n+1$ consecutive coordinates of $\mathbf{c}$ are zero. Then, the variation diminishing property, see Problem 4, implies that the spline

$$\sum_{i=0}^{m} c_i N_i^n$$

has at most m zeroes in (a_0, a_{m+n+1}) if $a_0 < a_{n-1}$ or $[a_0, a_{m+n+1})$ if $a_0 = \cdots = a_{n-1}$. Thus, at least one u_i lies outside the support of the corresponding B-spline N_i^n. If

$$c_r = \cdots = c_{r+n} = 0 \ ,$$

then one can consider the splines

$$\sum_{i=0}^{r-1} c_i N_i^n \quad \text{and} \quad \sum_{i=r+n+1}^{m} c_i N_i^n$$

instead, which are zero at $u_0, \ldots, u_{r-1}$ and $u_{r+n+1}, \ldots, u_m$, respectively. ◇

Remark 6: A solution of the system $NC = P$ above is not necessarily affinely invariant. This is only guaranteed if N is regular and if the rows sum to one, i.e., if

$$N\mathbf{e} = \mathbf{e} \ , \quad \text{where} \quad \mathbf{e} = [1 \ \ldots \ 1]^t \ .$$

Namely, the row condition $N\mathbf{e} = \mathbf{e}$ implies $\mathbf{e} = N^{-1}\mathbf{e}$, which means that the $\mathbf{c}_i$ given by $C = N^{-1}P$ are affine combinations of the $\mathbf{p}_i$. The row condition is satisfied if $a_n \leq u_i < a_{m+1}$ as illustrated in Figure 6.6.

Remark 7: The matrix N is **totally positive**,, see [Karlin '68, de Boor '76a].

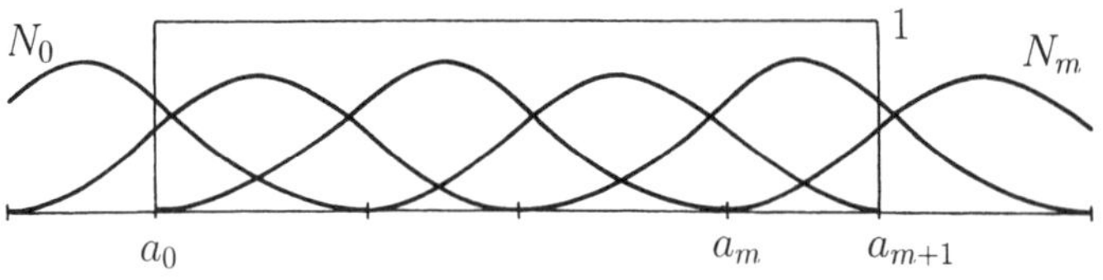

Figure 6.6: Where $N_0, \ldots, N_m$ sum to one.

6.8 Cubic spline interpolation

The collocation matrix N can easily be written down for cubic splines, $n = 3$, with the equidistant knots $a_i = i$. The B-spline $N_0^3(u)$ with its control polygon is shown in Figure 6.7.

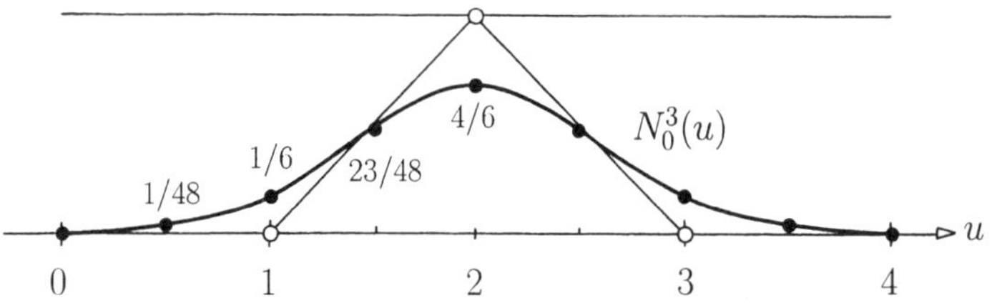

Figure 6.7: The cubic B-spline with knots $0, 1, 2, 3, 4$.

Choosing the interpolation abscissae

$$u_i = 2,\ 3,\ 4, \ldots,\ (m+1),\ (m+2)\ ,$$

the linear system (2) takes on the form

$$\frac{1}{6}\begin{bmatrix} 4 & 1 & & & & & \\ 1 & 4 & 1 & & & & \\ & & \cdot & & & & \\ & & & \cdot & & & \\ & & & & \cdot & & \\ & & & & 1 & 4 & 1 \\ & & & & & 1 & 4 \end{bmatrix} \begin{bmatrix} \mathbf{c}_0^t \\ \mathbf{c}_1^t \\ \cdot \\ \cdot \\ \cdot \\ \mathbf{c}_{m-1}^t \\ \mathbf{c}_m^t \end{bmatrix} = \begin{bmatrix} \mathbf{p}_0^t \\ \mathbf{p}_1^t \\ \cdot \\ \cdot \\ \cdot \\ \mathbf{p}_{m-1}^t \\ \mathbf{p}_m^t \end{bmatrix} .$$

However, the first and last row do not sum to one. Thus, a better choice are the abscissae

$$u_i = 3,\ (3 + \frac{1}{2}),\ 4,\ 5, \ldots, m,\ (m + \frac{1}{2}),\ (m+1)\ .$$

Then, the linear system (2) takes on the form

$$\frac{1}{6}\begin{bmatrix} 1 & 4 & 1 & & & & & \\ a & b & b & a & & & & \\ & 1 & 4 & 1 & & & & \\ & & & \cdot & & & & \\ & & & & \cdot & & & \\ & & & & & \cdot & & \\ & & & & & 1 & 4 & 1 \\ & & & & & a & b & b & a \\ & & & & & & 1 & 4 & 1 \end{bmatrix} \begin{bmatrix} \mathbf{c}_0^t \\ \mathbf{c}_1^t \\ \mathbf{c}_2^t \\ \cdot \\ \cdot \\ \cdot \\ \mathbf{c}_{m-2}^t \\ \mathbf{c}_{m-1}^t \\ \mathbf{c}_m^t \end{bmatrix} = \begin{bmatrix} \mathbf{p}_0^t \\ \mathbf{p}_1^t \\ \mathbf{p}_2^t \\ \cdot \\ \cdot \\ \cdot \\ \mathbf{p}_{m-2}^t \\ \mathbf{p}_{m-1}^t \\ \mathbf{p}_m^t \end{bmatrix},$$

where $a = 1/8$ and $b = 23/8$, as one can see from Figure 6.7.

Another possibility is to interpolate the points $\mathbf{p}_i$ at $u_i = i + 2$ for $i = 1, \ldots, m-1$, and given derivatives $\mathbf{a}$ and $\mathbf{b}$ at u_1 and u_{m+1}, respectively. The resulting linear system is

$$\frac{1}{6}\begin{bmatrix} -3 & 0 & 3 & & & \\ 1 & 4 & 1 & & & \\ & & \cdot & & & \\ & & & \cdot & & \\ & & & & \cdot & \\ & & & 1 & 4 & 1 \\ & & & -3 & 0 & 3 \end{bmatrix} \begin{bmatrix} \mathbf{c}_0^t \\ \mathbf{c}_1^t \\ \cdot \\ \cdot \\ \cdot \\ \mathbf{c}_{m-1}^t \\ \mathbf{c}_m^t \end{bmatrix} = \begin{bmatrix} \mathbf{a}^t \\ \mathbf{p}_1^t \\ \cdot \\ \cdot \\ \cdot \\ \mathbf{p}_{m-1}^t \\ \mathbf{b}^t \end{bmatrix}.$$

The solution is called a **clamped spline**. Figure 6.8 shows an example.

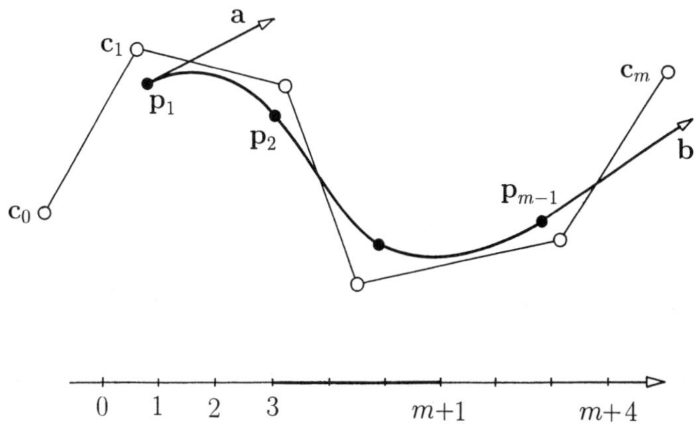

Figure 6.8: Clamped cubic spline.

In certain applications, one is interested in a **periodic spline** $\mathbf{s}(u) =$

$\sum_i \mathbf{c}_i N_i^3(u) = \mathbf{s}(u+m)$. Then, $\mathbf{c}_{i+m} = \mathbf{c}_i$, and one has to solve the cyclic linear system

$$\frac{1}{6}\begin{bmatrix} 4 & 1 & & & & 1 \\ 1 & 4 & 1 & & & \\ & & \cdot & & & \\ & & & \cdot & & \\ & & & & \cdot & \\ & & & 1 & 4 & 1 \\ 1 & & & & 1 & 4 \end{bmatrix} \begin{bmatrix} \mathbf{c}_1^t \\ \cdot \\ \cdot \\ \cdot \\ \mathbf{c}_m^t \end{bmatrix} = \begin{bmatrix} \mathbf{p}_1^t \\ \cdot \\ \cdot \\ \cdot \\ \mathbf{p}_m^t \end{bmatrix}.$$

An example is shown in Figure 6.9.

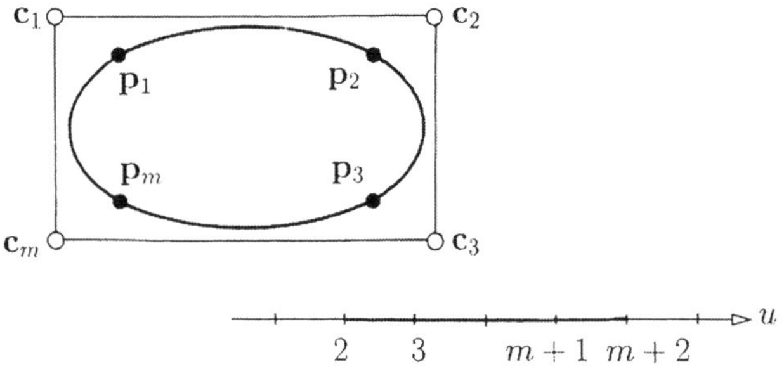

Figure 6.9: Periodic cubic spline.

6.9 Problems

1 Develop an intersection algorithm for spline curves by generalizing the algorithm given in 3.7.

2 Consider a planar cubic spline segment $\mathbf{s}(u) = \sum_{i=0}^{3} \mathbf{c}_i N_i^3(u)$ with the knots $0, 1, \ldots, 7$ for $u \in [3,4]$. For which positions of $\mathbf{c}_3$, relative to $\mathbf{c}_0, \mathbf{c}_1$ and $\mathbf{c}_2$, does the spline segment $\mathbf{s}[3,4]$ have a cusp, cf. also Problem 7 in 2.10.

3 Prove that the B-spline representation is variation diminishing in the sense of 3.8.

4 As a consequence of the variation diminishing property above, a spline function $s(u) = \sum_{i=0}^{m} c_i N_i^n$ is identically zero over some knot interval (a_i, a_{i+1}), $i \in \{0, \ldots, m+n\}$, or has at most $m+1$ zeros in I, where

$I = [a_0, a_{m+n+1})$ if $a_0 = \cdots = a_{n-1}$ and $I = (a_0, a_{m+n+1})$ otherwise [e.g., Schumaker '81, Thm 4.76].

5 Consider a polynomial curve with its nth and mth degree Bézier representation,

$$\mathbf{b}(u) = \sum_{i=0}^{n} \mathbf{b}_i B_i^n(u) = \sum_{i=0}^{m} \mathbf{c}_i B_i^m(u) \ ,$$

where $n < m$. Let $N_0^n, \ldots, N_n^n$ and $M_0^n, \ldots, M_m^n$ be the nth degree B-splines with the knots $0, 1, \ldots, n, m{+}1, m{+}2, \ldots, m{+}n{+}1$ and $0, 1, 2, \ldots, m{+}n{+}1$, respectively. Show that

$$\sum_{i=0}^{n} \mathbf{b}_i N_i^n = \sum_{i=0}^{m} \mathbf{c}_i M_i^n \ ,$$

i.e., a higher degree Bézier representation can be computed by knot insertion, see [Trump & Prautzsch '96].

6 Draw the Bézier points of the spline segments in Figures 6.8 and 6.9.

7 Let $a_0, \ldots, a_4$ be a knot sequence and let m_i be the number of knots which equal a_i. For each possible sequence $m_0, \ldots, m_4$, choose an underlying knot sequence $a_0, \ldots, a_4$ and draw the Bézier points of the corresponding cubic B-spline.

8 Derive a degree elevation algorithm for splines with the aid of symmetric polynomials, see also 3.11.

9 Let $s(u)$ be a spline function interpolating the values $p_0, \ldots, p_m$. If the rows of the collocation matrix N do not sum to one, then the spline interpolating $q_i = p_i + h$ differs from $s(u) + h$.

10 Use the identity

$$\begin{aligned} & (u_1 - u_2)[u_1 u_2 u_3 \ldots u_n]f \\ + \ & (u_2 - u_0)[u_0 u_2 u_3 \ldots u_n]f \\ + \ & (u_0 - u_1)[u_0 u_1 u_3 \ldots u_n]f = 0 \end{aligned}$$

for divided differences to derive the formula for knot deletion and knot insertion [Boehm '80].

7 Smooth curves

7.1 Contact of order r — 7.2 Arc length parametrization — 7.3 Gamma splines — 7.4 Gamma B-splines — 7.5 Nu-splines — 7.6 The Frenet frame — 7.7 Frenet frame continuity — 7.8 Osculants and symmetric polynomials — 7.9 Geometric meaning of the main theorem — 7.10 Splines with arbitrary connection matrices — 7.11 Knot insertion — 7.12 Basis splines — 7.13 Problems

There are several ways to define smoothness. Stärk's simple C^r condition establishes a very simple construction of a smooth curve continuation. More generally, a curve is said to be GC^r if it has an r times continuously differentiable parametrization. An even more general smoothness concept is based on the continuity of higher order geometric invariants. Piecewise polynomial curves with this general smoothness can be nicely studied using a geometric interpretation of symmetric polynomials.

7.1 Contact of order r

Two curves $\mathbf{p} = \mathbf{p}(s)$ and $\mathbf{q} = q(t)$, which are r times differentiable at $s = t = 0$, are said to have **contact of order** r or to have a general C^r joint at 0, short a GC^r **joint**, if $\dot{\mathbf{q}}(0) \neq \mathbf{o}$ and there exists a reparametrization $s(t)$ with $s(0) = 0$ such that $\mathbf{p}(s(t))$ and $\mathbf{q}(t)$ have identical derivatives at 0 up to order r. As before, we denote derivatives with respect to s and t by primes or dots, respectively.

Because of the chain and product rules, contact of order r at $s = t = 0$ means that

$$\begin{array}{lcccccl}
\mathbf{p} & & & & & = & \mathbf{q} \\
\mathbf{p}'\dot{s} & & & & & = & \dot{\mathbf{q}} \\
\mathbf{p}'\ddot{s} & + & \mathbf{p}''\dot{s}^2 & & & = & \ddot{\mathbf{q}} \\
\mathbf{p}'\dddot{s} & + & 3\mathbf{p}''\dot{s}\ddot{s} & + & \mathbf{p}'''\dot{s}^3 & = & \dddot{\mathbf{q}} \; . \\
 & & \vdots & & & & \vdots
\end{array}$$

In matrix notation these equations can be written as

$$[\mathbf{p}\,\mathbf{p}'\,\dots\,\mathbf{p}^{(r)}]\begin{bmatrix}1 & 0 & 0 & 0 & \cdots & 0\\ & \alpha & \beta & \gamma & \cdots & \\ & & \alpha^2 & 3\alpha\beta & \cdots & \\ & & & \alpha^3 & & \\ & & & & \ddots & \\ & & & & & \alpha^r\end{bmatrix} = [\mathbf{q}\,\dot{\mathbf{q}}\,\dots\,\overset{(r)}{\mathbf{q}}]$$

or, more succinctly, as

$$PC = Q \ ,$$

where $\alpha = \dot{s} \neq 0, \beta = \ddot{s}, \gamma = \dddot{s}, \dots$ The matrices P and Q are called **r-jets** of $\mathbf{p}$ and $\mathbf{q}$ at $s = 0$ and $t = 0$, respectively, while C is referred to as **chain rule connection matrix** of order r.

In particular, let $\mathbf{p}$ and $\mathbf{q}$ be polynomial curves of degree $n \geq r$ having the Bézier points $\mathbf{p}_i$ and $\mathbf{q}_i$, $i = 0, \dots, n$, respectively. As a consequence of the results in 2.4, the r-jets P and Q at $\mathbf{p}_n$, respectively $\mathbf{q}_0$, are determined by the last or first $r + 1$ Bézier points and vice versa. Hence, a contact of order r can be expressed in terms of Bézier points by the equation

$$[\mathbf{p}_n \ \cdots \ \mathbf{p}_{n-r}]\widetilde{C} = [\mathbf{q}_0 \ \cdots \ \mathbf{q}_r] \ ,$$

where $\widetilde{C}$ is a special upper triangular matrix with the diagonal elements $\widetilde{c_{ii}} = (-\alpha)^i \neq 0$, $i = 0, \dots, r$. Note that $\widetilde{C}$ depends on n.

Remark 1: For any given reparametrization $s(t)$ there exists an equivalent polynomial reparametrization, namely

$$s(t) = \dot{s}t + \frac{\ddot{s}}{2!}t^2 + \cdots + \frac{\overset{(r)}{s}}{r!}t^r \ .$$

In particular, any GC^1 joint can be transformed into a simple C^1 joint by an affine parameter transformation $s = \alpha t$.

Remark 2: Any GC^2 joint can be transformed into a simple C^2 joint by a quadratic transformation $s = \alpha t + (\beta/2)t^2$ and also by the projective transformation

$$s = \frac{\alpha^2 t}{\alpha - (\beta/2)t} \ ,$$

see [Degen '88].

Remark 3: If $s = t$, then $\ddot{s} = \cdots = \overset{(r)}{s} = 0$. Consequently, we obtain the simple C^r condition of Stärk illustrated in Figure 7.1 for $r = n = 3$ with the Bézier points $\mathbf{p}_i$ and $\mathbf{q}_j$ of $\mathbf{p}(s)$ and $\mathbf{q}(t)$ over $[-1, 0]$ and $[0, \alpha]$, respectively.

Remark 4: If $\beta = \ddot{s}$ changes, then $\mathbf{q}_2$ moves parallel to the tangent $\mathbf{q}_0\mathbf{q}_1$, as

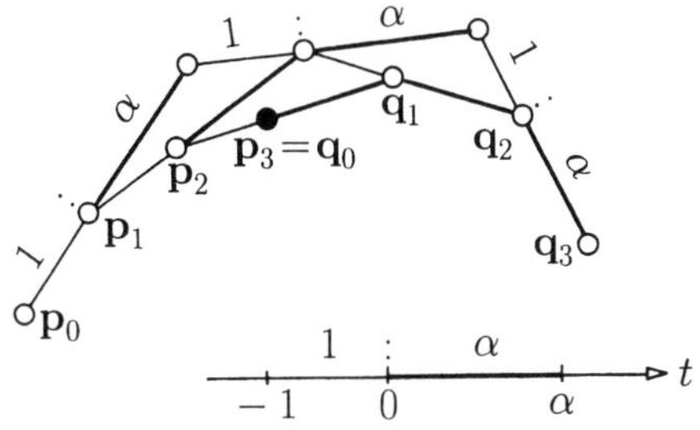

Figure 7.1: Stärk's C^3 contact.

illustrated in Figure 7.2. Thus, $\mathbf{p}$ and $\mathbf{q}$ have a GC^2 contact at $\mathbf{p}_n = \mathbf{q}_0$ if the distances g and h of $\mathbf{p}_{n-2}$ and $\mathbf{q}_2$ to the tangent $\mathbf{q}_0\mathbf{q}_1$ satisfy $g : h = 1 : \alpha^2$, see [Farin '82, Boehm '85].

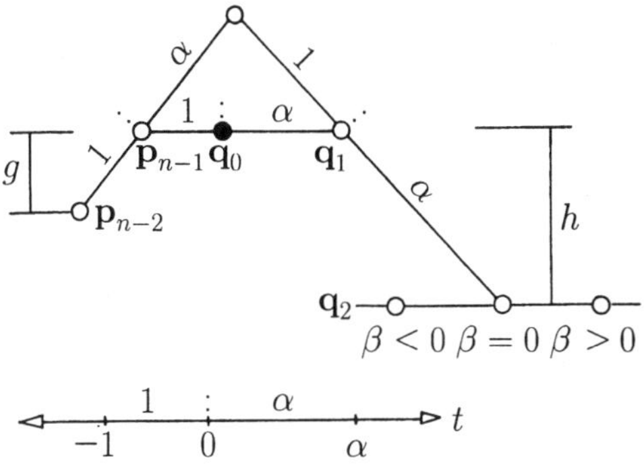

Figure 7.2: Farin's GC^2 contact.

Remark 5: If $\overset{(n)}{s}$ changes, then $\mathbf{q}_n$ moves parallel to the tangent $\mathbf{q}_0\mathbf{q}_1$.

7.2 Arc length parametrization

Let $\mathbf{x}(t)$ be a differentiable parametrization of a curve $\mathbf{x}$. Then

$$s(t) = \int_0^t \|\dot{\mathbf{x}}\| dt$$

is the **arc length** of the curve segment $\mathbf{x}[0, t]$ and

$$\|\mathbf{x}'\| = 1 \ ,$$

where $\mathbf{x}' = \frac{d}{ds}\mathbf{x}(t(s))$ denotes the first derivative with respect to the arc length and $t(s)$ the inverse to $s(t)$.

Furthermore, if $\mathbf{x}(t)$ is **regular**, which means $\dot{\mathbf{x}} \neq \mathbf{o}$, then the arc length $s(t)$, its inverse $t(s)$ and $\mathbf{x}(s) = \mathbf{x}(t(s))$ are as often differentiable as $\mathbf{x}(t)$ since $\dot{s} = \|\dot{\mathbf{x}}\|$. In other words, no parametrization is smoother than the arc length parametrization.

Hence, two regular curves $\mathbf{p}(s)$ and $\mathbf{q}(s)$ parametrized by their arc lengths have a contact of order r at $s = 0$ if and only if the composite curve

$$\mathbf{x}(s) = \begin{cases} \mathbf{p}(s) & \text{for} \quad s \leq 0 \\ \mathbf{q}(s) & \text{for} \quad s > 0 \end{cases}$$

is r times differentiable at $s = 0$. In other words,

the chain rule connection matrix is the identity matrix if both curves $\mathbf{p}(s)$ and $\mathbf{q}(t)$ are parametrized by their arc length.

7.3 Gamma-splines

Consider a piecewise cubic curve $\mathbf{s}(u)$ consisting of m segments with GC^2 contacts. Let $\mathbf{b}_{3i}, \ldots, \mathbf{b}_{3i+3}$ be the Bézier points of the ith segment, i.e.,

$$\mathbf{s}(u) = \sum_{j=0}^{3} \mathbf{b}_{3i+j} B_j^3(t) \ , \quad i = 0, \ldots, m-1 \ ,$$

for $u = a_i(1-t) + a_{i+1}t$, and $t \in [0,1]$.

In the following three sections on γ- and ν-splines, we assume that the knots a_i are chosen such that $\mathbf{s}(u)$ is everywhere continuously differentiable, which means $\mathbf{s}'(a_i-) = \mathbf{s}'(a_i+)$ or

$$(\mathbf{b}_{3i+1} - \mathbf{b}_{3i})\Delta_{i-1} = (\mathbf{b}_{3i} - \mathbf{b}_{3i-1})\Delta_i \ ,$$

where $\Delta_j = a_{j+1} - a_j$, as illustrated in Figure 3.8.

Since the segments of $\mathbf{s}$ have GC^2 contacts, the four points $\mathbf{b}_{3i-2}, \mathbf{b}_{3i-1}, \mathbf{b}_{3i+1}, \mathbf{b}_{3i+2}$ form a planar quadrilateral. This quadrilateral has outer vertices $\mathbf{c}_i$ and $\mathbf{d}_i$, as illustrated in Figure 7.3 for $i = 1$, and it is called the **generalized** or **γ-A-frame**.

A comparison with Figure 7.2 shows that there is a non-zero number γ such that

$$\varepsilon = \gamma\Delta_0 \quad \text{and} \quad \delta = \gamma\Delta_1 \ .$$

A piecewise cubic curve whose joints satisfy the above GC^2 condition with finite γ's is called a **γ-spline**. Any γ-spline is given by control points $\mathbf{c}_i$ with

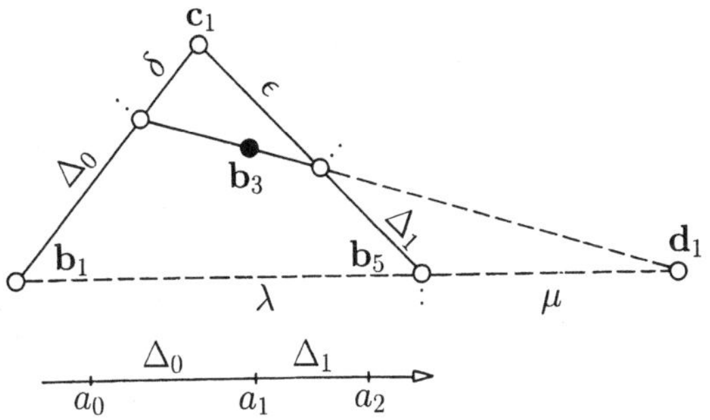

Figure 7.3: Generalized A-frame.

associated numbers $\gamma_i \neq 0$ and knots a_i [Boehm '85]. The construction of its Bézier representation is illustrated in Figure 7.4.

Note that γ may be negative.

Remark 6: The ratio $\lambda : \mu$ shown in Figure 7.3 is equal to $(\Delta_0^2 - \Delta_1^2) : \Delta_1^2$. The points $\mathbf{c}_1$ or $\mathbf{d}_1$ lie on the ideal line, i.e., they are infinitely far away, if $\gamma = \infty$ or $\mu = \infty$, respectively. Further the ratio $\lambda : \mu$ does not depend on γ.

Remark 7: In 7.7 it is shown that a GC^2 curve is curvature continuous.

Remark 8: A γ-spline for which all γ's equal 1 is a simple cubic C^2 spline, as discussed in 5.1, Figure 5.2.

7.4 Gamma B-splines

A γ-spline $\mathbf{s}(u)$ depends affinely on its control points $\mathbf{c}_i$. Therefore, it can be represented as an affine combination,

$$\mathbf{s}(u) = \sum_i \mathbf{c}_i M_i(u) \ ,$$

in which $M_i(u)$ is the basis γ-spline with the knots $a_i, \ldots, a_{i+3}$ and associated γ-values $\gamma_i, \ldots, \gamma_{i+3}$. The Bézier ordinates of M_i can easily be obtained by the construction 7.3 above with dimension one and $c_i = 1$ while all other control ordinates are zero. The Bézier abscissae subdivide the knot intervals uniformly, as shown in Remark 10 in 2.8.

The construction of the γ-B-spline M_2 is shown in Figure 7.5.

Remark 9: The γ-B-splines are non-negative only if all γ's are positive. If some γ_i is negative, there are γ-B-splines with negative values, and a γ-spline

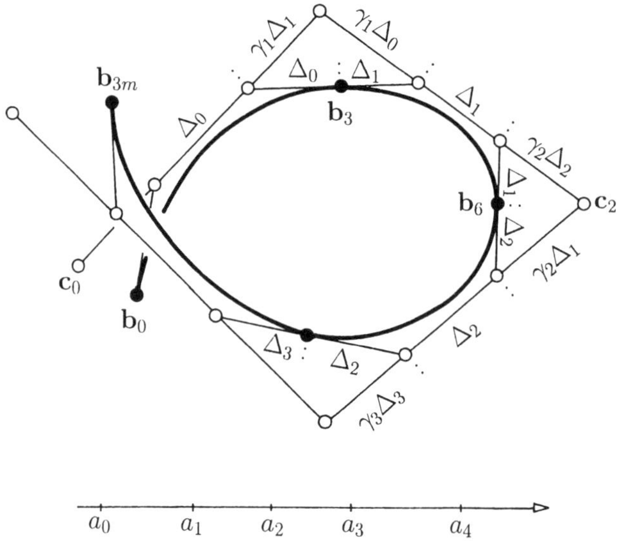

Figure 7.4: A γ-spline.

built with these γ-B-splines does, in general, not lie in the convex hull of its control points $\mathbf{c}_i$.

Remark 10: In the same way, **torsion continuous quartic splines** can be developed [Boehm '87].

7.5 Nu-splines

Since a γ-spline $\mathbf{s}(u)$ is GC^2, there are, according to 7.1, numbers ν_i such that

(1) $$\mathbf{s}''(a_i+) = \mathbf{s}''(a_i-) + \nu_i \mathbf{s}'(a_i-) \ .$$

In particular, for any j we have

$$M_j''(a_i+) = M_j''(a_i-) + \nu_i M_j'(a_i-) \ .$$

Using the Bézier representation of the M_j, it is easy to calculate the ν_i [Boehm '85]:

$$\nu_i = 2\left(\frac{1}{\Delta a_{i-1}} + \frac{1}{\Delta a_i}\right)\left(\frac{1}{\gamma_i} - 1\right) \ .$$

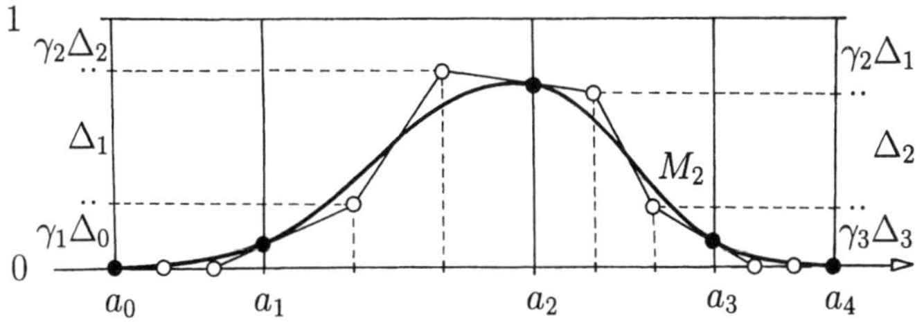

Figure 7.5: A γ-B-spline.

Piecewise cubic GC^2 splines satisfying condition (1) for given **tensions** ν_i at the knots a_i are called **ν-splines**. They were introduced by Nielson in 1974 to interpolate given points $\mathbf{s}(a_i)$ in the plane by more or less tighter curves, see [Nielson '74].

7.6 The Frenet frame

More generally, the geometric smoothness in spaces of higher dimension is based on the Frenet frame of a composite curve. For a derivation it is convenient to use the arc length parametrization $\mathbf{x}(s)$ of a regular curve in $\mathbb{R}^d$ and to assume that the derivatives up to order d are linearly independent, i.e., span $\mathbb{R}^d$. Consequently, $\mathbf{x}(s)$ also spans $\mathbb{R}^d$ since, otherwise, the derivatives would span a proper subspace only.

The **Frenet frame** of $\mathbf{x}(s)$ is a positively oriented orthogonal system $\mathbf{f}_1, \ldots, \mathbf{f}_d$ obtained from $\mathbf{x}', \ldots, \mathbf{x}^{(d)}$ by the Gram-Schmidt process such that $\mathbf{f}_i^t \mathbf{x}^{(i)} > 0$ for $i = 1, \ldots, d-1$. The vector $\mathbf{f}_1 = \mathbf{x}'$ is the **tangent vector** of $\mathbf{x}$. For $d = 3$, one calls $\mathbf{f}_2$ the **principal normal vector** and $\mathbf{f}_3$ the **binormal vector** of $\mathbf{x}$, see Figure 7.6 for an illustration.

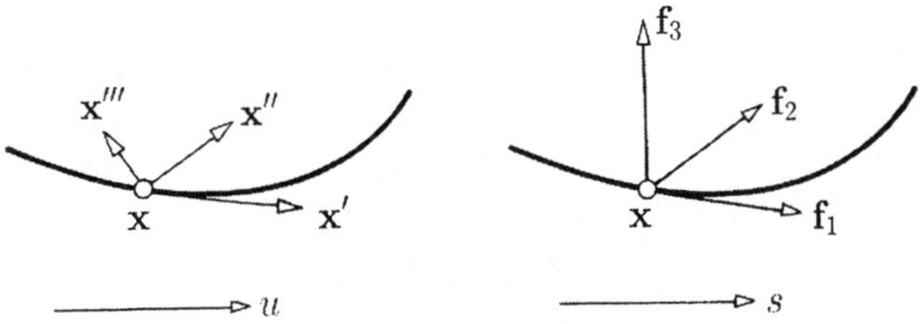

Figure 7.6: Frenet frame.

The local change of the Frenet frame as a function of the arc length s is given by the so-called **Frenet-Serret formulae**

$$[\mathbf{f}_1' \ldots \mathbf{f}_d'] = [\mathbf{f}_1 \ldots \mathbf{f}_d] \begin{bmatrix} 0 & -\kappa_1 & & & \\ \kappa_1 & 0 & -\kappa_2 & & \\ & \kappa_2 & 0 & & \\ & & & \ddots & \\ & & & & -\kappa_{d-1} \\ & & & \kappa_{d-1} & 0 \end{bmatrix} ,$$

with the quantities κ_i given by,

$$\begin{aligned} \kappa_1 &= \mathbf{f}_2^t \mathbf{x}'' , \\ \kappa_1 \kappa_2 &= \mathbf{f}_3^t \mathbf{x}''' , \\ \kappa_1 \kappa_2 \kappa_3 &= \mathbf{f}_4^t \mathbf{x}'''' , \\ & \text{etc.,} \end{aligned}$$

see, for example, [Carmo '76, Klingenberg '78], or, equivalently, by

$$\begin{aligned} \kappa_1 &= \mathrm{vol}_2[\mathbf{x}'\mathbf{x}''] , \\ \kappa_1^2 \kappa_2 &= \mathrm{vol}_3[\mathbf{x}'\mathbf{x}''\mathbf{x}'''] , \\ \kappa_1^3 \kappa_2^2 \kappa_3 &= \mathrm{vol}_4[\mathbf{x}'\mathbf{x}''\mathbf{x}'''\mathbf{x}''''] , \\ & \text{etc.} \end{aligned}$$

The non-zero quantities κ_i are called the **curvatures** and are **geometric invariants** of the curve $\mathbf{x}$. The first, κ_1, is the usual **curvature** and the second, κ_2, is the usual **torsion** of $\mathbf{x}$. For a geometric interpretation of κ_1 and κ_2, see Problems 1, 2 and 3.

7.7 Frenet frame continuity

A continuous curve $\mathbf{x}(s)$ in $\mathbb{R}^d$ is called **Frenet frame continuous** of order r, abbreviated by F^r, if the first r Frenet vectors $\mathbf{f}_1, \ldots, \mathbf{f}_r$ and the first $r-1$ curvatures $\kappa_1, \ldots, \kappa_{r-1}$ of $\mathbf{x}$ are continuous. Note that Frenet frame continuity of order r is only defined if $r \leq d$.

Let $\mathbf{x}_-', \ldots, \mathbf{x}_-^{(r)}$ and $\mathbf{x}_+', \ldots, \mathbf{x}_+^{(r)}$ be the left and right hand side derivatives of $\mathbf{x}$ at $s = s_0$. Then one has that

the curve $\mathbf{x}$*, parametrized by its arc length* s*, is Frenet frame* *continuous of order r at $s = s_0$ if and only if*

$$[\mathbf{x}'_- \ldots \mathbf{x}^{(r)}_-]C = [\mathbf{x}'_+ \ldots \mathbf{x}^{(r)}_+] \ , \tag{2}$$

where C is an arbitrary upper triangular connection matrix whose diagonal entries are all one.

Proof: Let $[\mathbf{f}_1 \ldots \mathbf{f}_r]_-$ and $[\mathbf{f}_1 \ldots \mathbf{f}_r]_+$ be the matrices of the left- and right-hand side Frenet vectors and let κ_i^-, and κ_i^+ be the left- and right-hand side curvatures of $\mathbf{x}$ at s_0. From 7.6 it follows that

$$[\mathbf{x}'_- \ldots \mathbf{x}^{(r)}_-] = [\mathbf{f}_1 \ldots \mathbf{f}_r]_- U_- \ ,$$

where U_- is an upper triangular matrix whose diagonal entries are

$$1, \kappa_1^-, \ldots, (\kappa_1^- \cdots \kappa_{r-1}^-) \ .$$

An analogous equation holds for the right-hand side derivatives. Hence, equation (2) holds if and only if

$$[\mathbf{f}_1 \ldots \mathbf{f}_r]_- U_- C U_+^{-1} = [\mathbf{f}_1 \ldots \mathbf{f}_r]_+ \ , \tag{3}$$

where $U_- C U_+^{-1}$ is an upper triangular matrix whose diagonal entries are

$$1, (\kappa_1^-/\kappa_1^+), \ldots, (\kappa_1^- \cdots \kappa_{r-1}^-/\kappa_1^+ \cdots \kappa_{r-1}^+) \ .$$

Since the Frenet vectors are orthogonal, (3) holds only if $U_- C U_+^{-1}$ is the identity matrix. Thus, (3) and (2) hold if and only if the curvatures and the Frenet vectors are continuous, which concludes the proof. ◇

According to 7.1, the r-jets of a curve with respect to different parametrizations are related by an upper triangular matrix whose diagonal entries are of the form $\alpha^0, \alpha^1, \ldots, \alpha^r$, where $\alpha \neq 0$. Hence, on reparametrizing $\mathbf{x}_+$ and $\mathbf{x}_-$, we can rephrase condition (2) in the following way. Let $\mathbf{x}(t)$ be any parametrization of the curve $\mathbf{x}$ with linearly independent derivatives up to order r to the left- and right-hand side of some $t = t_0$. Then we get:

The curve $\mathbf{x}(t)$ is Frenet frame continuous of order r at $t = t_0$ if and only if

$$[\dot{\mathbf{x}}_- \ldots \overset{(r)}{\mathbf{x}}_-]\overline{C} = [\dot{\mathbf{x}}_+ \ldots \overset{(r)}{\mathbf{x}}_+] \ ,$$

where $\overline{C}$ is an arbitrary upper triangular matrix whose diagonal entries are of the form $\alpha, \alpha^2, \ldots, \alpha^r$ with $\alpha > 0$.

Remark 11: Contact of order r implies Frenet frame continuity. The converse is in general only true for $r \leq 2$.

Remark 12: A planar curve in 3-space is torsion continuous.

Remark 13: Although the curvatures κ_i are only Euclidean invariants, Frenet frame continuity is affinely and even projectively invariant.

Remark 14: Contact of order r is invariant under projection. However, Frenet frame continuity of order r is invariant under projections only if the projected curve has linearly independent derivatives up to order r.

Remark 15: Consider two polynomial curves $\mathbf{p}(t)$ and $\mathbf{q}(t)$ in $\mathbb{R}^d$ such that

$$\mathbf{p}(0) = \mathbf{q}(0) \quad \text{and} \quad -\alpha\mathbf{p}'(0) = \mathbf{q}'(0) \ ,$$

with $\alpha > 0$. Let $\mathbf{p}_0, \ldots, \mathbf{p}_n$ and $\mathbf{q}_0, \ldots, \mathbf{q}_n$ be the respective Bézier points of $\mathbf{p}$ and $\mathbf{q}$ over $[0,1]$. Then $\mathbf{p}$ and $\mathbf{q}$ have the same Frenet frame and curvatures at $t = 0$ if the spaces spanned by $\mathbf{p}_0, \ldots, \mathbf{p}_{i-1}$ divide $\mathbf{p}_i$ and $\mathbf{q}_i$ in the ratios

$$1 : \alpha^i \quad \text{for even} \quad i \quad \text{and} \quad -1 : \alpha^i \quad \text{for odd} \quad i \ ,$$

see [Boehm '87]. Figure 7.7 gives an illustration for $d = 3$.

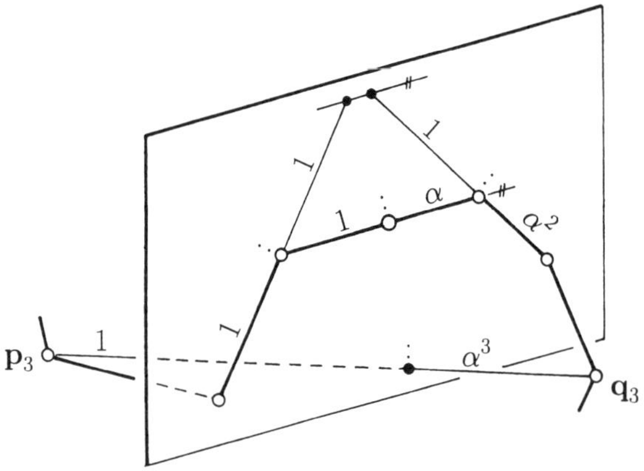

Figure 7.7: Frenet frame continuity due to Boehm.

7.8 Osculants and symmetric polynomials

For Frenet frame continuous piecewise polynomial curves there are also control polygons and a knot insertion algorithm as for simple C^r splines. Both can be obtained by a geometric derivation and interpretation of the main theorem 5.5, as shown in the sequel.

Let $\mathbf{p}(u)$ be a polynomial curve of degree n. Then

$$\mathbf{p}_a(u) = \mathbf{p}(u) + \frac{1}{n}(a-u)\mathbf{p}'(u)$$

is a curve of degree $n-1$ in u, called the **first osculant** of $\mathbf{p}$ at the knot a, see Figure 7.8 for an illustration.

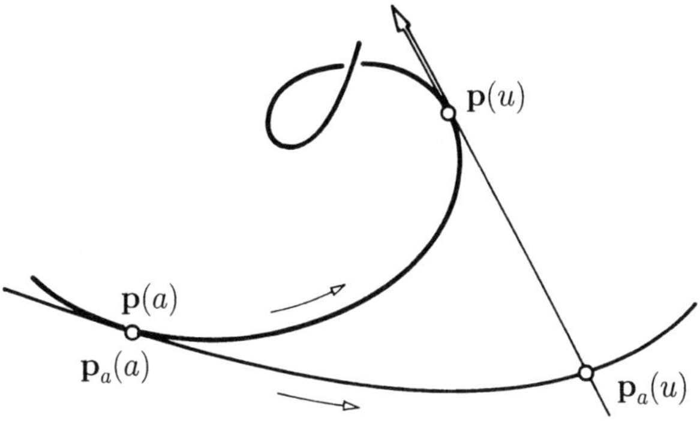

Figure 7.8: The first osculant of a curve.

The **second osculant** of $\mathbf{p}$ is obtained by forming the first osculant of $\mathbf{p}_a$. Repeating this process, one obtains further osculants of $\mathbf{p}$. The ith osculant of $\mathbf{p}$ at the knots $a_1, \ldots, a_i$ is written as

$$\mathbf{p}_{a_1 \ldots a_i} = (\mathbf{p}_{a_1})_{a_2 \ldots a_i} \ .$$

These osculants have the following properties, see Problem 7.

- *The* **diagonal** *of the* n*th osculant agrees with the curve,*

$$\mathbf{p}_{a \overset{n}{\cdots} a} = \mathbf{p}(a) \ .$$

- *Osculants are* **symmetric** *with respect to their knots*

$$\mathbf{p}_{ab}(u) = \mathbf{p}_{ba}(u) \ .$$

- *Osculants are* **affine** *in their knots. For* $a = (1-\alpha)c + \alpha d$ *one has*

$$\mathbf{p}_a = (1-\alpha) \cdot \mathbf{p}_c + \alpha \cdot \mathbf{p}_d \ .$$

Due to 3.1, these three properties characterize the nth osculant of $\mathbf{p}$ as the polar form of $\mathbf{p}$, which means that

$$\mathbf{p}_{u_1 \ldots u_n} = \mathbf{p}[u_1 \; \ldots \; u_n] \; .$$

Let the derivatives $\mathbf{p}^{(1)}, \ldots, \mathbf{p}^{(n)}$ be linearly independent, then $\mathbf{p}, \mathbf{p}^{(1)}, \ldots, \mathbf{p}^{(r)}$ span an r-dimensional space, denoted by $\mathcal{P}_u^r$, see also Problem 10. It is called the **rth osculating flat** of $\mathbf{p}$ at u. From 3.9 it follows that

$$\mathbf{p}_a^{(r)}(a) = \frac{n-r}{n}\mathbf{p}^{(r)}(a) \; .$$

Consequently, $\mathbf{p}$ and its first osculant have the same rth osculating flat at $u = a$ for $r = 0, \ldots, n-1$.

7.9 Geometric meaning of the main theorem

Let $\mathbf{s}(u) = \sum \mathbf{c}_i N_i^n$ be an ordinary spline with simple knots a_i, $i \in \mathbb{Z}$, and let $\mathbf{s}^j$ be the polynomial that agrees with $\mathbf{s}$ over the interval $[a_j, a_{j+1}]$. From the main theorem in 5.5 we recall that

$$\mathbf{c}_i = \mathbf{s}^j[a_{i+1} \; \ldots \; a_{i+n}] = \mathbf{s}^j_{a_{i+1} \ldots a_{i+n}}$$

for $j = i, \ldots, i+n$.

Further, let $\mathcal{S}_j^k$ be the kth osculating flat of $\mathbf{s}$ at a_j. Since the first osculant $\mathbf{s}^j_{a_j}$ spans $\mathcal{S}_j^{n-1}$, and since osculants are symmetric in their knots, we obtain the following interpretation of the main theorem.

> *The control point $\mathbf{c}_i$ lies in the osculating flats $\mathcal{S}_{i+1}^{n-1}, \ldots, \mathcal{S}_{i+n}^{n-1}$.*

Moreover,

> *any n consecutive osculating flats $\mathcal{S}_{i+1}^{n-1}, \ldots, \mathcal{S}_{i+n}^{n-1}$ intersect in precisely the control point $\mathbf{c}_i$ if all spaces $\mathcal{S}_j^n$ spanned by the segments $\mathbf{s}^j$ are n-dimensional.*

Namely, any $n+1$ consecutive control points $\mathbf{c}_{j-n}, \ldots, \mathbf{c}_j$ span $\mathcal{S}_j^n$. Hence, any $n+1$ consecutive control points are independent, which implies that the intersections $\mathcal{S}_{i+1}^{n-1} \cap \ldots \cap \mathcal{S}_{i+k}^{n-1}$ are spanned by $\mathbf{c}_{i-n+k}, \ldots, \mathbf{c}_i$. In particular, $n+1$ osculating flats $\mathcal{S}_i^{n-1}, \ldots, \mathcal{S}_{i+n}^{n-1}$ have an empty intersection.

Remark 16: The geometric meaning of the main theorem provided above holds also in case of multiple knots if we interpret a k-fold intersection $\mathcal{S}_j^{n-1} \cap \overset{k}{\ldots} \cap \mathcal{S}_j^{n-1}$ as the (n-k)th osculating flat $\mathcal{S}_j^{n-k}$.

Remark 17: Let $\mathbf{p}(u)$ be an n-dimensional polynomial curve of degree n and let $\mathcal{P}_u^k$ be its kth osculating flat at u. Since $\mathbf{p}(u)$ can be viewed as a spline with any arbitrary knot sequence with n-dimensional segments, any $m \leq n+1$, possibly coalescing, osculating flats $\mathcal{P}_{a_1}^{n-1}, \ldots, \mathcal{P}_{a_m}^{n-1}$ intersect in a space of dimension $n-m$. Otherwise, $n+1$ flats $\mathcal{P}_u^{n-1}$ would not have an empty intersection.

Remark 18: Because of Remark 16, there are at most m, possibly coalescing, osculating flats $\mathcal{P}_{a_1}^{n-1}, \ldots, \mathcal{P}_{a_m}^{n-1}$ whose intersection contains any given subspace of dimension $n-m$.

Remark 19: Moreover, it can be shown that the intersection of any osculating flat $\mathcal{P}_u^{n-r}$ with an m-dimensional subspace is of dimension $m-r$, except for at most finitely many n, see [Prautzsch '02].

Remark 20: The geometric approach to B-splines discussed above can be used in a more general form also for Tchebycheffian splines,
see [Pottmann '93, Mazure & Pottmann '96].

Remark 21: The Bézier points $\mathbf{b}_i$ of a cubic curve $\mathbf{p}(u)$ spanning $\mathbb{R}^3$ are given by the 3rd osculants

$$\begin{array}{lllll} \mathbf{b}_0 & = & \mathbf{p}_{000} & = & \mathcal{P}_0^0 \ , \\ \mathbf{b}_1 & = & \mathbf{p}_{001} & = & \mathcal{P}_0^1 \cap \mathcal{P}_1^2 \ , \\ \mathbf{b}_2 & = & \mathbf{p}_{011} & = & \mathcal{P}_0^2 \cap \mathcal{P}_1^1 \quad \text{and} \\ \mathbf{b}_3 & = & \mathbf{p}_{111} & = & \mathcal{P}_1^0 \ , \end{array}$$

see Figure 7.9 for an illustration.

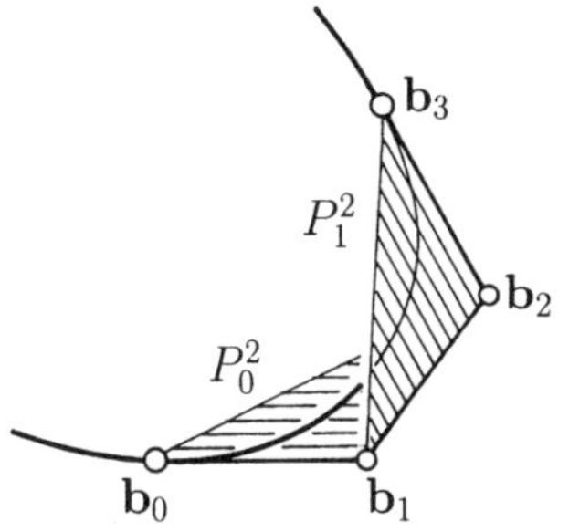

Figure 7.9: Osculating flats.

7.10 Splines with arbitrary connection matrices

Let a_i be simple knots such that $a_i < a_{i+1}$. Further, let $\mathbf{s}(u)$ be a continuous curve that is polynomial of degree n over each knot interval $[a_i, a_{i+1}]$ and whose left- and right-hand side derivatives up to order $n-1$ at the knots are related by arbitrary non-singular connection matrices. Thus, the curve $\mathbf{s}$ has a well-defined $(n-1)$th osculating flat at each knot a_i, denoted by $\mathcal{S}_i$ or $\mathcal{S}_{a_i}$, but it need not be Frenet frame continuous!

Further, we assume that the polynomial segments of $\mathbf{s}$ span n-dimensional spaces and introduce the notation

$$\mathbf{s}[a_{i+1} \ \dots \ a_{i+n}] = \mathcal{S}_{i+1} \cap \dots \cap \mathcal{S}_{i+n}$$

for the intersection of n osculating flats at consecutive knots.

Because of Remark 18 in 7.9, any two consecutive flats $\mathcal{S}_{j+1}$ and $\mathcal{S}_{j+2}$ intersect in an $(n-2)$-dimensional subspace of $\mathcal{S}_{j+2}$. Consequently, it follows by successive applications of Remark 19 in 7.9, that, in general, m consecutive osculating flats

$$\mathcal{S}_{j+1}, \dots, \mathcal{S}_{j+m}$$

intersect in an $(n-m)$-dimensional subspace of $\mathcal{S}_{j+m}$. Thus, in general, $\mathbf{s}[a_{i+1} \ \dots \ a_{i+n}]$ is a point. It may be at infinity, though, see Problem 9.

Since the properties of $\mathbf{s}$ assumed in the beginning also hold with any finer knot sequence, we can extend the definition of $\mathbf{s}[a_{i+1} \ \dots \ a_{i+n}]$ to any n consecutive knots $x_1, \dots, x_n$ of a refinement of the given knot sequence (a_i). This n-variate function $\mathbf{s}[x_1 \ \dots \ x_n]$ has the following three important properties, which follow directly from its definition.

- $\mathbf{s}[x_1 \ \dots \ x_n]$ *agrees with* $\mathbf{s}(u)$ *on its* **diagonal**, *i.e.*,

$$\mathbf{s}[u \ \dots \ u] = \mathbf{s}(u) \ .$$

- $\mathbf{s}[x_1 \ \dots \ x_n]$ *is* **symmetric** *in its variables, i.e., for any permutation* $(y_1, \dots, y_n)$ *of* $(x_1, \dots, x_n)$ *one has*

$$\mathbf{s}[y_1 \ \dots \ y_n] = \mathbf{s}[x_1 \ \dots \ x_n] \ .$$

- $\mathbf{s}[x_1 \ \dots \ x_n]$ *is a* **piecewise rational function** *of its variables. If, e.g.,* $x_2, \dots, x_n$ *are fixed, then* $\mathbf{s}[x\, x_2 \ \dots \ x_n]$ *lies on the line*

$$\mathcal{S}_{x_2} \cap \dots \cap \mathcal{S}_{x_n} \ .$$

Generalizing the main theorem in 5.5, we call the points $\mathbf{c}_i = \mathbf{s}[a_{i+1} \ \dots \ a_{i+n}]$ the **control points** of $\mathbf{s}$ [Seidel '92].

Remark 22: If the connection matrices are all totally positive, then any n consecutive flats $\mathcal{S}_{i+1}, \dots, \mathcal{S}_{i+n}$ intersect always only in a single point, see [Dyn & Micchelli '88].

7.11 Knot insertion

Consider a piecewise polynomial curve $\mathbf{s}(u)$ of degree n with the knots a_i, as in 7.10, such that any n consecutive flats $\mathcal{S}_{j+1}, \ldots, \mathcal{S}_{j+n}$ intersect in a single point. Let $\hat{a} \in [a_j, a_{j+1})$ be a new knot, and let

$$\hat{a}_i = \begin{cases} a_i & \text{for} \quad i \leq j \\ \hat{a} & \text{for} \quad i = j+1 \\ a_{i-1} & \text{for} \quad i \geq j+2 \end{cases}$$

be the knots of the refined knot sequence. Using the properties of the generalized osculant, $\mathbf{s}[x_1 \ldots x_n]$, we can compute the new control points $\hat{\mathbf{c}}_i = \mathbf{s}[\hat{a}_{i+1} \ldots \hat{a}_{i+n}]$ from the initial points $\mathbf{c}_i = \mathbf{s}[a_{i+1} \ldots a_{i+n}]$. Namely, it follows from 7.10

$$\hat{\mathbf{c}}_i = \begin{cases} \mathbf{c}_i & \text{for} \quad i \leq j-n \\ \mathbf{c}_{i-1}(1-\alpha_i) + \mathbf{c}_i\alpha_i & \text{for} \quad i = j-n+1, \ldots, j \\ \mathbf{c}_{i-1} & \text{for} \quad i \geq j+1 \end{cases} ,$$

where α_i is some piecewise rational function of $\hat{a}$, see Problem 6.

In particular, if we insert the knot $\hat{a}$ altogether n times, we obtain a **generalized de Boor algorithm** for the computation of $\mathbf{s}[\hat{a} \ldots \hat{a}] = \mathbf{s}(\hat{a})$. The only difference is that the weights α_i do not depend linearly on $\hat{a}$ in general.

This generalized de Boor algorithm works only if all intersections $\mathbf{s}[a_{j+1} \ldots a_{j+k}\ \hat{a} \overset{n-k}{\ldots} \hat{a}] = \mathcal{S}_{j+1} \cap \ldots \cap \mathcal{S}_{j+k} \cap$ (kth osculating flat at $\hat{a}$), needed for the computation of $\mathbf{s}[\hat{a} \ldots \hat{a}]$, are points. However, because of Remark 18, these intersections can fail to be points for at most a finite number of knots $\hat{a}$.

Remark 23: The control points $\mathbf{c}_i$, $i = 0, \ldots, m$, define the spline $\mathbf{s}(u)$ for all $u \in [a_n, a_{m+1}]$. Therefore, if one considers $\mathbf{s}$ only over $[a_n, a_{m+1}]$, it is convenient to assume n-fold end knots, i.e.,

$$a_0 = \cdots = a_{n-1} \qquad \text{and} \qquad a_{m+1} = \cdots = a_{m+n+1} .$$

7.12 Basis splines

So far, we have assumed that the curve $\mathbf{s}(u)$ has linearly independent derivatives up to order $n-1$. Even if this is not the case, control points and a knot insertion algorithm exist.

Let $\mathbf{a}$ be a sequence of simple knots $a_0, \ldots, a_m$, and let C be a sequence of $(n-1) \times (n-1)$ non-singular connection matrices $C_1, \ldots, C_{m-1}$. A continuous piecewise polynomial curve $\mathbf{s}(u)$ is called a spline of degree n over $\mathbf{a}$ and C if it is polynomial of degree n over each knot interval $(a_i, a_{i+1}), i = 0, \ldots, m-1$, and if at $u = a_1, \ldots, a_{m-1}$, the left and right hand derivatives are related by

$$[\mathbf{s}_-^{(1)} \ldots \mathbf{s}_-^{(n-1)}]C_i = [\mathbf{s}_+^{(1)} \ldots \mathbf{s}_+^{(n-1)}] \ .$$

Let $\mathbf{b}_0, \ldots, \mathbf{b}_{mn}$ be the Bézier points of some spline $\mathbf{s}_{\text{norm}}(u)$ of degree n over $\mathbf{a}$ and C and assume that the points $\mathbf{b}_0, \ldots, \mathbf{b}_n, \mathbf{b}_{2n}, \mathbf{b}_{3n}, \ldots, \mathbf{b}_{mn}$ span an mn-dimensional space.

The other Bézier points are determined by the connection conditions. Thus, $\mathbf{s}_{\text{norm}}$ spans the same mn-dimensional space, and any other spline $\mathbf{s}$ of degree n over $\mathbf{a}$ and C is an affine image of $\mathbf{s}_{\text{norm}}$. Therefore, $\mathbf{s}_{\text{norm}}$ is a **normal curve** for the space of all splines over $\mathbf{a}$ and C. It is also called a universal spline in [Seidel '92]. The image of the control polygon of $\mathbf{s}_{\text{norm}}$ is the control polygon of $\mathbf{s}$. This shows that for any spline $\mathbf{s}$ over $\mathbf{a}$ and C, there is a control polygon with a knot insertion algorithm and shows also that the weights α_i in 7.11 depend only on $\mathbf{a}$, C and $\hat{a}$, but not on the particular spline.

Furthermore, the splines over $\mathbf{a}$ and C with control points $0, \ldots, 0, 1, 0, \ldots, 0$ form a basis for the space of all splines over $\mathbf{a}$ and C. The generalized de Boor algorithm implies that these basis splines have the same supports as the ordinary B-splines over $\mathbf{a}$.

Remark 24: If it is impossible to define some control points of a normal spline over $\mathbf{a}$ and C by intersecting n osculating flats, one can move the knots a_i slightly such that the osculating flats at all n consecutive knots intersect in exactly one point, see Remark 18 in 7.9. See also Problems 13, 14 and 15.

7.13 Problems

1 Consider a curve $\mathbf{x}(t)$ with curvature $\kappa(t) \neq 0$. Show that the circles contacting $\mathbf{x}$ with order 2 have radius $1/\kappa$.

2 Consider a curve $\mathbf{x}(s)$ parametrized by its arc length s. Its tangent vector $\mathbf{t}(s)$ describes a curve on the unit sphere. Thus, the arc length of $\mathbf{t}$, here denoted by $\alpha(s)$, represents the angle by which $\mathbf{t}$ turns along the curve. Show that the curvature κ of $\mathbf{x}$ equals the angular velocity of $\mathbf{t}$ with respect to s, i.e., $\alpha' = \kappa$.

3 Show that the torsion of a curve in $\mathbb{R}^3$ equals the angular velocity of the binormal with respect to s.

4 Convert the ν- into the γ-spline representation, i.e., express the γ_i in terms of the ν_i, see 7.5.

5 A γ- or ν-spline over the knots $a_0, \ldots, a_{m+2}$ with corresponding ν-values $\nu_1, \ldots, \nu_{m+1}$ and γ-values $\gamma_1, \ldots, \gamma_{m+1}$ also is a γ-spline over a finer knot sequence $\hat{a}_0 = a_0, \ldots, \hat{a}_i = a_i, \hat{a}_{i+1}, \hat{a}_{i+2} = a_{i+1}, \ldots, \hat{a}_{m+3} = a_{m+2}$, with corresponding ν-values $\hat{\nu}_1, \ldots, \hat{\nu}_{m+2}$ and γ-values $\hat{\gamma}_1, \ldots, \hat{\gamma}_{m+2}$. Express the $\hat{\nu}_i$ and $\hat{\gamma}_i$ in terms of the ν_i and γ_i, respectively.

6 In 7.3 it is shown how to obtain the spline control polygon of a γ-spline $\mathbf{s}(u)$ from its Bézier polygon. Subdivide the Bézier polygon of one segment of $\mathbf{s}(u)$ and construct the corresponding spline control polygon. Express the new control points as affine combinations of the old ones, cf. [Boehm '85].

7 Verify that osculants are affine and symmetric in their knots.

8 Consider a periodic γ-spline $\mathbf{s}(u) = \mathbf{s}(u+4)$ over the knots $0, 1, 2, 3, 4$ with control points $[1\ 1]^t$, $[-1\ 1]^t$, $[-1\ -1]^t$, $[1\ -1]^t$ and equal γ-values $\gamma = \gamma_1 = \gamma_2 = \gamma_3 = \gamma_4$. For which γ does $\mathbf{s}(u)$ interpolate the circle with radius $\rho < \sqrt{2}$ around the origin at $u = 1, 2, 3, 4$? For which γ does $\mathbf{s}(u)$ interpolate the circle also at $u = 1/2, 3/2, 5/2, 7/2$?

9 Come up with two cubics $\mathbf{p}(u)$ and $\mathbf{q}(u)$ with the same Frenet frame at $u = 0$ such that the osculating planes of $\mathbf{p}$ and $\mathbf{q}$ at $u = 1$ are different and parallel.

10 Let $\mathbf{p}(u)$ be a polynomial curve of degree n spanning $\mathbb{R}^n$. Show that every ith osculating flat of $\mathbf{p}$ has dimension i for $i = 0, 1, \ldots, n$.

11 Come up with a cubic spline $\mathbf{s}$ with simple knots and planar polynomial segments such that each control point forms the intersection of 3 consecutive osculating planes $\mathcal{S}_j^2$.

12 Let $\mathbf{p}(u)$ be a polynomial curve of degree n spanning $\mathbb{R}^n$, and let $\mathcal{P}_u^k$ be its kth osculating flat at u. Show that any intersection $\mathcal{P}_a^{n-i} \cap \mathcal{P}_b^{n-j}$ converges to $\mathcal{P}_a^{n-i+j}$, as b converges to a.

13 Consider the cubic splines over $\mathbb{Z}$ with connection matrices $C_i = \begin{bmatrix} -1 & 0 \\ 0 & 1 \end{bmatrix}$ for $i = 1, \ldots, m$ and $C_i = \begin{bmatrix} 1 & 0 \\ 0 & 1 \end{bmatrix}$ for all other knots i. Show that three osculating planes of such a spline at consecutive knots can intersect in a line.

14 Find basis functions for the space of the cubic splines in Problem 13. Try to have as many ordinary B-splines in your basis as possible.

15 Construct a minimally supported basis for the space of the cubic splines in Problem 13.

8 Uniform subdivision

8.1 Uniform B-splines — 8.2 Uniform subdivision — 8.3 Repeated subdivision — 8.4 The subdivision matrix — 8.5 Derivatives — 8.6 Stationary subdivision — 8.7 Convergence theorems — 8.8 Computing the difference scheme — 8.9 The four-point scheme — 8.10 Analyzing the four-point scheme — 8.11 Problems

Splines are particularly simple if their knots are all simple and evenly spaced. In this case, the B-splines are translates of each other. As a consequence, there are simple efficient knot insertion algorithms to convert a B-spline representation to a B-spline representation over a finer and also evenly spaced knot sequence. Moreover, these algorithms are the prototypes for the class of the so-called **stationary subdivision algorithms**.

8.1 Uniform B-splines

B-splines over the knot sequence $\mathbb{Z}$ can be defined as in 5.3 or, more beneficially for our purpose here, by a convolution derived from the derivative formula given in 5.7. Let

$$N^0(u) = \begin{cases} 1 & \text{if} \quad u \in [0,1) \\ 0 & \text{otherwise} \end{cases}$$

be the piecewise constant B-spline with the knots 0 and 1, see Figure 5.4. Using the derivative formula of B-splines, the nth degree B-spline $N^n(u)$ with the knots $0, 1, 2, \ldots, n+1$ is obtained by the recursion formula

$$N^j(u) = \int_0^u (N^{j-1}(v) - N^{j-1}(v-1))dv \ , \qquad j = 1, \ldots, n \ ,$$

which can also be written as

$$N^j(u) = \int_0^1 N^{j-1}(u-t)dt$$

or

$$\begin{aligned} N^j(u) &= \int_{\mathbb{R}} N^{j-1}(u-t)N^0(t)dt \\ &= N^{j-1} * N^0 \end{aligned}$$

and is known as the **convolution** of N^{j-1} with N^0. Figure 8.1 illustrates this recursive construction. In the sequel we will consider the **translates** $N^n(u-i)$ and denote these B-splines as in Chapters 5 and 6 by

$$N_i^n(u) = N^n(u-i) \ .$$

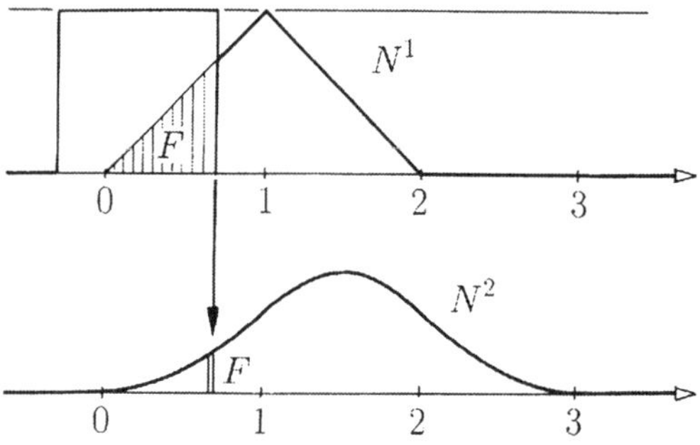

Figure 8.1: Construction of the uniform B-spline $N^2(u)$.

8.2 Uniform subdivision

An affine combination

$$\mathbf{s}^n(u) = \sum \mathbf{c}_i N_i^n(u)$$

of points $\mathbf{c}_i$ with uniform B-splines as weights is called a **uniform spline** over $\mathbb{Z}$. **Uniform subdivision** means to compute the B-spline representation of $\mathbf{s}^n(u)$ over the finer knot sequence $\frac{1}{2}\mathbb{Z}$. For this representation, we need the **scaled B-spline** $M^n(u) = N^n(2u)$ and its translates

$$M_i^n(u) = M^n(u - i/2) = N^n(2u - i) \ ,$$

see Figure 8.2 for an illustration.

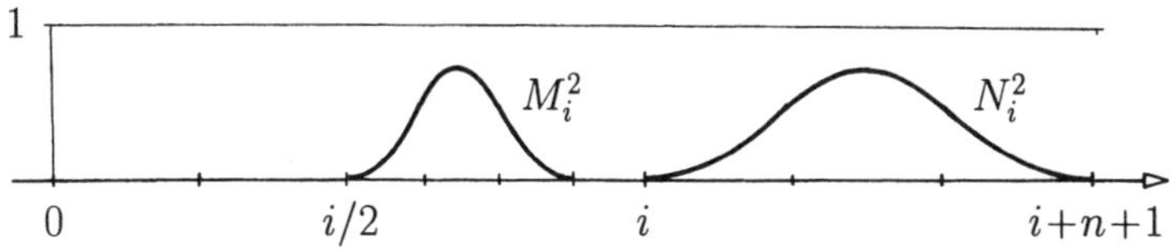

Figure 8.2: Scaled uniform B-spline.

Constructing the finer representation

$$\mathbf{s}^n(u) = \sum \mathbf{b}_i^n M_i^n(u)$$

is rather simple. For $n = 0$, one has

$$\begin{aligned} \mathbf{s}^n(u) &= \sum \mathbf{c}_i N_i^0(u) \\ &= \sum \mathbf{c}_i (M_{2i}^0(u) + M_{2i+1}^0(u)) \end{aligned}$$

and, therefore,

$$\mathbf{b}_{2i}^0 = \mathbf{b}_{2i+1}^0 = \mathbf{c}_i \ .$$

For $n = j + 1 > 0$, the recursion formula for uniform B-splinesimplies that

$$\begin{aligned} \mathbf{s}^n(u) &= \sum \mathbf{c}_i N_i^{j+1}(u) \\ &= \int_{\mathbb{R}} \sum \mathbf{c}_i N_i^j(u-t) N^0(t) dt \\ &= \int_{\mathbb{R}} \sum \mathbf{b}_i^j M_i^j(u-t)[M_0^0(t) + M_1^0(t)] dt \\ &= \frac{1}{2} \sum \mathbf{b}_i^j [M_i^{j+1}(u) + M_{i+1}^{j+1}(u)] \\ &= \sum \frac{1}{2}(\mathbf{b}_{i-1}^j + \mathbf{b}_i^j) M_i^{j+1}(u) \end{aligned}$$

and, therefore,

$$\mathbf{b}_i^{j+1} = \frac{1}{2}(\mathbf{b}_{i-1}^j + \mathbf{b}_i^j) \ .$$

This recursive computation of the $\mathbf{b}_i^n$ is the **algorithm of Lane and Riesenfeld** [Lane & Riesenfeld '80].

Given a control polygon, first **double** *all control points and then* *construct the polygons connecting the* **midpoints** n *times repeatedly.*

Figure 8.3 shows the corresponding construction for $n = 3$. Solid dots mark given points and their doubles, empty circles midpoints and small empty circles points constructed in the preceding step. The first m steps of this algorithm are the same for all uniform splines of degree n,where $m > n$.

Remark 1: The construction for $n = 2$ bears **Chaikin's** name [Chaikin '74] but has already been investigated by de Rham [Rham '47].

Remark 2: One has

$$\mathbf{s}^{n+1}(u) = \int_{u-1}^{u} \mathbf{s}^n(t)dt = \mathbf{s}^n * N^0 = \mathbf{s}^0 * N^n \ .$$

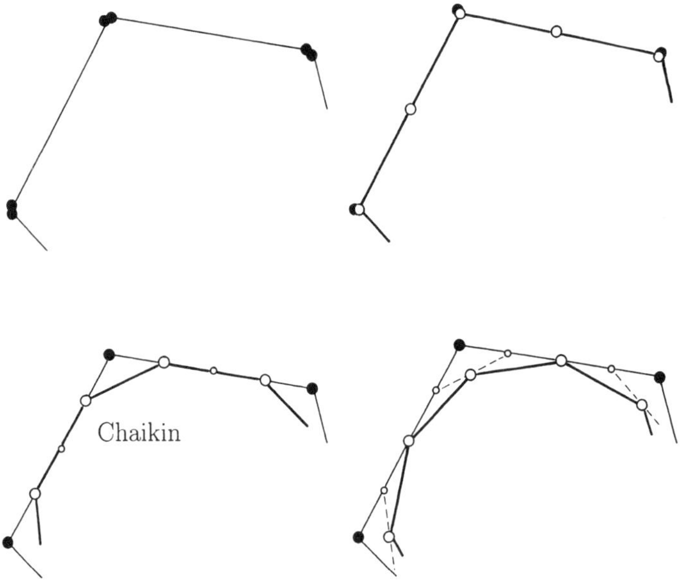

Figure 8.3: Uniform subdivision.

8.3 Repeated subdivision

The uniform subdivision algorithm can be described in matrix notation. Let

$$C = [\dots \quad \mathbf{c}_{-1}\,\mathbf{c}_0\,\mathbf{c}_1 \quad \dots]$$

and

$$B_n = [\quad \dots \quad \mathbf{b}^n_{-1}\, \mathbf{b}^n_0\, \mathbf{b}^n_1 \quad \dots \quad]$$

be the matrices formed by the control points of a uniform spline $\mathbf{s}^n$ of degree n over $\mathbb{Z}$ and $\frac{1}{2}\mathbb{Z}$, respectively. Then the Lane-Riesenfeld algorithm, see 8.2, can be written as

$$B_0 = CD \quad \text{and} \quad B_{j+1} = B_j M \ , \quad j = 0, \dots, n-1 \ ,$$

with the biinfinite matrices

$$D = \begin{bmatrix} \ddots & & & & \\ & 1 & 1 & & \\ & & & 1 & 1 \\ & & & & & \ddots \end{bmatrix}$$

and

$$M = 1/2 \begin{bmatrix} \ddots & & & \\ \ddots & 1 & & \\ & 1 & 1 & \\ & & 1 & \ddots \\ & & & \ddots \end{bmatrix} .$$

Multiplication by D means to double each control point, and multiplication by M means to compute the midpoints of any two consecutive control points. So, the matrix

$$S_n = DM^n$$

represents the subdivision operator for uniform splines of degree n. The subdivision process can be repeated. Applying $S_n = DM^n$ twice gives the control polygon of the spline $\mathbf{s}^n$ over $\frac{1}{4}\mathbb{Z}$. This means

$$\begin{aligned} \mathbf{s}^n(u) &= \sum \mathbf{c}_i N_i^n(u) \\ &= \sum \mathbf{b}_i^n N_i^n(2u) \\ &= \sum \mathbf{a}_i^n N_i^n(4u) \ , \end{aligned}$$

where

$$[\dots \mathbf{a}_{-1}\, \mathbf{a}_0\ \mathbf{a}_1 \dots] = CS_nS_n \ ,$$

and so on.

Remark 3: It is shown in 6.3 that the control polygons

$$C_k = CS_n^k$$

of the spline $\mathbf{s}^n$ over $2^{-n}\mathbb{Z}$ converge to $\mathbf{s}^n$ as k tends to infinity.

8.4 The subdivision matrix

The subdivision matrix $S_1 = D \cdot M$ for piecewise linear splines can be read off directly from Figure 8.3,

$$S_1 = \frac{1}{2} \begin{bmatrix} \cdot & \cdot & \cdot & & & & \\ & & 1 & 2 & 1 & & \\ & & & & 1 & 2 & 1 \\ & & & & \cdot & \cdot & \cdot \end{bmatrix} .$$

Similarly, the subdivision matrix S_2 for piecewise quadratic splines (Chaikin's algorithm) is obtained as

$$S_2 = \frac{1}{4} \begin{bmatrix} \cdot & \cdot & \cdot & \cdot & & & & \\ & & 1 & 3 & 3 & 1 & & \\ & & & & 1 & 3 & 3 & 1 \\ & & & & \cdot & \cdot & \cdot & \cdot \end{bmatrix} .$$

In general, the matrix S_n is of the form

$$S_n = \begin{bmatrix} \cdots & & & & & & \\ & a_0 \; a_1 & \cdots & a_{n+1} & & & \\ & & a_0 \; a_1 & \cdots & a_{n+1} & & \\ & & & & & & \cdots \end{bmatrix} ,$$

where a_i represents the binomial coefficient

$$a_i = \binom{n+1}{i} .$$

Another derivation of S_n is given in Remark 5 of 8.8.

Multiplying a control polygon C by the subdivision matrix S_n leads to the so-called **refinement equations**

$$\mathbf{b}_{2i}^n = \sum_j \mathbf{c}_j a_{2i-2j} \quad \text{and} \quad \mathbf{b}_{2i+1}^n = \sum_j \mathbf{c}_j a_{2i+1-2j} ,$$

which can be combined into one equation for the new control points,

$$\mathbf{b}_i^n = \sum_j \mathbf{c}_j a_{i-2j} .$$

8.5 Derivatives

The derivative of a spline $\mathbf{s}(u) = \sum \mathbf{c}_i N_i^n(u)$ over $\mathbf{Z}$ has a particular simple form. Specializing the general formula (1) in 5.6 gives

$$\mathbf{s}'(u) = \sum \nabla \mathbf{c}_i N_i^{n-1}(u) ,$$

where $\nabla \mathbf{c}_i = \mathbf{c}_i - \mathbf{c}_{i-1}$ denotes the backward difference as before. Thus, the derivative $\mathbf{s}'$ is controlled by the **difference polygon** $\nabla C = [\ldots \nabla \mathbf{c}_i \ldots]$ associated with the control polygon $C = [\ldots \mathbf{c}_i \ldots]$ of $\mathbf{s}$. Furthermore, let

$$C_k = [\ldots \mathbf{c}_i^k \ldots] = C(S_n)^k$$

be the polygons obtained from C by k-fold subdivision. Then, the representations of $\mathbf{s}$ and its derivative, both over $2^{-k}\mathbf{Z}$, are

$$\mathbf{s}(u) = \sum \mathbf{c}_i^k N_i^n(2^k u)$$

and

$$\mathbf{s}'(u) = \sum 2^k (\nabla \mathbf{c}_i^k) N_i^{n-1}(2^k u) .$$

Hence, subdividing $\mathbf{s}'$ gives the (by 2^{-k} divided) difference polygons

$$2^k \nabla C_k = (\nabla C)(S_{n-1})^k .$$

Thus, $\frac{1}{2}S_{n-1}$ maps the difference polygons ∇C_k onto the next difference polygons ∇C_{k+1}. Therefore, $\frac{1}{2}S_{n-1}$ is referred to as the matrix of the **difference scheme** associated with the subdivision scheme represented by S_n.

8.6 Stationary subdivision

More generally than in 8.4, let S be a biinfinite subdivision matrix of the form

$$S = \begin{bmatrix} \cdots & & & & & & \\ \cdots & \alpha_{-1} & \alpha_0 & \alpha_1 & \cdots & & \\ & & \cdots & \alpha_{-1} & \alpha_0 & \alpha_1 & \cdots \\ & & & & & & \cdots \end{bmatrix} ,$$

where the entries α_i are arbitrary numbers such that $\sum_i \alpha_{2i} = \sum_i \alpha_{2i+1} = 1$ and only finitely many α_i are non-zero.

Repeated subdivision with S gives a sequence of control polygons

$$C_k = [\ldots \mathbf{c}_i^k \ldots] = CS^k .$$

This sequence is said to converge uniformly to a continuous curve $\mathbf{c}(u)$ if

$$\sup_i \| \mathbf{c}_i^k - \mathbf{c}(2^k i) \| \underset{k\to\infty}{\longrightarrow} 0 .$$

Uniform convergence of the polygons c_k to $\mathbf{c}(u)$, over each compact interval, implies that the piecewise constant splines

$$\mathbf{c}_k(u) = \sum_i \mathbf{c}_i^k N_i^1(u)$$

also converge uniformly to $\mathbf{c}(u)$, which is uniformly continuous over compact intervals.

Further, an important necessary and sufficient criterion for uniform convergence is the following [Dyn et al. '91][Micchelli & Prautzsch '87].

The polygons c_k converge uniformly to a uniformly continuous curve $\mathbf{c}(u)$ if and only if the difference polygons ∇C_k converge uniformly to zero.

A proof is given in 15.3, see also Problems 3 and 4.

8.7 Convergence theorems

Let $C_k = [\dots \mathbf{c}_i^k \dots], k = 0, 1, \dots$, be arbitrary polygons, not necessarily obtained by subdivision, and assume that the second divided difference polygons $2^k \nabla^2 C_k$ converge uniformly to zero. Due to 8.6, this means that the first divided difference polygons $2^k \nabla C_k$ converge uniformly to a uniformly continuous curve, say $\mathbf{d}(u)$. Therefore, the first difference polygons ∇C_k converge uniformly to zero and the polygons C_k to a uniformly continuous curve $\mathbf{c}$.

This fact implies that the piecewise constant splines

$$\mathbf{d}_k(u) = \sum 2^k \nabla \mathbf{c}_i^k N_i^0(2^k u)$$

and the piecewise linear splines

$$\mathbf{c}_k(u) = \sum \mathbf{c}_i^k N_i^1(2^k u)$$

converge uniformly to $\mathbf{d}(u)$ and $\mathbf{c}(u)$, respectively. Since

$$\mathbf{c}_k(u) = \mathbf{c}_{-1}^k + \int_0^u \mathbf{d}_k(t)\, dt ,$$

and since integration commutes with the limit of a uniformly converging sequence, it follows altogether

$$\mathbf{c}(u) = \mathbf{c}(0) + \int_0^u \mathbf{d}(t)\, dt \ .$$

Consequently, $\mathbf{c}$ is differentiable and has the derivative $\mathbf{c}'(u) = \mathbf{d}(u)$. Iterated application of this result finally gives the following general theorem.

If the polygons $2^{kr}\nabla^{r+1}C_k$ converge uniformly to zero, as k tends to infinity, then $\mathbf{c}$ is a C^r curve and the polygons $2^{ki}\nabla^i C_k$ converge uniformly to the derivatives $\mathbf{c}^{(i)}$ of some C^r curve $\mathbf{c}$ for $i = 0, \ldots, r$.

8.8 Computing the difference scheme

In order to exploit the results of 8.7, one needs to compute the difference polygons ∇C_k. Let $C = [\ldots \mathbf{c}_i \ldots]$ be a control polygon and S be the subdivision matrix as in 8.6. As explained in 8.4, the vertices $\mathbf{b}_i$ of the subdivided control polygon $B = CS$ are computed by the **refinement equation**

$$\mathbf{b}_i = \sum_j \mathbf{c}_j \alpha_{i-2j} \ .$$

Multiplication by the monomial z^i and summation over all i gives the Laurent polynomial

$$\begin{aligned}\sum_i \mathbf{b}_i z^i &= \sum_i \sum_j \mathbf{c}_j \alpha_{i-2j} z^{2j} z^{i-2j} \\ &= \sum_j \mathbf{c}_j z^{2j} \sum_k \alpha_k z^k \ ,\end{aligned}$$

which we abbreviate by

$$\mathbf{b}(z) = \mathbf{c}(z^2) \cdot \alpha(z) \ .$$

The factor

$$\alpha(z) = \sum_k \alpha_k z^k$$

does not depend on C and is called the **characteristic polynomial**, or **symbol**, of the subdivision scheme S. Every subdivision scheme has a unique characteristic polynomial and vice versa.

Representing subdivision schemes by characteristic polynomials allows for an easy derivation of their associated difference schemes. Multiplying the differences $\nabla \mathbf{b}_i$ by z^i and summing over all i results in the Laurent polynomial

$$\nabla \mathbf{b}(z) = \sum_i \nabla \mathbf{b}_i z^i = \mathbf{b}(z)(1-z) \ .$$

Substituting the above equation for $\mathbf{b}(z)$ gives

$$\begin{aligned} \nabla \mathbf{b}(z) &= \mathbf{c}(z^2)\alpha(z)(1-z) \\ &= \nabla \mathbf{c}(z^2)\alpha(z)\frac{1-z}{1-z^2} \\ &= \nabla \mathbf{c}(z^2)\frac{\alpha(z)}{1+z} \ . \end{aligned}$$

The assumption $\sum \alpha_{2i} = \sum \alpha_{2i+1} = 1$ made in 8.5 is equivalent to

$$\alpha(-1) = 0 \quad \text{and} \quad \alpha(1) = 2 \ .$$

Therefore, $(1+z)$ is a factor of $\alpha(z)$, which implies that

$$\beta(z) = \frac{\alpha(z)}{1+z}$$

is the **characteristic polynomial of the difference scheme** associated with S.

Remark 4: The subdivision matrix D for piecewise constant splines over $\mathbb{Z}$ is given in 8.3. It has the characteristic polynomial

$$\sigma_0(z) = (1+z) \ .$$

Remark 5: The matrix $\frac{1}{2}S_{n-1}$ defined in 8.4 represents the difference scheme underlying the subdivision algorithm for splines of degree n over $\mathbb{Z}$, see 8.5. Hence, the characteristic polynomial $\sigma_n(z)$ of S_n is given by

$$\begin{aligned} \sigma_n &= \frac{1}{2}(1+z)\sigma_{n-1}(z) \\ &= \frac{1}{4}(1+z)^2\sigma_{n-2}(z) \\ &\ \ \vdots \\ &= 2^{-n}(1+z)^n\sigma_0(z) = 2^{-n}(1+z)^{n+1} \ , \end{aligned}$$

which, again, proves the identity $a_i = \binom{n+1}{i}$ in 8.4.

8.9 The four-point scheme

To provide an example, we apply the results above to the so-called **four-point scheme** by Dyn, Gregory and Levin [Dyn et al. '87]. Given a polygon $P_0 = [\dots \mathbf{p}_i^0 \dots]$ they construct a sequence of polygons $P_k = [\dots \mathbf{p}_i^k \dots]$ by the rules

$$\begin{aligned} \mathbf{p}_{2i-1}^{k+1} &= \mathbf{p}_{i-2}^k \\ \mathbf{p}_{2i}^{k+1} &= -\omega\mathbf{p}_{i-3}^k + (1/2+\omega)\mathbf{p}_{i-2}^k + (1/2+\omega)\mathbf{p}_{i-1}^k - \omega\mathbf{p}_i^k \ , \end{aligned}$$

where ω is a free design parameter. The corresponding construction is illustrated in Figure 8.4.

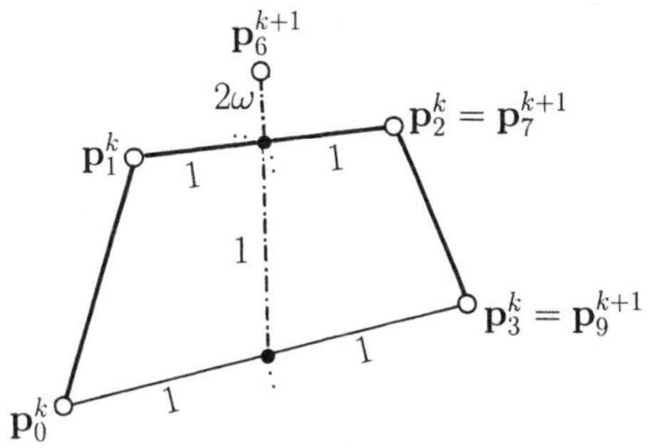

Figure 8.4: The four-point scheme - construction of $\mathbf{p}_6^{k+1}$.

Every polygon P_{k+1} obtained by the four-point scheme interpolates the preceding polygon P_k. Subdivision algorithms with this property are also referred to as **iterative interpolation schemes**.

Remark 6: Let $\mathbf{p}(u)$ be any cubic curve and choose the polygon

$$P_0 = [\dots \mathbf{p}_i^0 \dots]$$

such that $\mathbf{p}_i^0 = \mathbf{p}(i)$, $i \in \mathbb{Z}$. If $\omega = 1/16$, then all polygons P_k obtained by the four-point scheme also lie on this cubic, i.e.,

$$\mathbf{p}_{i+1}^k = \mathbf{p}(1 + i/2^k) \ .$$

Thus, the four-point scheme has cubic precision for $\omega = 1/16$.

Remark 7: Based on polynomial interpolation of degree $2k-1$ at equidistant abscissae, one can build $2k$-point schemes with polynomial precision of degree $2k-1$, see [Kobbelt '94].

8.10 Analyzing the four-point scheme

From the definition of the four-point scheme one can easily read off its characteristic polynomial,

$$\begin{aligned}\alpha(z) &= -\omega + (1/2+\omega)z^2 + z^3 + (1/2+\omega)z^4 - \omega z^6 \\ &= (1+z)\beta(z)\ ,\end{aligned}$$

where

$$\beta(z) = -\omega + \omega z + 1/2z^2 + 1/2z^3 + \omega z^4 - \omega z^5$$

is the characteristic polynomial of the difference scheme. Thus, it follows that

$$\begin{aligned}\|\nabla \mathbf{p}_{2i}^{k+1}\| &= \| - \omega \nabla \mathbf{p}_i^k + 1/2 \nabla \mathbf{p}_{i+1}^k + \omega \nabla \mathbf{p}_{i+2}^k \| \\ &\le (1/2 + 2|\omega|) \sup_i \|\nabla \mathbf{p}_i^k\|\end{aligned}$$

and, similarly,

$$\|\nabla \mathbf{p}_{2i+1}^{k+1}\| \le (1/2 + 2|\omega|) \sup_i \|\nabla \mathbf{p}_i^k\|\ .$$

Hence, for $|\omega| < 1/4$, the difference polygons ∇P_k converge to zero and the polygons P_k to a continuous curve. Further, according to 8.7 differentiability depends on the second differences $2^k \nabla^2 P_k$. Due to 8.8, the associated subdivision scheme has the characteristic polynomial

$$\gamma(z) = \frac{2\beta(z)}{1+z} = -2\omega + 4\omega z + (1-4\omega)z^2 + 4\omega z^3 - 2\omega z^4\ .$$

Again, one can show that the differences $2^k \nabla^2 P_k$ go to zero if $0 < \omega < 1/8$. Hence, the four-point scheme produces C^1 interpolants in this case.

Remark 8: One can show [Dyn et al. '91] that the four-point scheme produces C^1 interpolants also if $0 < \omega < (\sqrt{5}-1)/8 \approx 0.15$. However, the four-point scheme does not produce C^2 curves in general.

Remark 9: Kobbelt's $2k$-point schemes produce C^{k-1} interpolants, see [Kobbelt '94].

8.11 Problems

1 The uniform B-spline N^n can be obtained by n-fold convolution of N^0 with itself,

$$N^n = N^0 * \overset{n}{\cdots} * N^0\ .$$

2 Use the recursion formula for uniform B-splines in 8.1 to show that

$$\sum_i N_i^n = 1, \ \sum_i (i + \frac{n+1}{2}) N_i^n(u) = u, \quad \text{and} \quad \int_{\mathbb{R}} N_i^n = 1 \ .$$

3 Use the notation of 8.6 to show that $\|\mathbf{c}_{2i}^{k+1} - \mathbf{c}_i^k\|$ and $\|\mathbf{c}_{2i+1}^{k+1} - 1/2(\mathbf{c}_i^k + \mathbf{c}_{i+1}^k)\|$ are bounded by some multiple of $\sup_i \|\mathbf{c}_i^k\| \cdot \sum_i |\alpha_i|$.

4 Conclude from Problem 3 that the piecewise linear curves $\sum \mathbf{c}_i^k N_i^1(2^{-k}u)$ converge uniformly if the differences $\nabla \mathbf{c}_i^k$ converge uniformly to zero.

5 Let S_k be the matrix of the subdivision algorithm for uniform splines of degree k, as given in 8.4 and consider the sequence of polygons

$$C_k = C_{k-1} S_k, \ k = 1, 2, 3, \ \dots \ ,$$

where $C_0 = [\dots \mathbf{c}_i \dots]$ is an arbitrary control polygon. Show that the polygons C_k converge to a C^∞ curve [Rvachev '90].

6 Consider the uniform splines $\mathbf{s}_n = \sum \mathbf{c}_i N_i^n(u)$ of degree n over $\mathbb{Z}$. For any integer $r \in \mathbf{N}$, let

$$\mathbf{s}_n = \sum \mathbf{b}_i^n N_i^n(ru)$$

be the corresponding representations over $\frac{1}{r}\mathbb{Z}$. Show that the control points $\mathbf{b}_i^n$ can be computed by the recursion

$$\begin{aligned} \mathbf{b}_{ri}^0 &= \dots = \mathbf{b}_{ri+r-1}^0 = \mathbf{c}_i \\ \mathbf{b}_i^{n+1} &= \frac{1}{r}(\mathbf{b}_{i-r+1}^n + \cdots + \mathbf{b}_i^n) \ , \end{aligned}$$

7 Consider two subdivision matrices R and S with characteristic polynomials $\alpha(z)$ and $(1+z)\alpha(z)/2$, respectively. Given a polygon C, let the polygons CR^k converge to a curve $\mathbf{c}(u)$. Show that the polygons CS^k converge to the curve $\mathbf{d}(u) = \int_{u-1}^{u} \mathbf{c}(u) du$.

Part II

Surfaces

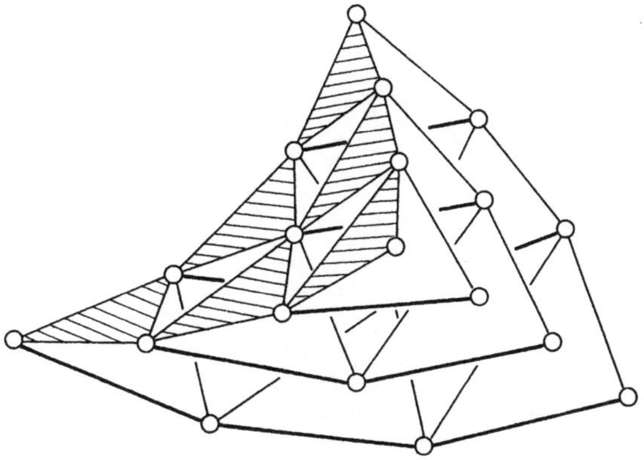

9 Tensor product surfaces

9.1 Tensor products — 9.2 Tensor product Bézier surfaces — 9.3 Tensor product polar forms — 9.4 Conversion to and from monomial form — 9.5 The de Casteljau algorithm — 9.6 Derivatives — 9.7 Simple C^r joints — 9.8 Piecewise bicubic C^1 interpolation — 9.9 Surfaces of arbitrary topology — 9.10 Singular parametrization — 9.11 Bicubic C^1 splines of arbitrary topology — 9.12 Problems

The easiest way to build a surface is to sweep a curve through space such that its control points move along some curves. The control points of these control curves control the surface. The surface representation by these control points has properties analogous to those of a univariate curve (e.g., Bézier or B-spline) representation. This is due to the fact that one can deal with these surfaces by applying just curve algorithms. Similarly, one can build multidimensional volumes by sweeping a surface or volume through space such that its control points move along curves. Again, one obtains control nets having properties analogous to those of the underlying curve representations.

9.1 Tensor products

To show how to build surfaces from curves, let

$$A_0(u), \ldots, A_m(u) \quad \text{and} \quad B_0(v), \ldots, B_n(v)$$

be two independent sequences of functions and consider a space curve,

$$\mathbf{p}(u) = \sum_{i=0}^{m} \mathbf{a}_i A_i(u) \ ,$$

where each control point lies itself on a curve, e.g.,

$$\mathbf{a}_i = \mathbf{a}_i(v) = \sum_{j=0}^{n} \mathbf{b}_{ij} B_j(v) \ .$$

The surface, $\mathbf{p}$ defined as

$$\mathbf{p}(u,v) = \sum_i \sum_j \mathbf{b}_{ij} A_i(u) B_j(v) \ .$$

is called a **tensor product surface**. The products $A_i B_j$ are linearly independent and, therefore, form a basis.

Remark 1: The product of two functions is a special tensor product function,

$$\sum_i a_i A_i(u) \cdot \sum_j b_j B_j(v) = \sum_i \sum_j a_i b_j A_i(u) B_j(v) \ .$$

Remark 2: There are tensor product functions that are not the product of two univariate functions as for example the function $uv + (1-u)(1-v)$, see Figure 9.1.

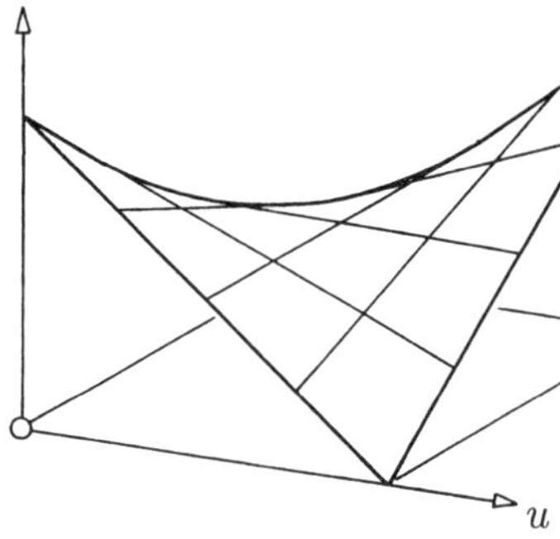

Figure 9.1: The function $uv + (1-u)(1-v)$.

Remark 3: Let $A_i(u)$ and $B_j(v)$ be the **Lagrange polynomials**, see 4.2, associated with the knots $u_0, \ldots, u_m$ and $v_0, \ldots, v_n$ and defined by $A_i(u_k) = \delta_{i,k}$ and $B_j(v_l) = \delta_{j,l}$ as illustrated in Figure 9.2. Then the tensor product surface

$$\mathbf{p}(u,v) = \sum_i \sum_j \mathbf{p}_{ij} A_i(u) B_j(v)$$

is a **Lagrange surface** interpolating the control points $\mathbf{p}_{ij}$. It is illustrated in Figure 9.3.

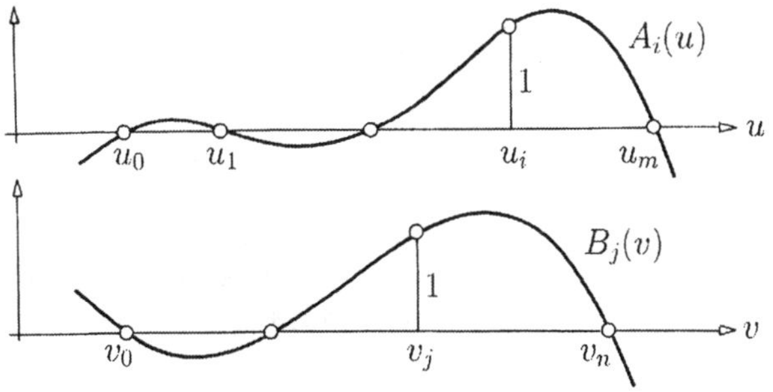

Figure 9.2: Lagrange polynomials.

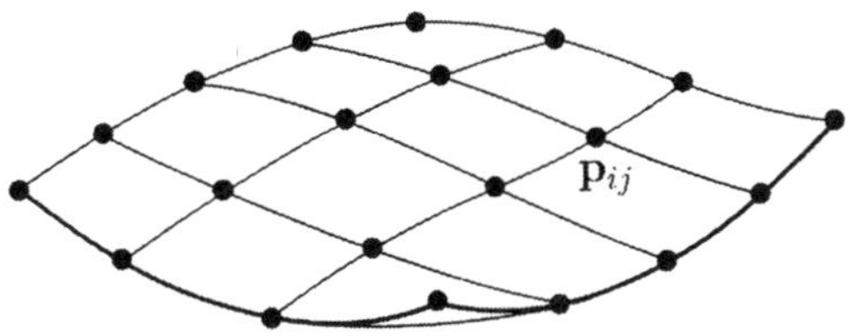

Figure 9.3: Interpolating surface.

9.2 Tensor product Bézier surfaces

We will discuss how to extend Bézier techniques for curves to tensor product surfaces. The extension of B-spline or other curve techniques to tensor product surfaces is straightforward then.

A polynomial surface $\mathbf{b}(\mathbf{u}) = \mathbf{b}(u, v)$ is said to be of degree $\mathbf{m} = (m, n)$ if it is of degree m and n in u and v, respectively. Any affine transformations

$$u = c_1(1-s) + d_1 s \quad \text{and} \quad v = c_2(1-t) + d_2 t$$

leave the degree of $\mathbf{b}(\mathbf{u})$ unchanged, i.e., the reparametrized polynomial

$$\mathbf{b}(s,t) = \mathbf{b}(\mathbf{u}(s,t))$$

is also of degree $\mathbf{m}$ in $\mathbf{s} = (s,t)$. The polynomial $\mathbf{b}(s,t)$ can be viewed as a polynomial of degree m in s with polynomial coefficients of degree n in t. Hence, $\mathbf{b}(s,t)$ has a **Bézier representation,**

$$\mathbf{b}(s,t) = \sum_i \mathbf{b}_i(t) B_i^m(s) = \sum_i \sum_j \mathbf{b}_{ij} B_i^m(s) B_j^n(t) \ ,$$

which we abbreviate by

$$\mathbf{b}(\mathbf{s}) = \sum_{\mathbf{i}} \mathbf{b_i} B_{\mathbf{i}}^{\mathbf{m}}(\mathbf{s}) \ ,$$

where $\mathbf{i} = (i, j)$.

The coefficients $\mathbf{b_i}$ are called the **Bézier points** of $\mathbf{b}(\mathbf{u})$ over the interval $[\mathbf{c}, \mathbf{d}] = [c_1, d_1] \times [c_2, d_2]$. They define the vertices of the **Bézier net** of $\mathbf{b}$, as illustrated in Figure 9.4. The parameter $\mathbf{s}$ is called the **local** and $\mathbf{u}$ the **global parameter** of $\mathbf{b}$.

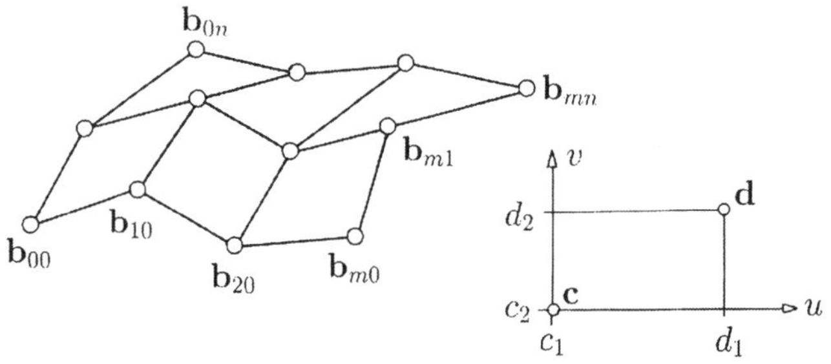

Figure 9.4: A Bézier net.

As for curves the properties of the univariate Bernstein polynomials in 2.1 are passed on to the Bézier representation of a surface $\mathbf{b}(\mathbf{u})$.

- *The* **symmetry** *property of the Bernstein polynomials implies that*

$$\mathbf{b}(\mathbf{u}) = \sum_{i,j} \mathbf{b}_{ij} B_{ij}^{\mathbf{m}}(\mathbf{s}) = \sum_{i,j} \mathbf{b}_{m-i,j} B_{i,j}^{\mathbf{m}}(\bar{\mathbf{s}}) \ ,$$

  where $\bar{\mathbf{s}} = (1 - s, t)$, *etc.*

Since a curve passes through its first and last Bézier point, it follows that

- *the boundary of the Bézier net represents the four* **boundary curves** *of the patch* $\mathbf{b}[\mathbf{c}, \mathbf{d}]$.

For example

$$\mathbf{b}(c_1, v) = \sum_j \mathbf{b}_{0j} B_j^n(v) \ .$$

In particular, this implies that

- *the four* **corners** *of the patch and the corner Bézier points coincide.*

This means

$$\mathbf{b}(c_1, d_1) = \mathbf{b}_{00}, \quad \mathbf{b}(c_1, d_2) = \mathbf{b}_{0n} \ , \quad \text{etc.}$$

Since the Bernstein polynomials sum to one,

- *their products form a* **partition of unity***,*

$$\sum_{\mathbf{i}} B_{\mathbf{i}}^{\mathbf{m}} = \sum_j 1 \cdot B_j^n = 1 \ .$$

- *Thus,* $\mathbf{b}(\mathbf{u})$ *is an* **affine combination** *of its Bézier points and the Bézier representation is* **affinely invariant**.

Since the Bernstein polynomials are non-negative in $[0, 1]$, it follows that

- *for every* $\mathbf{b}(\mathbf{u})$ *is a* **convex combination** *of the* $\mathbf{b}_{\mathbf{i}}$ *for every* $\mathbf{u} \in [\mathbf{c}, \mathbf{d}]$.

Hence,

- *the patch* $\mathbf{b}[\mathbf{c}, \mathbf{d}]$ *lies in the* **convex hull** *of its Bézier points.*

Remark 4: Using the convex hull property separately for each coordinate a bounding box is obtained for the surface patch $\mathbf{b}[\mathbf{c}, \mathbf{d}]$,

$$\mathbf{b}(\mathbf{u}) \in [\min_{\mathbf{i}} \mathbf{b}_{\mathbf{i}}, \max_{\mathbf{i}} \mathbf{b}_{\mathbf{i}}] \quad \text{for} \qquad \mathbf{u} \in [\mathbf{c}, \mathbf{d}] \ ,$$

which is illustrated in Figure 9.5.

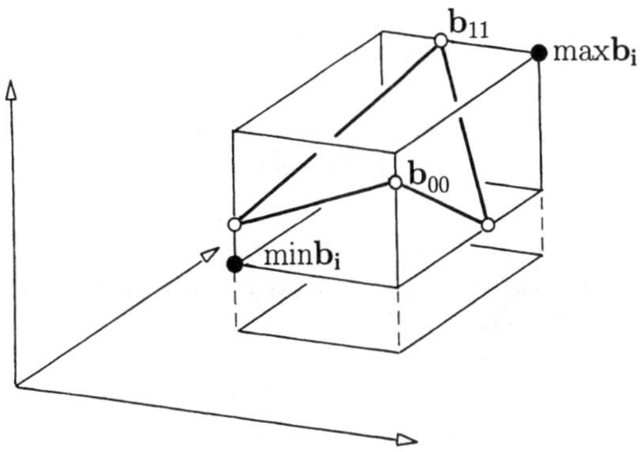

Figure 9.5: A bounding box.

9.3 Tensor product polar forms

Let $A_0(u), \ldots, A_m(u)$ and $B_0(v), \ldots, B_n(v)$ be bases for the space of all polynomials up to degree m and n, respectively. Further, let

$$A_i[u_1 \ldots u_m] \quad \text{and} \quad B_j[v_1 \ldots v_n]$$

be the corresponding polar forms. Then, any tensor product surface

$$\mathbf{b}(u, v) = \sum_i \sum_j \mathbf{b}_{ij} A_i(u) B_j(v)$$

has the **tensor product polar form**

$$\mathbf{b}[u_1 \ldots u_m, v_1 \ldots v_n] = \sum_i \sum_j \mathbf{b}_{ij} A_i[u_1 \ldots u_m] B_j[v_1 \ldots v_n] \ .$$

This polar form has the following three properties.

- $\mathbf{b}[u_1 \ldots u_m, v_1 \ldots v_n]$ *agrees with* $\mathbf{b}(u, v)$ *on its* **diagonal**,

which means $\mathbf{b}[u \ldots u, v \ldots v] = \mathbf{b}(u, v)$.

- $\mathbf{b}[u_1 \ldots u_m, v_1 \ldots v_n]$ *is* **symmetric** *in the variables* u_i *and symmetric in the variables* v_j,

which means that

$$\mathbf{b}[s_1 \ldots s_m, t_1 \ldots t_n] = \mathbf{b}[u_1 \ldots u_m, v_1 \ldots v_n]$$

for any permutations $(s_1, \ldots, s_m)$ and $(t_1, \ldots, t_n)$ of $(u_1, \ldots, u_m)$ and $(v_1, \ldots, v_n)$, respectively.

- $\mathbf{b}[u_1 \ldots u_m, v_1 \ldots v_n]$ *is* **affine** *in each variable.*

The Bézier points of $\mathbf{b}(u, v)$ over some interval $[a, b] \times [c, d]$ can easily be obtained by the main theorem 3.2. For any fixed u, the polynomial $\mathbf{b}(v) = \mathbf{b}(u, v)$ has the Bézier points

$$\mathbf{b}_j(u) = \mathbf{b}[u \overset{m}{\ldots} u, a \overset{n-j}{\ldots} a b \overset{j}{\ldots} b] \ , \ j = 0, \ldots, n \ ,$$

and, for each j, these polynomials $\mathbf{b}_j(u)$ have the Bézier points

$$\mathbf{b}_{ij} = \mathbf{b}[a \overset{n-i}{\ldots} a b \overset{i}{\ldots} b, c \overset{m-j}{\ldots} c d \overset{j}{\ldots} d] \ .$$

Thus, we have proved the following form of the **main theorem**

A tensor product polynomial $\mathbf{b}(u,v)$ *with tensor product polar form* $\mathbf{b}[u_1 \dots u_m, v_1 \dots v_n]$ *has the Bézier points*

$$\mathbf{b}_{ij} = \mathbf{b}[a \overset{n-i}{\dots} ab \overset{i}{\dots} b, c \overset{m-j}{\dots} cd \overset{j}{\dots} d]$$

over any interval $[a,b] \times [c,d]$.

Any tensor product polar form $\mathbf{b}[u_1 \dots u_m, v_1 \dots v_n]$ can be computed by the generalized de Casteljau algorithm from the points

$$\mathbf{b}_j(u) = \mathbf{b}[u_1 \dots u_m, c \overset{m-j}{\dots} cd \overset{j}{\dots} d] \ , \ j = 0, \dots, m \ ,$$

and these from the Bézier points $\mathbf{b}_{ij}$. Since the Bézier points are unique, any tensor product polynomial $\mathbf{b}(u,v)$ of degree $\leq (m,n)$ has a unique tensor product polar form $\mathbf{b}[u_1 \dots u_m, v_1 \dots v_n]$.

9.4 Conversion to and from monomial form

The monomial form of a polynomial tensor product surface

$$\mathbf{b}(u,v) = \sum_{k=0}^{m} \sum_{l=0}^{n} \mathbf{a}_{kl} \binom{m}{k} \binom{n}{l} u^k v^l$$

can be written more concisely with bold vector notation as

$$\mathbf{b}(\mathbf{u}) = \sum_{\mathbf{k}=0}^{\mathbf{m}} \mathbf{a}_{\mathbf{k}} \binom{\mathbf{m}}{\mathbf{k}} \mathbf{u}^{\mathbf{k}} \ ,$$

where $\mathbf{u} = (u,v), \mathbf{k} = (k,l)$ and $\mathbf{m} = (m,n)$.

The conversion of the monomial form to the Bézier representation of $\mathbf{b}(\mathbf{u})$ over $[0,1]^2$ is straightforward. Applying the conversion formula in 2.8 for univariate polynomials twice gives

$$\mathbf{b}(\mathbf{u}) = \sum_{\mathbf{i}=0}^{\mathbf{m}} \mathbf{b}_{\mathbf{i}} B_{\mathbf{i}}^{\mathbf{m}}(\mathbf{u}) \ ,$$

where

$$\mathbf{b}_{\mathbf{i}} = \sum_{\mathbf{k}=0}^{\mathbf{i}} \binom{\mathbf{i}}{\mathbf{k}} \mathbf{a}_{\mathbf{k}} \ .$$

Similarly, by applying the conversion formula in 2.9 twice, we obtain the converse relation

$$\mathbf{a_k} = \sum_{\mathbf{i}=\mathbf{0}}^{\mathbf{k}} (-1)^{k+l-i-j} \binom{\mathbf{k}}{\mathbf{i}} \mathbf{b_i} \ .$$

Remark 5: If $\mathbf{b}(\mathbf{u})$ is a bilinear polynomial, i.e., a biaffine map, then $\mathbf{a_k} = \mathbf{0}$ for all $\mathbf{k} \geq (2,2)$. Hence, $\mathbf{b}(\mathbf{u})$ has the Bézier points

$$\begin{aligned} \mathbf{b}_{ij} &= \mathbf{a}_{00} + i\mathbf{a}_{10} + j\mathbf{a}_{01} + ij\mathbf{a}_{11} \\ &= \mathbf{b}(i/m, j/n) \ . \end{aligned}$$

This property is referred to as the **bilinear precision** of the Bézier representation.

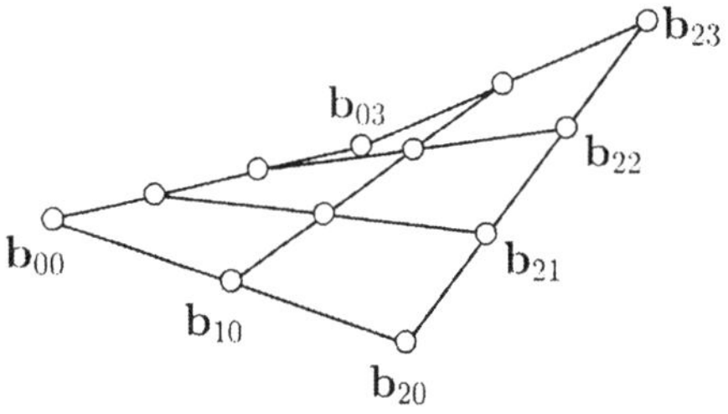

Figure 9.6: Uniform Bézier net on a bilinear interpolant.

9.5 The de Casteljau algorithm

A bivariate polynomial surface in Bézier representation,

$$\mathbf{b}(\mathbf{u}) = \sum_{\mathbf{i}} \mathbf{b_i} B_{\mathbf{i}}^{\mathbf{m}}(\mathbf{s}) \ ,$$

can be evaluated at any $\mathbf{s} = (s,t)$ by $(m+2)$ or $(n+2)$ applications of de Casteljau's algorithm for curves. This leads to the following surface algorithm

Use de Casteljau's curve algorithm to compute

1. *the points* $\mathbf{b}_i = \sum_{j=0}^{n} \mathbf{b}_{ij} B_j^n(t)$ *and*
2. *the surface point* $\mathbf{b}(\mathbf{u}) = \sum_{i=0}^{m} \mathbf{b}_i B_i^m(s)$.

Remark 6: For example, consider the polynomial $b(s,t) = \sum b_{ij} B_{ij}^{3,2}(s,t)$ whose Bézier matrix $[b_{ij}]$ appears in the upper left corner of Figure 9.7. This

9.6. Derivatives

Figure illustrates the algorithm above by showing the $8 = 4 \cdot 2$ de Casteljau steps for $t = 1/2$ at the left and the three steps for $s = 2/3$ at the bottom. Each arrow corresponds to one step.

$$\begin{bmatrix} 0 & 0 & 0 & 6 \\ 18 & 2 & 0 & 8 \\ 4 & 0 & 4 & 18 \end{bmatrix} \dashrightarrow \begin{bmatrix} * & * & * \\ * & * & * \\ * & * & * \end{bmatrix} \dashrightarrow \begin{bmatrix} * & * \\ * & * \\ * & * \end{bmatrix} \dashrightarrow \begin{bmatrix} * \\ * \\ * \end{bmatrix}$$

$$\downarrow \downarrow \downarrow \downarrow \qquad \downarrow \downarrow \downarrow \qquad \downarrow \downarrow \qquad \downarrow$$

$$\begin{bmatrix} 9 & 1 & 0 & 7 \\ 11 & 1 & 2 & 3 \end{bmatrix} \dashrightarrow \begin{bmatrix} \cdots \\ \cdots \end{bmatrix} \dashrightarrow \begin{bmatrix} \cdot & \cdot \\ \cdot & \cdot \end{bmatrix} \dashrightarrow \begin{bmatrix} * \\ * \end{bmatrix}$$

$$\downarrow \downarrow \downarrow \downarrow \qquad \downarrow \downarrow \downarrow \qquad \downarrow \downarrow \qquad \downarrow$$

$$\begin{bmatrix} 10 & 1 & 1 & 10 \end{bmatrix} \longrightarrow \begin{bmatrix} 4 & 1 & 7 \end{bmatrix} \longrightarrow \begin{bmatrix} 2 & 5 \end{bmatrix} \longrightarrow \begin{bmatrix} 4 \end{bmatrix}$$

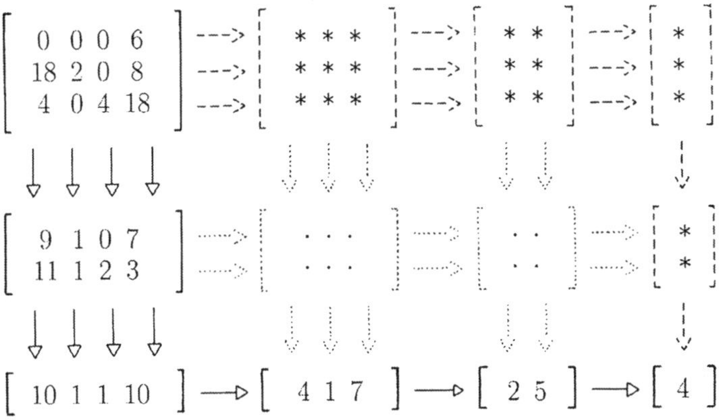

Figure 9.7: Evaluating a tensor product polynomial by de Casteljau's algorithm.

Remark 7: One can also first apply de Casteljau's algorithm to the Bézier rows for $s = 2/3$ and then to the columns for $t = 1/2$, as indicated by the dashed matrices and arrows. Moreover, one can alternate arbitrarily between row and column operations, as indicated by the dotted matrices and arrows and, as illustrated by Figure 9.8 for a parametric surface.

Remark 8: The subdivision property of de Casteljau's algorithm described in 3.3 shows that the Bézier points of $\mathbf{b}(s,t)$ over $[0,1] \times [0,t]$ and $[0,1] \times [t,1]$ are certain intermediate points computed by the above algorithm.

Remark 9: The polynomial $\mathbf{b}(u,v)$ defined in Figure 9.7 has the Bézier matrices

$$\begin{bmatrix} 0 & 0 & 0 & 6 \\ 9 & 1 & 0 & 7 \\ 10 & 1 & 1 & 10 \end{bmatrix} \quad \text{and} \quad \begin{bmatrix} 10 & 1 & 1 & 10 \\ 11 & 1 & 2 & 13 \\ 4 & 0 & 4 & 18 \end{bmatrix}$$

over $[0,1] \times [0,1/2]$ and $[0,1] \times [1/2,1]$, respectively.

9.6 Derivatives

The derivatives of a tensor product surface can also be obtained by the underlying curve schemes. The first partial derivative of a surface

$$\mathbf{b}(\mathbf{s}) = \sum \mathbf{b_i} B_i^m(\mathbf{s})$$

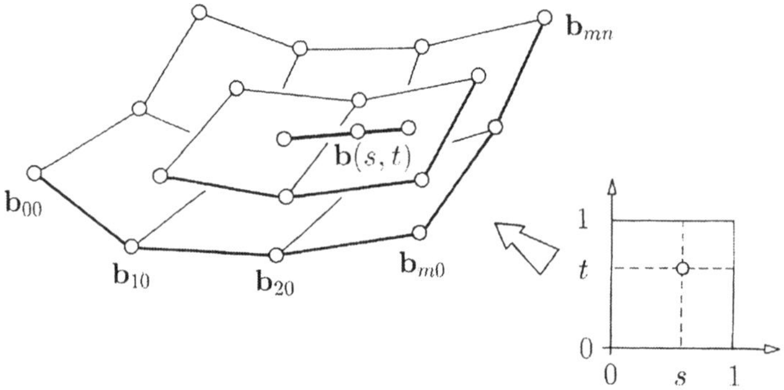

Figure 9.8: Evaluation by using de Casteljau's algorithm alternately for s and t.

has the Bézier representation

$$\mathbf{b}_s = \frac{\partial}{\partial s}\mathbf{b} = m \sum \Delta^{10}\mathbf{b_i} B_{\mathbf{i}}^{m-1,n}(\mathbf{s}) \ ,$$

where $\Delta^{10}\mathbf{b_i} = \mathbf{b}_{i+1,j} - \mathbf{b}_{i,j}$, see Figure 9.9.

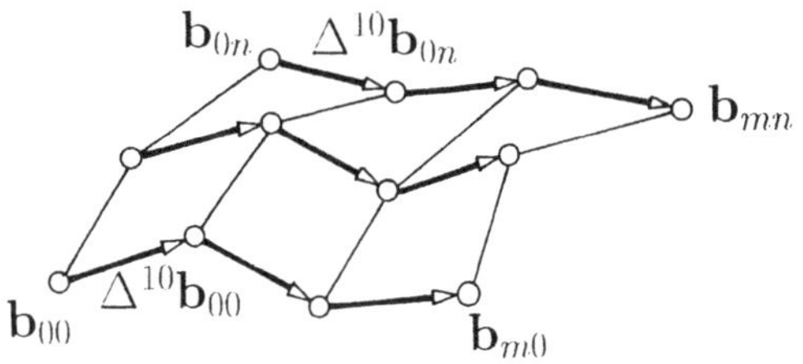

Figure 9.9: The Bézier points $\Delta^{10}\mathbf{b_i}$ of the derivative $\mathbf{b}_u$.

Repeated differentiation leads to the general formula

$$\frac{\partial^{q+r}}{\partial s^q \partial t^r}\mathbf{b} = \frac{\mathbf{m}!}{(\mathbf{m}-\mathbf{q})!} \sum \Delta^{\mathbf{q}}\mathbf{b_i} B_{\mathbf{i}}^{\mathbf{m}-\mathbf{q}} \ ,$$

where $\Delta^{qr} = \Delta^{10}\Delta^{q-1,r} = \Delta^{01}\Delta^{q,r-1}$ and $\mathbf{m}! = m!n!$ and $\mathbf{q} = (q,r)$.

It follows that $\mathbf{b}_{00}, \mathbf{b}_{10}, \mathbf{b}_{01}$ span the **tangent plane** of $\mathbf{b}$ at $(s,t) = (0,0)$, and that the **corner twist** is given by

$$\begin{aligned} \mathbf{b}_{st}(0,0) &= mn\Delta^{11}\mathbf{b}_{00} \\ &= mn(\mathbf{b}_{11} - \mathbf{b}_{10} - \mathbf{b}_{01} + \mathbf{b}_{00}) \ , \end{aligned}$$

see Figure 9.10.

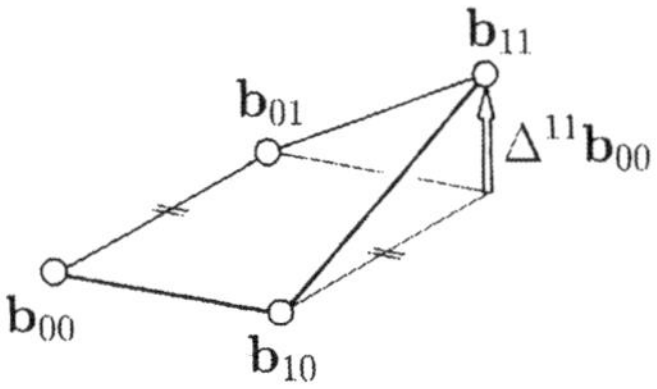

Figure 9.10: The corner twist $\Delta^{11}\mathbf{b}_{00}$.

Remark 10: Because of the symmetry of the Bernstein polynomials, one obtains analogous results for the three other corners of $[0, 1]^2$.

9.7 Simple $\mathbf{C^r}$ joints

Stärk's theorem in 3.10 leads also to a simple C^r joint of two polynomial patches $\mathbf{b}(\mathbf{u})$ and $\mathbf{c}(\mathbf{u})$ given by their Bézier points $\mathbf{b_0}, \ldots, \mathbf{b_m}$ and $\mathbf{c_0}, \ldots, \mathbf{c_m}$ over $[u_0, u_1] \times [v_0, v_1]$ and $[u_1, u_2] \times [v_0, v_1]$, respectively.

The derivatives of $\mathbf{b}$ *and* $\mathbf{c}$ *along the common parameter line* $u = u_1$ *agree up to order* r *if and only if the points*

$$\mathbf{b}_{m-r,j}, \ldots, \mathbf{b}_{mj} = \mathbf{c}_{0j}, \ldots, \mathbf{c}_{rj}$$

form the composite Bézier polygon of some curve of degree r *over* $[u_0, u_1, u_2]$ *for all* $j = 0, \ldots, n$.

Figure 9.11 shows such a simple C^1 joint for $m = n = 3$, where $\Delta_i = \Delta u_i$.

9.8 Piecewise bicubic $\mathbf{C^1}$ interpolation

Interpolation schemes for curves can be extended in a straightforward manner to tensor product schemes. We describe the principle for the example of the cubic interpolation scheme discussed in 4.5.

Given $(m+1)\times(n+1)$ interpolation points $\mathbf{p}_{ij}$ with corresponding parameter values (u_i, v_j), for $i = 0, \ldots, m$ and $j = 0, \ldots, n$, we construct a piecewise bicubic C^1 surface $\mathbf{s}(u, v)$ such that $\mathbf{s}(u_i, v_j) = \mathbf{p}_{ij}$. More exactly, we construct for all (i, j) the Bézier points $\mathbf{b}_{3i,3j}, \ldots, \mathbf{b}_{3i+3,3j+3}$ defining the bicubic segment of $\mathbf{s}$ over the interval $[u_i, u_{i+1}] \times [v_j, v_{j+1}]$.

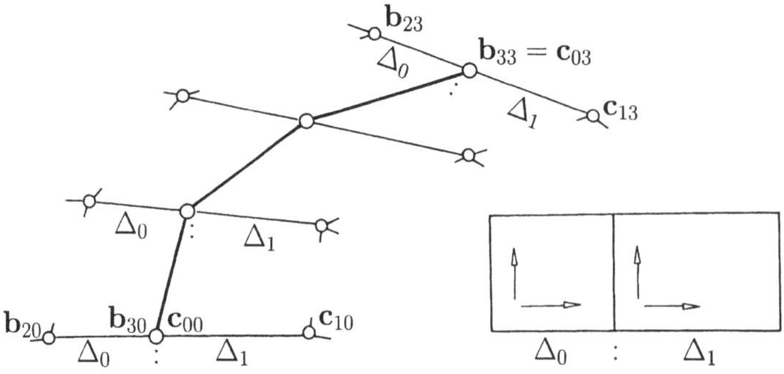

Figure 9.11: A simple C^1 joint of two bicubic patches.

Let $P = [\mathbf{p}_{ij}]$ be the $(m+1) \times (n+1)$ matrix formed by the interpolation points. Note that the entries are coordinate columns rather than scalars, in general. Further, let S and T be the $(m+1) \times (3m+1)$ and $(n+1) \times (3n+1)$ matrices of two linear interpolation schemes over the abscissae $u_0, \ldots, u_m$ and $v_0, \ldots, v_n$, respectively, as described in 4.5, Remark 9.

Then, the tensor product interpolation scheme based on S and T is as follows.

1 Interpolate every column of P by computing $A = S^t P$.

2 Interpolate every row of A by computing $B^t = AT$.

The desired Bézier points are the entries of $B = [\mathbf{b}_{ij}] = S^t PT$. See Figure 9.12 for an illustration.

Row and column interpolation are interchangeable. Namely, row interpolation gives $C^t = PT$ and subsequent column interpolation $B = S^t C = S^t PT$.

Obviously, $\mathbf{s}(u_i, v_j) = \mathbf{b}_{3i,3j} = \mathbf{p}_{i,j}$, and if S and T generate C^1 interpolants, the interpolant $\mathbf{s}(u, v)$ is differentiable in u, and since steps *1* and *2* commute, it is also differentiable in v.

Remark 11: Similarly, the curve approximation scheme in 4.6 can be used to obtain an approximating surface.

9.9 Surfaces of arbitrary topology

Tensor product interpolation schemes are excellent if one has to interpolate the vertices of a regular quadrilateral net. However, not every surface can be decomposed into quadrilateral patches such that the patch boundaries form a regular quadrilateral net, as shown in Figure 9.13.

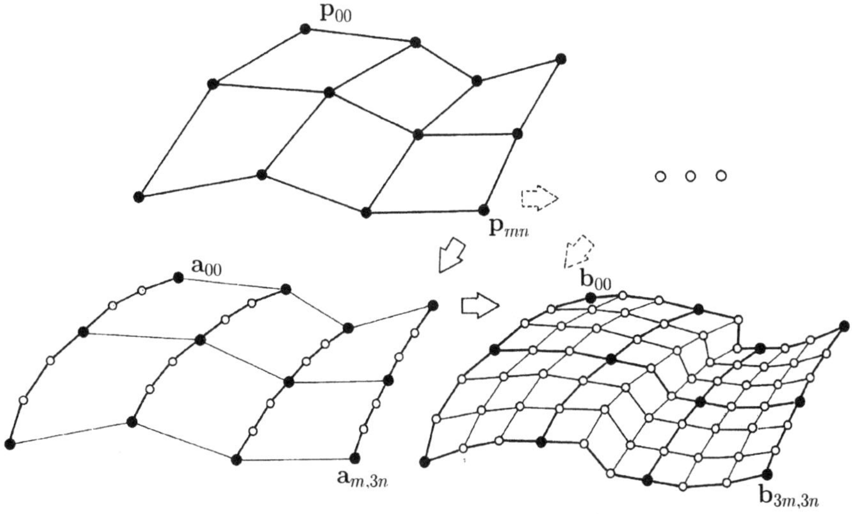

Figure 9.12: Tensor product interpolation scheme.

All one can do, in general, is to refine such a net so as to obtain at least a non-regular quadrilateral net, as illustrated in Figure 9.14.

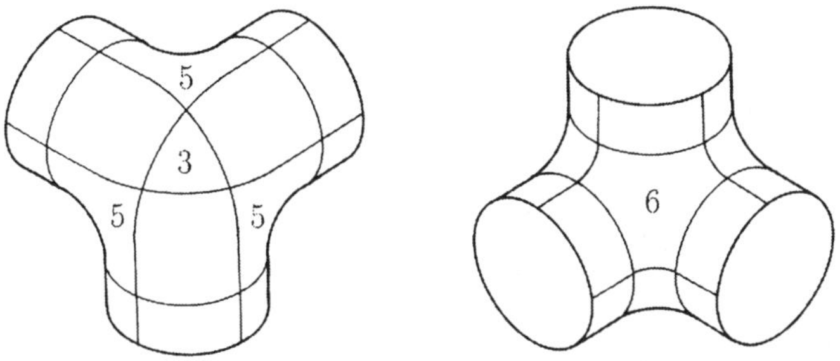

Figure 9.13: Non-regular net.

Still, it is difficult to construct a C^1 surface with regular quadrilateral patches arranged in an arbitrary net. The C^1 constructions given in 9.7 allow us only to join exactly four patches at any interior vertex. One way to overcome this problem is described in Chapters 13 and 14. Another, simpler possibility is to use singular parametrizations as described in the rest of this chapter.

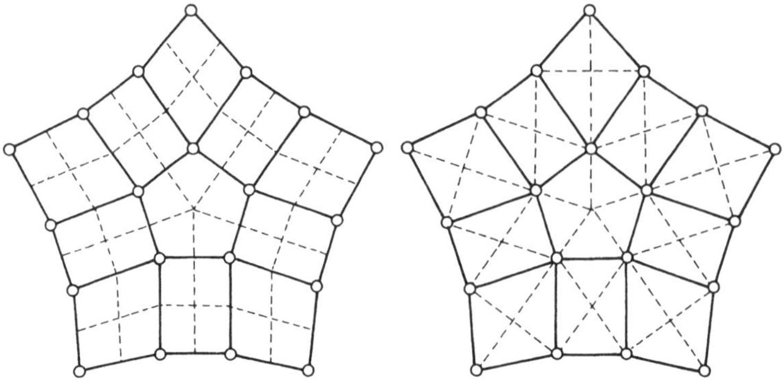

Figure 9.14: Two possible conversions to a quadrilateral net.

9.10 Singular parametrization

Consider a polynomial patch

$$\mathbf{b}(u,v) = \sum \mathbf{b}_{ij} B_{ij}^{mn}(u,v) \ ,$$

with a singularity at $(u,v) = \mathbf{o}$, i.e., with $\mathbf{b}_{00} = \mathbf{b}_{10} = \mathbf{b}_{01} = \mathbf{b}_{11}$ as illustrated for $m = n = 3$ in Figure 9.15.

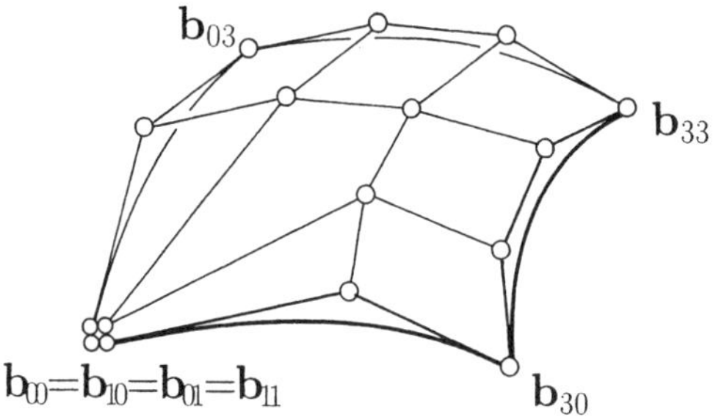

Figure 9.15: Bézier net with singularity.

The partial derivatives $\mathbf{b}_u$ and $\mathbf{b}_v$ are zero at $(u,v) = (0,0)$, i.e., the parametrization $\mathbf{b}$ is singular at $(0,0)$, and the Taylor expansion of $\mathbf{b}$ around $(0,0)$ is of the form

$$\mathbf{b}(u,v) = \sum \mathbf{a}_{ij} u^i v^j \ ,$$

where $\mathbf{a}_{10} = \mathbf{a}_{01} = \mathbf{a}_{11} = \mathbf{o}$. In general, there is no tangent plane for $\mathbf{b}$ at $(0,0)$. However, if $\mathbf{b}_{00}, \mathbf{b}_{20}, \mathbf{b}_{02}$ are independent and coplanar with $\mathbf{b}_{21}$

and $\mathbf{b}_{12}$, if $\mathbf{b}_{21}$ lies on the same side of the tangent $\mathbf{b}_{00}\mathbf{b}_{20}$ as $\mathbf{b}_{02}$ and if $\mathbf{b}_{12}$ lies on the same side of the tangent $\mathbf{b}_{00}\mathbf{b}_{02}$ as $\mathbf{b}_{20}$, then $\mathbf{b}(u, v)$ has a regular parametrization in a neighborhood of $(0, 0)$. In particular, $\mathbf{b}(u, v)$ has a well-defined and continuous tangent plane under these conditions [Reif '93, Theorem 3.3].

9.11 Bicubic C^1 splines of arbitrary topology

It is possible to build arbitrary patchworks with regular and degenerate bicubic patches such that adjacent patches have simple C^1 joints [Reif '93]. We describe how to build such spline surfaces, which can be used to interpolate the vertices of arbitrary quadrilateral nets.

Any piecewise bicubic surface with simple C^1 joints is completely determined by the inner Bézier points of each patch. These are marked by ∘ in Figure 9.16. The boundary Bézier points marked by □ and • , respectively, can be computed as the midpoints of two adjacent boundary Bézier points respectively.

The inner Bézier points next to a vertex surrounded by three or more than four patches must coincide in order to obtain C^1 joints along all patch boundary curves emanating from this so-called **extraordinary vertex**. Moreover, the patches around an extraordinary vertex have a common tangent plane at this vertex only if the interior Bézier points connected by dashed lines in Figure 9.16 are all coplanar and if they satisfy the conditions given in 9.10.

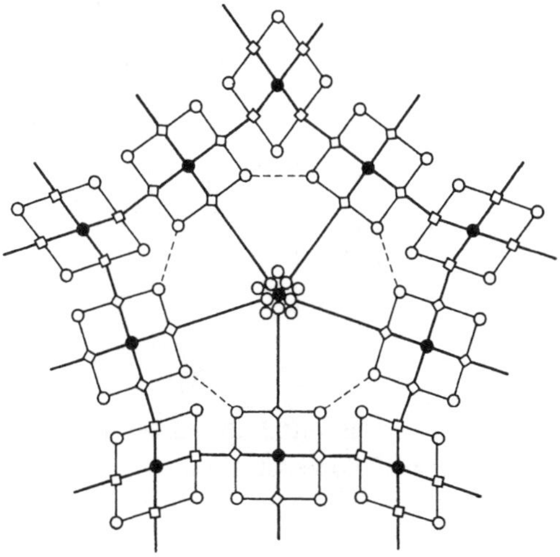

Figure 9.16: A bicubic C^1 spline.

Remark 12: The coplanarity condition above can, in general, only be satisfied if each patch has at most one extraordinary vertex. This assumption is always satisfied if one subdivides each patch into four subpatches.

Remark 13: The inner Bézier points $\circ$ determine the spline surface completely. However, they must meet certain restrictions for the spline to be a C^1 surface. Therefore, Reif calls these points **quasi control points**.

Remark 14: Reif has also developed a projection mapping arbitrary control nets onto quasi control nets satisfying the conditions above, see Problem 6. A more general method is presented in 14.6.

9.12 Problems

1 Generalize the curve generation algorithm in 3.5 to tensor product surfaces.

2 Generalize the curve generation algorithm based on forward differences in 3.6 to tensor product surfaces.

3 Generalize the curve-curve intersection algorithm in 3.7 to tensor product surfaces.

4 Implement an interpolation scheme based on the bicubic C^1 splines in 9.11 to interpolate the vertices of arbitrary quadrilateral nets.

5 Subdividing each patch of the bicubic C^1 spline in 9.10 into four subpatches over $[0, 1/2]^2, \ldots, [1/2, 1]^2$ corresponds to a refinement of the quasi control net, see Remark 13. Find this refinement construction.

6 Consider $2n$ vectors $\mathbf{c}_1, \ldots, \mathbf{c}_{2n}$ spanning $\mathbb{R}^3$ and let $c_i = \cos(i\frac{\pi}{n})$. Show that the vectors $\mathbf{d}_1, \ldots, \mathbf{d}_{2n}$, given by

$$[\mathbf{d}_1 \ldots \mathbf{d}_{2n}] = [\mathbf{c}_1 \ldots \mathbf{c}_{2n}] \begin{bmatrix} c_0 & c_1 & \cdots & c_{2n-1} \\ c_{-1} & c_0 & \cdots & c_{2n-2} \\ \vdots & \vdots & \ddots & \vdots \\ c_{1-2n} & c_{2-2n} & \cdots & c_0 \end{bmatrix},$$

are coplanar.

7 Extend the Hermite interpolation, as discussed in 4.4 for curves, to tensor product surfaces in Bézier representation.

10 Bézier representation of triangular patches

10.1 Bernstein polynomials — 10.2 Bézier triangles — 10.3 Linear precision — 10.4 The de Casteljau algorithm — 10.5 Derivatives — 10.6 Convexity — 10.7 Limitations of the convexity property — 10.8 Problems

The Bézier representation of triangular patches generalizes the univariate Bézier representation naturally. Except for the variation diminishing property, basically all properties of the Bézier representation of curves have a surface equivalent. The Bézier representation over triangles can be generalized further to Bézier representations over multi-dimensional simplices, see Chapter 19.

10.1 Bernstein polynomials

Because of their symmetry, Bernstein polynomials are best described by barycentric coordinates. Let A be a triangle in $\mathbb{R}^2$, and let A also denote the matrix $[\mathbf{a}_0, \mathbf{a}_1, \mathbf{a}_2]$ formed by the three vertices $\mathbf{a}_i$. Further, let $\mathbb{u} = [u\, v\, w]^t$ be the barycentric coordinate vector of a point $\mathbf{x}$ with respect to A. Then we can write this point as $\mathbf{x} = A\mathbb{u} = \mathbf{a}_0 u + \mathbf{a}_1 v + \mathbf{a}_2 w$, see 1.2.

Proceeding as in 2.1, we compute the trinomial expansion

$$1 = (u + v + w)^n = \sum_{i,j,k} \frac{n!}{i!j!k!} u^i v^j w^k \ ,$$

where $0 \le i, j, k$ and $i+j+k = n$. This leads to the **Bernstein polynomials** of degree n,

$$B^n_{ijk}(u, v, w) = \frac{n!}{i!j!k!} u^i v^j w^k \ ,$$

which we abbreviate to

$$B_{\mathbb{i}} = B_{\mathbb{i}}^n(\mathbb{u}) = \binom{n}{\mathbb{i}} \mathbb{u}^{\mathbb{i}} ,$$

where $\mathbb{i} = (i, j, k) \in \{0, 1, \ldots, n\}^3$ and $|\mathbb{i}| = i + j + k = n$. Examples are illustrated in Figure 10.1 while the canonical listing of the cubic Bernstein polynomials is shown in Figure 10.2.

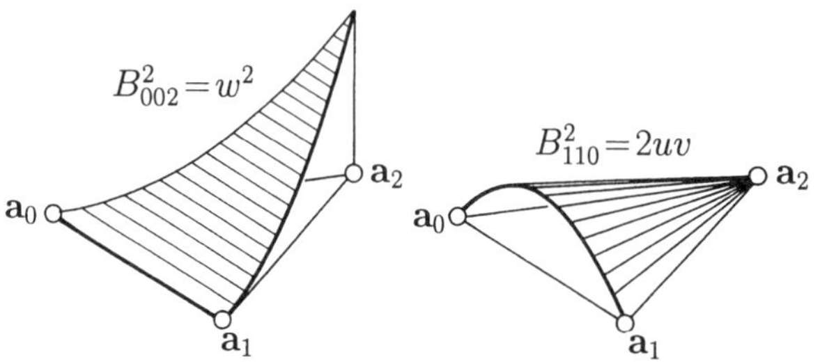

Figure 10.1: Two quadratic Bernstein polynomials.

$$\begin{array}{ccccccc} & & & B_{003} & & & \\ & & B_{102} & & B_{012} & & \\ & B_{201} & & B_{111} & & B_{021} & \\ B_{300} & & B_{210} & & B_{120} & & B_{030} \end{array} \qquad \begin{array}{ccccccc} & & & www & & & \\ & & 3uw^2 & & 3vw^2 & & \\ & 3u^2w & & 6uvw & & 3v^2w & \\ uuu & & 3u^2v & & 3uv^2 & & vvv \end{array}$$

Figure 10.2: The cubic Bernstein polynomials in canonical ordering.

Note that there are only two independent variables. Here, $\mathbb{u}$ represents the **local parameter** with respect to A and $\mathbf{x}$ the **global parameter**.

These definitions can be generalized. Multivariate Bernstein polynomials over a d-dimensional simplex A are defined completely analogously as

$$B_{\mathbb{i}}^n(\mathbb{u}) = \binom{n}{\mathbb{i}} \mathbb{u}^{\mathbb{i}} = \frac{n!}{i_0! \ldots i_d!} u_0^{i_0} \cdots u_d^{i_d} ,$$

where $\mathbb{i} = (i_0, \ldots, i_d) \in \{0, 1, \ldots, n\}^{d+1}$, $|\mathbb{i}| = i_0 + \cdots + i_d = n$, and $\mathbb{u} = (u_0, \ldots, u_d)$ is the barycentric coordinate vector of a point $\mathbf{x}$ with respect to A. In the following we list properties of Bernstein polynomials for arbitrary dimension d, see also 2.1.

- *The Bernstein polynomials of degree n are* **linearly independent**.

Namely, dividing

$$\sum b_{\mathbb{i}} \mathbb{u}^{\mathbb{i}} = 0$$

by u_0^n gives

$$\sum b_{\mathbb{i}} v_1^{i_1} \dots v_d^{i_d} = 0 \ ,$$

where $v_k = u_k/u_0$. Since the monomials are linearly independent, all $b_{\mathbb{i}}$ are zero, which concludes the proof.

There are $\binom{n+d}{d}$ Bernstein polynomials of degree n. Hence,

- *they form a* **basis** *for all d-variate polynomials of total degree $\leq n$.*
- *They are* **symmetric**, *i.e., $B_{\mathbb{i}}^n(\mathbb{u}) = B_{\pi(\mathbb{i})}^n(\pi(\mathbb{u}))$ for any permutation π.*
- *They have only* **roots** *at the faces of the simplex A. In particular,*

$$B_{\mathbb{i}}^n(\mathbb{e}_k) = \begin{cases} 1 & \text{for} \quad i_k = n \\ 0 & \quad i_k < n \ , \end{cases}$$

where $\mathbb{e}_0, \dots, \mathbb{e}_d$ represent the columns of the $(d+1)\times(d+1)$ identity matrix .

- *They form a* **partition of unity**,

$$\sum_{\mathbb{i}} B_{\mathbb{i}}^n(\mathbb{u}) \equiv 1 \ .$$

- *They are* **positive** *for $\mathbb{u} > \mathbb{0}$,*

which is the reason why one uses Bernstein polynomials mainly over the reference simplex A.

- *They satisfy the* **recursion formula**

$$B_{\mathbb{i}}^n(\mathbb{u}) = u_0 B_{\mathbb{i}-\mathbb{e}_0}^{n-1} + \cdots + u_d B_{\mathbb{i}-\mathbb{e}_d}^{n-1} \ ,$$

where $B_{\mathbb{o}}^0 = 1$ and $B_{\mathbb{i}}^n = 0$ if $\mathbb{i}$ has negative coordinates and $|\mathbb{i} - \mathbb{e}_j| = n - 1$.

10.2 Bézier simplices

Since the Bernstein polynomials form a basis, every polynomial surface $\mathbf{b}(\mathbf{x})$ has a unique **Bézier representation**,

$$\mathbf{b}(\mathbf{x}) = \sum_{\mathbb{i}} \mathbf{b}_{\mathbb{i}} B_{\mathbb{i}}^n(\mathbb{u}) \ ,$$

with respect to the reference simplex A. The coefficients $\mathbf{b}_{\mathbb{i}}$ are called the **Bézier points** of $\mathbf{b}$. They are the vertices of the **Bézier net** of $\mathbf{b}(\mathbf{x})$ over the simplex A, see Figure 10.3, where $d = 2$ and $n = 3$.

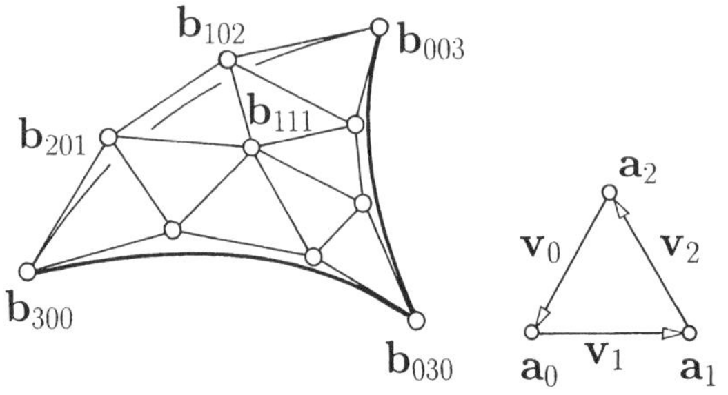

Figure 10.3: The Bézier net of a cubic surface.

The properties of Bernstein polynomials in 10.1 are passed on to the Bézier representation of a surface.

The symmetry property of the Bernstein polynomials implies that

- *The points* $\mathbf{b}_{\pi(\mathbb{i})}$ *are the Bézier points of* $\mathbf{b}$ *with respect to* $\pi^{-1}(\mathbf{a}_0 \ldots \mathbf{a}_d)$*, for any permutation* π.

- *The Bézier points of* $\mathbf{b}(\mathbf{x})$ *restricted to any lower dimensional* **face** *of the simplex* A *form the corresponding "face" of the Bézier net. For example, it holds, for any* $\delta < d$*, that*

$$\mathbf{b}(\mathbf{x}) = \sum \mathbf{b}_{i_0 \ldots i_\delta 0 \ldots 0} B^n_{i_0 \ldots i_\delta}(\mathbb{v}) \;,$$

where $\mathbb{v}$ *are the barycentric coordinates of* $\mathbf{x}$ *with respect to* $\mathbf{a}_0 \ldots \mathbf{a}_\delta$.

In particular, for the vertices and the edges of A, it follows that

$$\mathbf{b}(\mathbf{a}_0) = \mathbf{b}_{n0 \ldots 0}\,, \ldots, \mathbf{b}(\mathbf{a}_d) = \mathbf{b}_{0 \ldots 0n} \;,$$

and that

$$n(\mathbf{b}_{n-1,1,0,\ldots,0} - \mathbf{b}_{n,0,\ldots,0})$$

is the directional derivative with respect to $\mathbf{a}_1 - \mathbf{a}_0$ of $\mathbf{b}$ at $\mathbf{a}_0$.

Since the Bernstein polynomials sum to one,

- $\mathbf{b}(\mathbf{x})$ *is an* **affine combination** *of its Bézier points.*

Consequently,

- *the Bézier representation is* **affinely invariant**.

Since the Bernstein polynomials are non-negative over the reference simplex,

- $\mathbf{b}(\mathbf{x})$ *is a* **convex combination** *for every* $\mathbb{u} \geq \mathbb{o}$.

Hence,

- *the surface* $\mathbf{b}(A)$ *lies in the* **convex hull** *of its Bézier points.*

10.3 Linear precision

A linear polynomial

$$\mathbf{b}(\mathbb{u}) = u_0\mathbf{p}_0 + \cdots + u_d\mathbf{p}_d = P\mathbb{u} \ , \quad 1 = u_0 + \cdots + u_d \ ,$$

has also a Bézier representation of degree $n > 1$. It can be obtained as follows

$$\begin{aligned} \mathbf{b}(\mathbb{u}) &= [u_0\mathbf{p}_0 + \cdots + u_d\mathbf{p}_d] \sum_{\mathbb{j}} B_{\mathbb{j}}^{n-1}(\mathbb{u}) \\ &= \sum_{\mathbb{i}} \mathbf{b}_{\mathbb{i}} B_{\mathbb{i}}^{n}(\mathbb{u}) \ , \end{aligned}$$

where the points

$$\mathbf{b}_{\mathbb{i}} = [i_0\mathbf{p}_0 + \cdots + i_d\mathbf{p}_d]/n = P\mathbb{i}/n$$

are the vertices of a **uniform refinement** of the simplex $\mathbf{p}_0 \ldots \mathbf{p}_d$, see Figure 10.4.

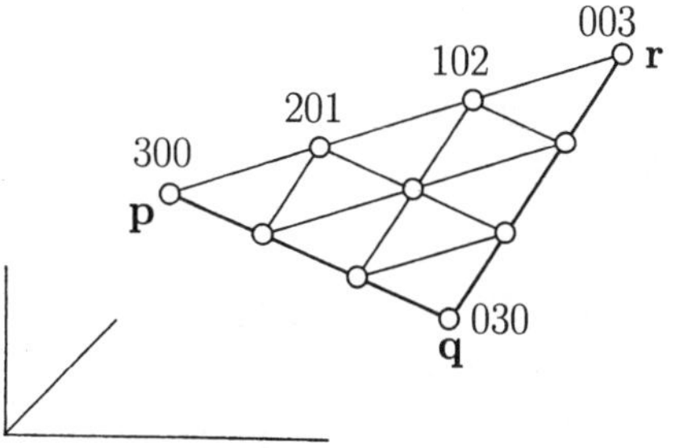

Figure 10.4: Uniform partition of the triangle **pqr**.

Remark 1: Sampling any linear polynomial $\mathbf{b}(\mathbb{u})$ at the grid points $\mathbf{b}_{\mathbb{i}}$, results in the polynomial's Bézier points, i.e.,

$$\mathbf{b}(\mathbb{i}/n) = \mathbf{b}_{\mathbb{i}} \quad \text{or} \quad \mathbf{b}(\mathbb{u}) = \sum_{\mathbb{i}} \mathbf{b}(\mathbb{i}/n) B_{\mathbb{i}}^n(\mathbb{u}) \ .$$

This property is referred to as the **linear precision** of the Bézier representation.

Remark 2: As a consequence of the linear precision, a functional surface

$$\mathbf{b}(\mathbf{x}) = \begin{bmatrix} \mathbf{x} \\ b(\mathbf{x}) \end{bmatrix} , \quad \text{where} \quad b(\mathbf{x}) = \sum b_{\mathbb{i}} B_{\mathbb{i}}^n ,$$

has the Bézier points $[\mathbf{a}_{\mathbb{i}}^t \; b_{\mathbb{i}}]^t$, where $n\mathbf{a}_{\mathbb{i}} = [\mathbf{a}_0 \dots \mathbf{a}d]\mathbb{i}$, as illustrated in Figure 10.5. The $b_{\mathbb{i}}$ are called **Bézier ordinates** and the corresponding points $\mathbf{a}_{\mathbb{i}}$ the **Bézier abscissae** of $b(\mathbf{x})$.

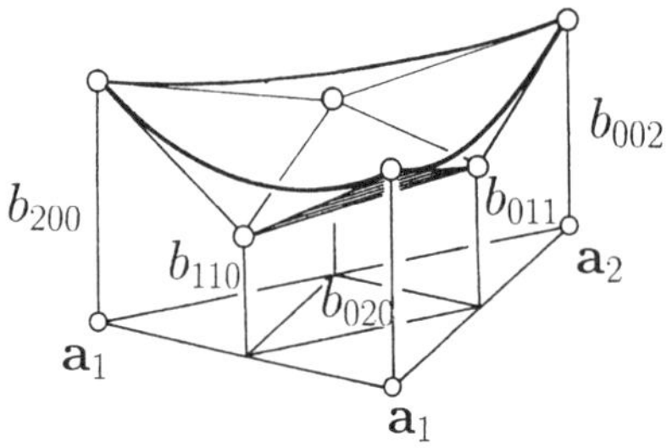

Figure 10.5: A quadratic function with its Bézier polyhedron.

10.4 The de Casteljau algorithm

A Bézier simplex $\mathbf{b} = \sum \mathbf{b}_{\mathbb{i}} B_{\mathbb{i}}^n$ can be evaluated by a generalization of de Casteljau's algorithm. Using the recurrence relation of the Bernstein polynomials repeatedly as in the case of curves, we first obtain

$$\begin{aligned} \mathbf{b}(\mathbf{x}) &= \sum_{|\mathbb{i}|=n} \mathbf{b}_{\mathbb{i}} B_{\mathbb{i}}^n(\mathbb{u}) \\ &= \sum_{|\mathbb{i}|=n-1} \mathbf{b}_{\mathbb{i}} B_{\mathbb{i}}^{n-1}(\mathbb{u}) \end{aligned}$$

and after $n-2$ further steps

$$\mathbf{b}(\mathbf{x}) = \sum_{|\mathbb{i}|=0} \mathbf{b}_{\mathbb{i}} B_{\mathbb{i}}^0(\mathbb{u}) = \mathbf{b}_{000} \ ,$$

where

$$\mathbf{b}_{\mathbb{i}} = [\mathbf{b}_{\mathbb{i}+\mathbb{e}_1} + \cdots + \mathbf{b}_{\mathbb{i}+\mathbb{e}_d}]\mathbb{u} \ .$$

An example is illustrated in Figure 10.6.

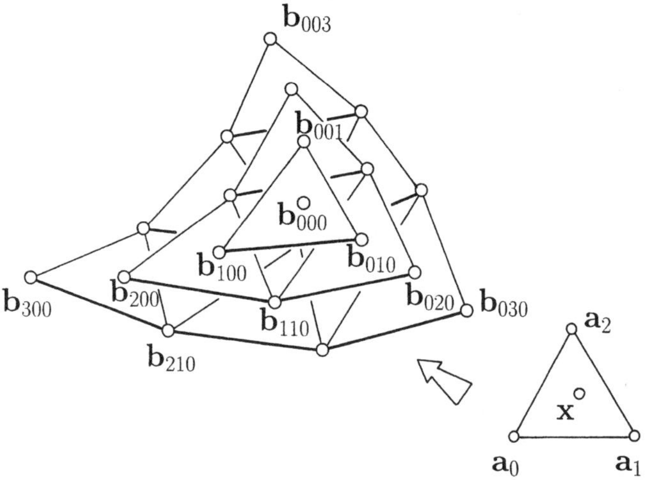

Figure 10.6: The de Casteljau construction.

The intermediate points $\mathbf{b}_{\mathbb{i}}$, $|\mathbb{i}| = n, \ldots, 0$, of the de Casteljau algorithm in their canonical ordering form a tetrahedral array. Further, if $\mathbf{x}$ lies in the simplex A, then the de Casteljau algorithm consists only of convex combinations, which accounts for the numerical stability of this algorithm.

10.5 Derivatives

The derivatives of a Bernstein polynomial are simple to compute. For a moment, let $u_0, \ldots, u_d$ be independent variables. Then, we obtain

$$\frac{\partial}{\partial u_j} B_{\mathbb{i}}^n = n B_{\mathbb{i}-\mathbb{e}_j}^{n-1} \ , \quad \text{etc.,}$$

where $B_{\mathbb{j}} = 0$ if $\mathbb{j}$ has a negative coordinate. Next, consider a line

$$\mathbf{x}(t) = \mathbf{p} + t\mathbf{v} \ ,$$

where $\mathbf{v} = \mathbf{a}_0 v_0 + \cdots + \mathbf{a}_d v_d$ and $v_0 + \cdots + v_d = 0$. This line in the domain corresponds to a curve $\mathbf{b}(\mathbf{x}(t))$ on a given surface

$$\mathbf{b}(\mathbf{x}) = \mathbf{b}(\mathbb{u}) = \sum_{\mathbb{i}} \mathbf{b}_{\mathbb{i}} B^n_{\mathbb{i}}(\mathbb{u}) \ .$$

Its **derivative** with respect to t at $t = 0$ is given by

$$\begin{aligned} D_{\mathbf{v}}\mathbf{b}(\mathbf{p}) &= \frac{\mathrm{d}}{\mathrm{d}t}\mathbf{b}\left(\mathbf{x}(t)\right)\Big|_{t=0} \\ &= v_0 \frac{\partial}{\partial u_0}\mathbf{b} + \cdots + v_d \frac{\partial}{\partial u_d}\mathbf{b} \\ &= n \sum_{\mathbb{j}} \mathbf{c}_{\mathbb{j}} B^{n-1}_{\mathbb{j}} \ , \end{aligned}$$

where

$$\mathbf{c}_{\mathbb{j}} = v_0 \mathbf{b}_{\mathbb{j}+\mathbb{e}_1} + \cdots + v_d \mathbf{b}_{\mathbb{j}+\mathbb{e}_d} \ ,$$

which we abbreviate by $\mathbf{c}^{\mathbb{j}} = \Delta_{\mathbf{v}} \mathbf{b}_{\mathbb{j}}$ as illustrated in Figure 10.7.

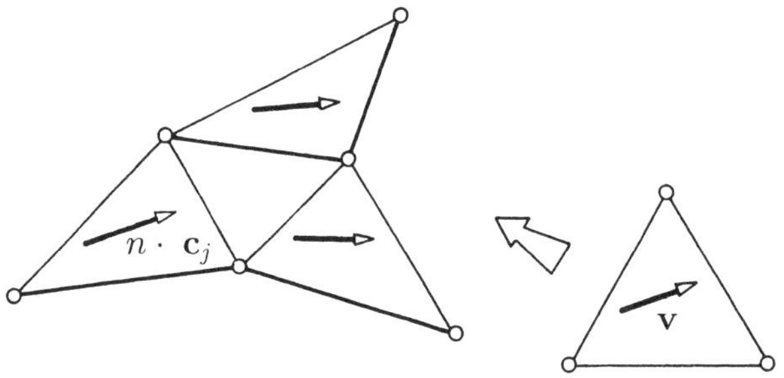

Figure 10.7: The differences $\mathbf{c}_{\mathbb{j}}$.

Similarly, one can compute higher derivatives. An rth directional derivative $D_{\mathbf{v}_1} \ldots D_{\mathbf{v}_r}\mathbf{b}$ has the Bézier coefficients $\Delta^{\mathbf{v}_1} \ldots \Delta^{\mathbf{v}_r}\mathbf{b}_{\mathbb{j}}$, where $|\mathbb{j}| = n - r$. The difference operator $\Delta^{\mathbf{v}}$ commutes with the steps of the de Casteljau algorithm since the computation of affine combinations of affine combinations is commutative, see 2.6.

Hence, we can compute an rth derivative also by first computing $n - r$ steps of the de Casteljau algorithm and then r differencing steps. In particular, it follows that the points $\mathbf{b}_{10\ldots0}, \ldots, \mathbf{b}_{0\ldots01}$ computed in the next to last step of the de Casteljau algorithm span the tangent plane of $\mathbf{b}$ at $\mathbf{x}$.

Remark 3: If $d = 2$, we can view the Bézier net of a polynomial $\mathbf{b}(\mathbf{x}) = \sum \mathbf{b}_{\mathbb{i}} B^n_{\mathbb{i}}(\mathbb{u})$ as a piecewise linear function $\mathbf{p}(\mathbf{x})$ over $\mathbf{a}_0\mathbf{a}_1\mathbf{a}_2$. Then,

the directional derivative $D_{\mathbf{v}}\mathbf{p}(\mathbf{x})$ of the Bézier net contains the

Bézier points of $D_{\mathbf{v}}\mathbf{b}(\mathbf{x})$.

This fact is illustrated in Figure 10.8 for a functional surface.

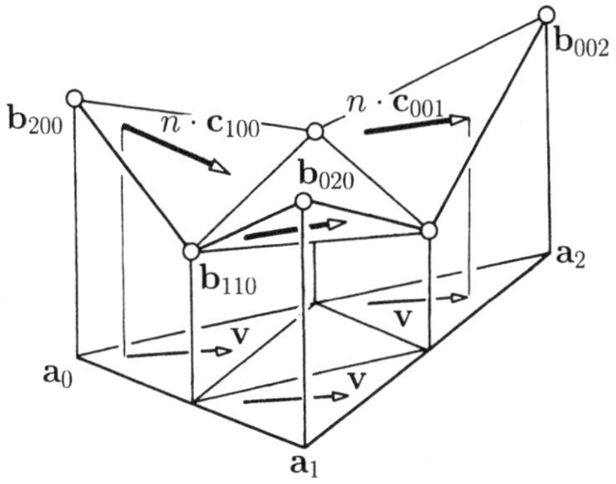

Figure 10.8: A Bézier net with its derivative.

10.6 Convexity

In the following we restrict ourselves to bivariate functions. Hence, we set $d = 2$ and $\mathbb{u} = [u, v, w]^t$.

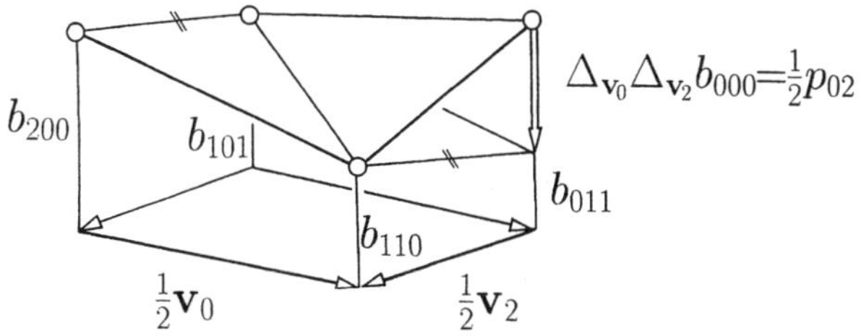

Figure 10.9: Convexity of a Bézier net.

Given the Bézier representation of a polynomial over some triangle $\mathbf{a}_1\mathbf{a}_2\mathbf{a}_3$, there is a piecewise linear polynomial $p(\mathbf{x})$ that interpolates the Bézier ordinates $b_{\mathbb{i}}$ at the corresponding abscissae $A\mathbb{i}/n$. We call it the **Bézier polyhedron** of $b(\mathbf{x})$ over A and show that

a polynomial $b(\mathbf{x})$ *is convex if its Bézier polyhedron* $p(\mathbf{x})$ *is convex.*

The converse does not hold in general, see 3.13 Problem 11.

For a proof, let

$$\mathbf{v}_0 = \mathbf{a}_2 - \mathbf{a}_1 \ , \quad \mathbf{v}_1 = \mathbf{a}_0 - \mathbf{a}_2 \ , \quad \mathbf{v}_2 = \mathbf{a}_1 - \mathbf{a}_0$$

as illustrated in Figure 10.3, and let

$$b_{\mu\nu} = D_{\mathbf{v}_\mu} D_{\mathbf{v}_\nu} b$$

denote the second derivative of b with respect to the directions $\mathbf{v}_\mu$ and $\mathbf{v}_\nu$, and let $p_{\mu\nu}(\mathbf{x})$ denote its Bézier polyhedron.

The Bézier polyhedron $p(\mathbf{x})$ is convex if and only if every pair of its triangles with a common edge form a convex function. This is true exactly if

$$p_{01} \le 0, p_{12} \le 0 \quad \text{and} \quad p_{20} \le 0 \ ,$$

see Figure 10.9. Because of the convex hull property, this implies $b_{01} \le 0, b_{12} \le 0$, and $b_{20} \le 0$, and furthermore,

$$\begin{aligned} b_{00} &= -b_{01} - b_{02} \ \ge \ 0 \ , \\ b_{11} &= -b_{12} - b_{10} \ \ge \ 0 \ , \\ b_{22} &= -b_{20} - b_{21} \ \ge \ 0 \ . \end{aligned}$$

Since we can write any direction $\mathbf{v} \in \mathbb{R}^2$ as

$$\mathbf{v} = \alpha \mathbf{v}_\mu + \beta \mathbf{v}_\nu \ , \quad \text{where} \quad \alpha\beta \le 0 \quad \text{and} \quad \mu, \nu \in \{0,1,2\} \ ,$$

we can express any second derivative $D_{\mathbf{v}} D_{\mathbf{v}} b$ as a sum of three non-negative terms

$$D_{\mathbf{v}} D_{\mathbf{v}} b = \alpha^2 b_{\mu\mu} + 2\alpha\beta b_{\mu\nu} + \beta^2 b_{\nu\nu} \ge 0 \ ,$$

which proves the convexity of b. $\diamond$

Remark 4: A convex Bézier polygon in the plane represents a convex curve segment. For triangular Bézier nets in space, the analogue is not true, in general, as is shown by Figure 10.10.

10.7 Limitations of the convexity property

Functional polynomials with a convex Bézier polyhedron are convex, as we have seen. The converse, however, is not true in general. Consider the quadratic polynomial

$$b = 3B_{200} - B_{101} + 3B_{002} \ ,$$

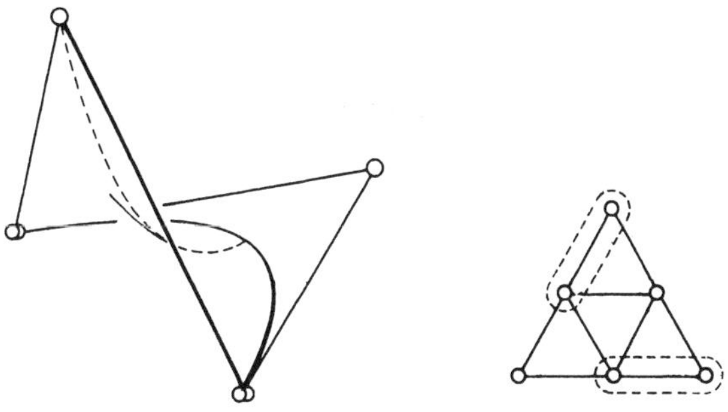

Figure 10.10: A non-convex quadratic patch with convex Bézier net.

which is shown with its Bézier polyhedron in Figure 10.11.

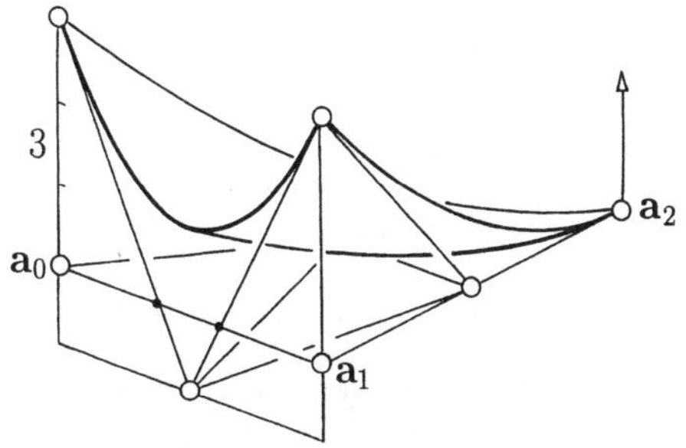

Figure 10.11: A convex polynomial patch with non-convex Bézier net.

Obviously, the Bézier polyhedron is not convex, but b is convex. Namely, the second partial derivatives are $b_{00} = b_{11} = 6$ and $b_{01} = 2$. Thus, for any $\mathbf{v} = \alpha\mathbf{v}_0 + \beta\mathbf{v}_1 \neq \mathbf{o}$ it follows that

$$\begin{aligned} D_{\mathbf{v}}D_{\mathbf{v}}b &= \alpha^2 b_{00} + 2\alpha\beta b_{01} + \beta^2 b_{11} \\ &= 4(\alpha^2+\beta^2) + 2(\alpha+\beta)^2 > 0 \ , \end{aligned}$$

which means that b is even strictly convex.

The Bézier polyhedra of all higher degree representations of b are also **not** convex since for any nth degree representation of the constant polynomial $b_{01} = 2$ all Bézier ordinates equal 2. This result is striking since the nth degree Bézier polyhedra of b converge to b if n goes to infinity, see 11.9. Thus, we have a sequence of non-convex functions with a strictly convex limit.

Another negative result is due to Grandine [Grandine '89]. Consider a non-convex quadrilateral **abdc**, as the ones illustrated in Figure 10.12, and let b and c be polynomials with C^1 contact along the line **ad**. Then, if b and c have convex Bézier nets over **adc** and **abd**, respectively, they must be linear over **ad**.

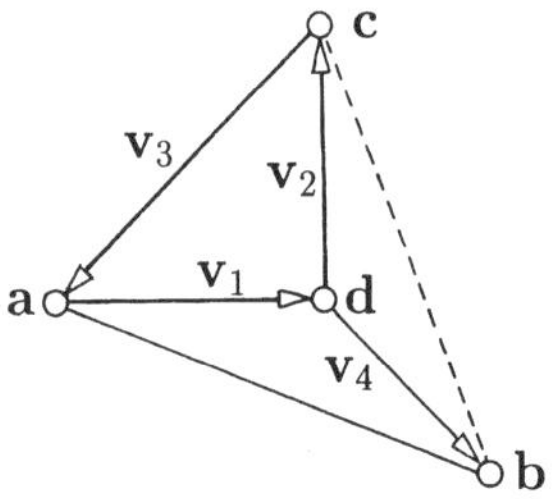

Figure 10.12: Non-convex quadrilateral domain.

Because of the C^1 contact, we have $b_{14} = c_{14} (= D_{\mathbf{v}_1} D_{\mathbf{v}_4} c)$ along **ad**. Since there are positive constants α, and β such that $\mathbf{v}_1 = \alpha \mathbf{v}_2 + \beta \mathbf{v}_4$, the convexity of the Bézier polyhedra implies that

$$b_{11} = \alpha b_{12} + \beta b_{14} \leq 0 \qquad \text{over the line} \quad \mathbf{ad}$$

and also $b_{11} \geq 0$. Thus, $b_{11} = 0$, which concludes the proof.

This result has a surprising consequence. Let b have a convex Bézier polyhedron over **abc**. Then, its Bézier polyhedra over the three triangles **abd**, **bcd**, and **adc** cannot all be convex unless b is linear over the three lines **ad**, **bd** and **cd**. Since b is convex, this implies that b is entirely linear. Hence,

subdivision, as described in 11.3, does not preserve convexity.

However, see Problem 1 in 11.12.

10.8 Problems

1 A Bernstein polynomial $B^n_{\mathbb{i}}(\mathbb{u})$ has only one maximum over the reference simplex, namely at $\mathbb{u} = \mathbb{i}/n$.

2 The **Bernstein operator** $\mathcal{B}$ assigns to a function $f(\mathbb{u})$ over the reference triangle a polynomial by

$$\mathcal{B}[f] = \sum f(\mathbb{i}/n) B^n_{\mathbb{i}}(\mathbb{u}) \ .$$

Show that the Bernstein operator maps polynomials of degree $m \leq n$ to polynomials of degree m.

3 Show that the monomial $u^\alpha v^\beta$ has the Bézier representation

$$u^\alpha v^\beta = \frac{\alpha!\beta!\gamma!}{n!} \sum_{\mathbf{i}} \binom{i}{\alpha}\binom{j}{\beta} B^n_{\mathbf{i}}(u, v, w) \ ,$$

where $w = 1 - u - v$ and $\gamma = n - \alpha - \beta$ and $\alpha, \beta, \gamma \geq 0$.

4 Compute the Bézier representation over the triangle $\begin{bmatrix}1\\0\end{bmatrix} \begin{bmatrix}0\\1\end{bmatrix} \begin{bmatrix}0\\0\end{bmatrix}$ of the polynomial

$$\mathbf{b}(x, y) = \sum_{0 \leq i+j \leq n} \mathbf{a}_{ij} x^i y^j \ .$$

5 Let $\Delta_{\alpha\beta}$ be the difference operator defined recursively by

$$\begin{aligned} \Delta^{\alpha\beta}\mathbf{b}_{\mathbf{i}} &= \Delta^{\alpha-1,\beta}\mathbf{b}_{\mathbf{i}+\mathbf{e}_1-\mathbf{e}_3} - \Delta^{\alpha-1,\beta}\mathbf{b}_{\mathbf{i}} \\ &= \Delta^{\alpha,\beta-1}\mathbf{b}_{\mathbf{i}+\mathbf{e}_2-\mathbf{e}_3} - \Delta^{\alpha,\beta-1}\mathbf{b}_{\mathbf{i}} \\ \text{and} \quad \Delta^{00}\mathbf{b}_{\mathbf{i}} &= \mathbf{b}_{\mathbf{i}} \ . \end{aligned}$$

Show that

$$\Delta^{\alpha\beta}\mathbf{b}_{00n} = \sum_{i=0}^{\alpha}\sum_{j=0}^{\beta} \binom{\alpha}{i}\binom{\beta}{j}(-1)^{\alpha+\beta-i-j}\mathbf{b}_{ijk} \ ,$$

where $k = n - i - j$.

6 Use a Taylor expansion and Problem 5 to show that

$$\sum \mathbf{b}_{\mathbf{i}} B^n_{\mathbf{i}}(\mathbf{u}) = \sum_{0 \leq \alpha+\beta \leq n} \sum_{\mathbf{i}} \binom{\alpha}{i}\binom{\beta}{j}(-1)^{k-\gamma}\mathbf{b}_{\mathbf{i}} \frac{n!}{\alpha!\beta!\gamma!} u^\alpha v^\beta \ ,$$

where $\gamma = n - \alpha - \beta$.

11 Bézier techniques for triangular patches

11.1 Symmetric polynomials — 11.2 The main theorem — 11.3 Subdivision and reparametrization — 11.4 Convergence under subdivision — 11.5 Surface generation — 11.6 The symmetric polynomial of the derivative — 11.7 Simple C^r joints — 11.8 Degree elevation — 11.9 Convergence under degree elevation — 11.10 Conversion to tensor product Bézier representation — 11.11 Conversion to triangular Bézier representation — 11.12 Problems

Just as univariate polynomials, multivariate polynomials correspond uniquely to symmetric multiaffine polynomials. With symmetric polynomials, it is a simple task to derive algorithms which evaluate, degree elevate, reparametrize, or subdivide a triangular surface in Bézier representation. The generalization of the techniques described for univariate polynomials in Chapter 3 is straightforward.

11.1 Symmetric polynomials

Every polynomial surface $\mathbf{b}(\mathbf{x})$ of total degree $\le n$ can be associated with a unique n-affine **symmetric polynomial** $\mathbf{b}[\mathbf{x}_1 \ldots \mathbf{x}_n]$ over $\mathbb{R}^2$ having the following three properties.

- $\mathbf{b}[\mathbf{x}_1 \ldots \mathbf{x}_n]$ *agrees with* $\mathbf{b}(\mathbf{x})$ *on its* **diagonal**, *i.e.,*

$$\mathbf{b}[\mathbf{x} \ldots \mathbf{x}] = \mathbf{b}(\mathbf{x}) \ .$$

- $\mathbf{b}[\mathbf{x}_1 \ldots \mathbf{x}_n]$ *is* **symmetric** *in its variables, i.e., for any permutation* $(\mathbf{y}_1 \cdots \mathbf{y}_n)$ *of* $(\mathbf{x}_1 \ldots \mathbf{x}_n)$, *we obtain*

$$\mathbf{b}[\mathbf{y}_1 \cdots \mathbf{y}_n] = \mathbf{b}[\mathbf{x}_1 \ldots \mathbf{x}_n] \ .$$

- $\mathbf{b}[\mathbf{x}_1 \ldots \mathbf{x}_n]$ *is* **affine** *in each variable, i.e.,*

$$\mathbf{b}[(\alpha\mathbf{x}+(1-\alpha)\mathbf{y})\,\mathbf{x}_2\,\dots\,\mathbf{x}_n] = \alpha\mathbf{b}[\mathbf{x}\,\mathbf{x}_2\,\dots\,\mathbf{x}_n] + (1-\alpha)\mathbf{b}[\mathbf{y}\,\mathbf{x}_2\,\dots\,\mathbf{x}_n]\ .$$

The symmetric polynomial $\mathbf{b}[\mathbf{x}_1\,\dots\,\mathbf{x}_n]$ is also referred to as the **polar form** [Casteljau '85] or **blossom** [Ramshaw '87] of $\mathbf{b}(\mathbf{x})$.

To show that such symmetric polynomials exist for all polynomials, it suffices to consider basis polynomials and to derive explicit representations of their symmetric forms. Any linear combination

$$\mathbf{b}(\mathbf{x}) = \sum_{\mathbb{i}} \mathbf{c}_{\mathbb{i}}\, C_{\mathbb{i}}(\mathbf{x})$$

of nth degree polynomials $C_{\mathbb{i}}(\mathbf{x})$ with polar forms $C_{\mathbb{i}}[\mathbf{x}_1\,\dots\,\mathbf{x}_n]$, where $\mathbb{i} = (i,j,k) \geq \mathbb{o}$ and $|\mathbb{i}| = n$, has the symmetric polynomial

$$\mathbf{b}[\mathbf{x}_1\,\dots\,\mathbf{x}_n] = \sum_{\mathbb{i}} \mathbf{c}_{\mathbb{i}}\, C_{\mathbb{i}}[\mathbf{x}_1\,\dots\,\mathbf{x}_n]\ ,$$

which clearly satisfies the three properties above.

Note that the diagonal $\mathbf{b}[\mathbf{x}\,\dots\,\mathbf{x}]$ can be of lower degree than n, although $\mathbf{b}[\mathbf{x}_1\,\dots\,\mathbf{x}_n]$ depends on n variables.

In case the $C_{\mathbb{i}}$ are the weighted monomials $A^n_{ij}(x,y) = \binom{n}{\mathbb{i}} x^i y^j$, we obtain the **elementary symmetric polynomials**

$$A^n_{ij}[\mathbf{x}_1\,\dots\,\mathbf{x}_n] = \sum_{\substack{\alpha<\dots<\beta\\ \gamma<\dots<\delta}} x_\alpha \overset{i}{\dots} x_\beta\, y_\gamma \overset{j}{\dots} y_\delta\ ,$$

where $\mathbf{x}_\alpha = (x_\alpha, y_\alpha)$ and $\alpha,\dots,\beta,\gamma,\dots,\delta$ are $i+j$ distinct integers between 1 and n. Obviously, the three characterizing properties of the polar form are satisfied.

In case the $C_{\mathbb{i}}$ are the Bernstein polynomials $B^n_{\mathbb{i}}(\mathbb{u}) = \binom{n}{\mathbb{i}}\mathbb{u}^{\mathbb{i}}$, we obtain

$$B^n_{\mathbb{i}}[\mathbb{u}_1\,\dots\,\mathbb{u}_n] = \sum_{\substack{\alpha<\dots<\beta\\ \gamma<\dots<\delta\\ \varepsilon<\dots<\varphi}} u_\alpha \overset{i}{\cdots} u_\beta\, v_\gamma \overset{j}{\cdots} v_\delta\, w_\varepsilon \overset{k}{\cdots} w_\varphi\ ,$$

where $u_\alpha, v_\alpha, w_\alpha$ are the coordinates of $\mathbb{u}_\alpha$ and $(\alpha,\dots,\beta,\gamma,\dots,\delta,\varepsilon,\dots,\varphi)$ is a permutation of $(1,\dots,n)$. Again, one can easily check the three properties above.

Remark 1: The symmetric polynomials $B^n_{\mathbb{i}}[\mathbb{u}_1\,\dots\,\mathbb{u}_n]$ satisfy the recursion

$$B^n_{\mathbb{i}}[\mathbb{u}_1\,\dots\,\mathbb{u}_n] = u_1 B^{n-1}_{\mathbb{i}-\mathbb{e}_1}[\mathbb{u}_2\,\dots\,\mathbb{u}_n] + v_1 B^{n-1}_{\mathbb{i}-\mathbb{e}_2}[\mathbb{u}_2\,\dots\,\mathbb{u}_n] + w_1 B^{n-1}_{\mathbb{i}-\mathbb{e}_3}[\mathbb{u}_2\,\dots\,\mathbb{u}_n].$$

Remark 2: The barycentric coordinate vector $\mathbb{u}$ and the affine coordinate vector $\mathbf{x}$ are related by two transformations, given by

$$\mathbf{x} = \mathbf{x}(\mathbb{u}) = [\mathbf{a}_0\mathbf{a}_1\mathbf{a}_2]\mathbb{u} \quad \text{and} \quad \mathbb{u} = \mathbb{u}(\mathbf{x}) = \mathbb{p} + \mathbb{v}\mathbf{x} \ .$$

Since these transformations are affine, one can transform a polar form $\mathbf{a}[\mathbf{x}_1 \dots \mathbf{x}_n]$, given by affine coordinates, to the corresponding polar form $\mathbf{b}[\mathbb{u}_1 \dots \mathbb{u}_n] = \mathbf{a}[A\mathbb{u}_1 \dots A\mathbb{u}_n]$ and vice versa, i.e., $\mathbf{a}[\mathbf{x}_1 \dots \mathbf{x}_n] = \mathbf{b}[\mathbb{u}(\mathbf{x}_1) \dots \mathbb{u}(\mathbf{x}_n)]$.

11.2 The main theorem

The uniqueness of the symmetric polynomials and their relationship to the Bézier representation is given by the following extension of the **main theorem**.

> *For every polynomial surface* $\mathbf{b}(\mathbf{x})$ *of degree* $\le n$*, there exists only one symmetric* n*-variate multiaffine polynomial* $\mathbf{b}[\mathbf{x}_1 \dots \mathbf{x}_n]$ *with diagonal* $\mathbf{b}[\mathbf{x} \dots \mathbf{x}] = \mathbf{b}(\mathbf{x})$*, and the points*
>
> $$\mathbf{b}^0_{\mathbb{i}} = \mathbf{b}[\mathbf{p} \overset{i}{\dots} \mathbf{p}\, \mathbf{q} \overset{j}{\dots} \mathbf{q}\, \mathbf{r} \overset{k}{\dots} \mathbf{r}]$$
>
> *are the Bézier points of* $\mathbf{b}(\mathbf{x})$ *over* $\mathbf{pqr}$.

Proof: Consider the points

$$\mathbf{b}^l_{\mathbb{i}} = \mathbf{b}[\mathbf{p} \overset{i}{\dots} \mathbf{p}\, \mathbf{q} \overset{j}{\dots} \mathbf{q}\, \mathbf{r} \overset{k}{\dots} \mathbf{r}\, \mathbf{x}_1 \overset{l}{\dots} \mathbf{x}_l] \ , \qquad i+j+k+l = n \ .$$

Since $\mathbf{b}^n_{\mathbb{o}} = \mathbf{b}[\mathbf{x}_1 \dots \mathbf{x}_n]$ is symmetric and multiaffine, it can be computed from the points $\mathbf{b}^0_{\mathbb{i}}$ by the recursion formula

$$(1) \qquad \mathbf{b}^l_{\mathbb{i}} = u_l \mathbf{b}^{l-1}_{\mathbb{i}+\mathbb{e}_1} + v_l \mathbf{b}^{l-1}_{\mathbb{i}+\mathbb{e}_2} + w_l \mathbf{b}^{l-1}_{\mathbb{i}+\mathbb{e}_3} \ ,$$

where u_l, v_l, w_l are the barycentric coordinates of $\mathbf{x}_l$ with respect to $\mathbf{pq}$, see Figure 11.1, where the points $\mathbf{b}[\mathbf{x}_1\mathbf{x}_2\mathbf{x}_3]$ are labelled by their arguments $\mathbf{x}_1\mathbf{x}_2\mathbf{x}_3$. Thus, different symmetric multiaffine maps must differ at some argument $[\mathbf{p} \overset{i}{\dots} \mathbf{p}\, \mathbf{q} \overset{j}{\dots} \mathbf{q}\, \mathbf{r} \overset{k}{\dots} \mathbf{r}]$.

If all $\mathbf{x}_l$ equal $\mathbf{x}$, then the recursion formula above reduces to de Casteljau's algorithm for the computation of $\mathbf{b}(\mathbf{x})$. Consequently, since the Bézier representation is unique, the points $\mathbf{b}^0_{\mathbb{i}}$ are the Bézier points of $\mathbf{b}(\mathbf{x})$ over $\mathbf{pqr}$ and, furthermore, there can be only one symmetric n-affine polynomial with the diagonal $\mathbf{b}(\mathbf{x})$. $\diamond$

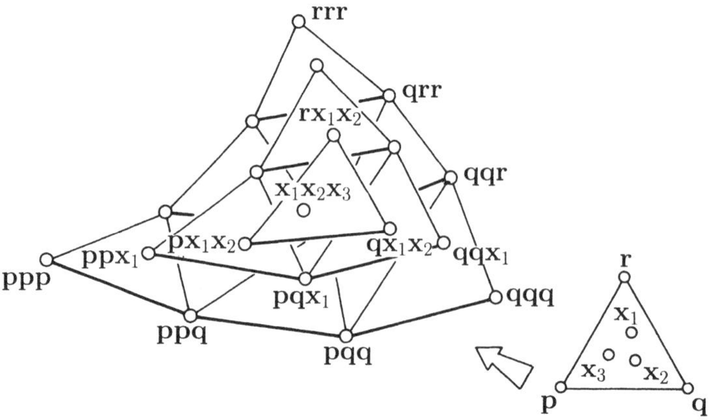

Figure 11.1: The generalized de Casteljau algorithm.

11.3 Subdivision and reparametrization

Recursion formula (1), which is illustrated in Figure 11.1, reveals another important property of de Casteljau's algorithm. The computation of $\mathbf{b}(\mathbf{x})$ generates also the Bézier points

$$\mathbf{b}[\mathbf{p} \overset{i}{\ldots} \mathbf{p}\,\mathbf{q} \overset{j}{\ldots} \mathbf{q}\,\mathbf{x} \overset{k}{\ldots} \mathbf{x}], \quad \mathbf{b}[\mathbf{p} \overset{i}{\ldots} \mathbf{p}\,\mathbf{x} \overset{j}{\ldots} \mathbf{x}\,\mathbf{r} \overset{k}{\ldots} \mathbf{r}]\ ,$$

and

$$\mathbf{b}[\mathbf{x} \overset{i}{\ldots} \mathbf{x}\,\mathbf{q} \overset{j}{\ldots} \mathbf{q}\,\mathbf{r} \overset{k}{\ldots} \mathbf{r}]$$

of $\mathbf{b}$ over $\mathbf{pqx}$, $\mathbf{pxr}$, and $\mathbf{xqr}$, respectively. Figure 11.2 shows an example for $n = 3$.

The Bézier nets of $\mathbf{b}(\mathbf{x})$ over $\mathbf{pqx}$, $\mathbf{pxr}$, and $\mathbf{xqr}$ form one connected net. It is folded if $\mathbf{x}$ lies outside $\mathbf{pqr}$. The computation of this composed net will be referred to as the **subdivision** of the Bézier net over $\mathbf{pqr}$ in $\mathbf{x}$.

One can compute the Bézier net of a polynomial surface $\mathbf{b}$ over a second triangle $\mathbf{xyz}$ by repeated subdivision from the net over $\mathbf{pq}$, see [Prautzsch '84a, Boehm et al. '84]. First one subdivides the net over $\mathbf{pq}$ in $\mathbf{x}$, then one subdivides the net over $\mathbf{xqr}$ in $\mathbf{y}$, and, finally, one subdivides the net over $\mathbf{xyr}$ in $\mathbf{z}$, see Figure 11.3.

A permutation of $\mathbf{pq}$ and $\mathbf{xyz}$ results in a different construction. If possible, one should subdivide at interior points in order to avoid non-convex combinations.

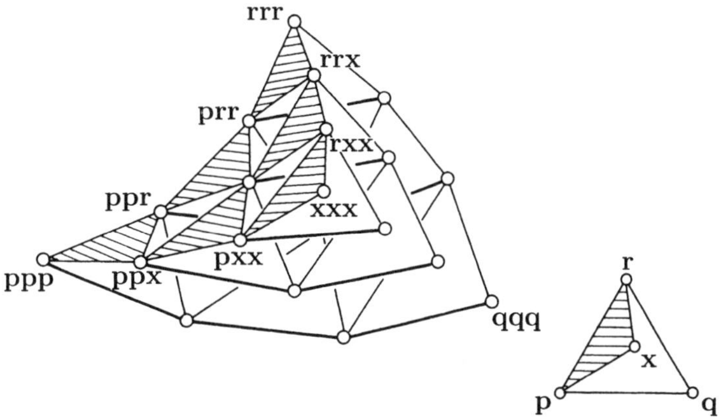

Figure 11.2: Subdividing a Bézier net.

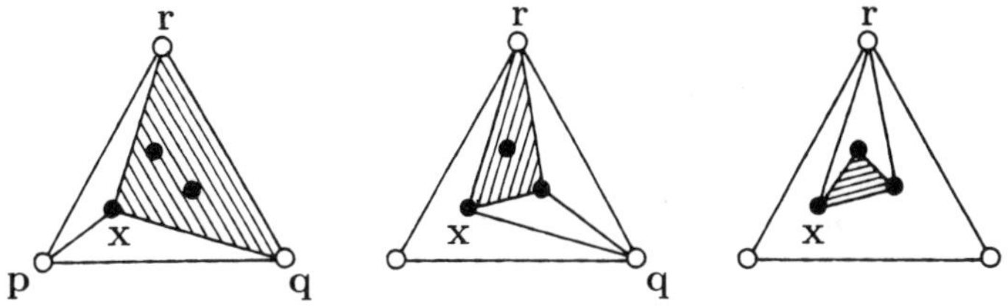

Figure 11.3: Reparametrization by repeated subdivision.

Figure 11.4 shows a situation, where it is impossible to avoid non-convex combinations with the above construction, no matter how one permutes **pq** and **xyz**.

Remark 3: The construction requires one to compute $3 \cdot \binom{n+3}{3} = O(n^3)$ affine combinations.

Remark 4: Every single Bézier point $\mathbf{b}[\mathbf{x} \overset{i}{\ldots} \mathbf{x}\,\mathbf{y} \overset{j}{\ldots} \mathbf{y}\,\mathbf{z} \overset{k}{\ldots} \mathbf{z}]$ of **b** over **xyz** can also be computed by the generalized de Casteljau algorithm, see Figure 11.1. The affine combinations computed by this algorithm are convex if **x**, **y** and **z** all lie in the triangle **pq**.

Remark 5: To compute the Bézier net over **xyz** by $\binom{n+2}{2}$ applications of the generalized de Casteljau algorithm, one needs to compute $\binom{n+2}{2} \cdot \binom{n+3}{3} = O(n^5)$ affine combinations.

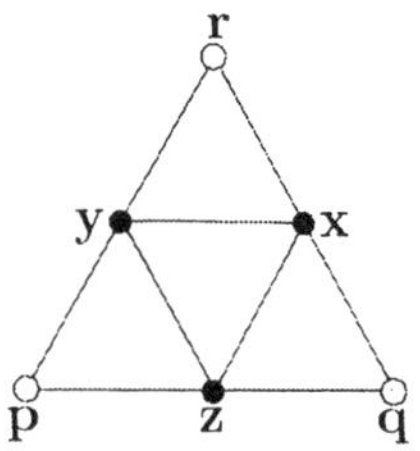

Figure 11.4: Special reference triangles.

11.4 Convergence under subdivision

The Bézier net of $\mathbf{b}(\mathbf{x})$ over a triangle $\mathbf{pq}$ is a good approximation of the patch $\mathbf{b}$ if the triangle is sufficiently small. To make this statement precise, let $\mathbf{pq}$ be any triangle in some fixed bounded region and let h be its diameter. Furthermore, let

$$\mathbf{i} = \mathbf{p}\frac{i}{n} + \mathbf{q}\frac{j}{n} + \mathbf{r}\frac{k}{n}$$

represent the point with barycentric coordinates $\mathbb{i}/n$. Then,

there is a constant M not depending on $\mathbf{pq}$ such that

$$\max_{\mathbb{i}} \|\mathbf{b}(\mathbf{i}) - \mathbf{b}_{\mathbb{i}}\| \leq Mh^2 \ .$$

For a proof, let D be the differential of $\mathbf{b}[\mathbf{x}\,\mathbf{i} \dots \mathbf{i}] = \cdots = \mathbf{b}[\mathbf{i} \dots \mathbf{i}\,\mathbf{x}]$ at $\mathbf{x} = \mathbf{i}$. Expanding the symmetric polynomial $\mathbf{b}[\mathbf{x}_1 \dots \mathbf{x}_n]$ around $[\mathbf{i} \dots \mathbf{i}]$, we obtain

$$\begin{aligned} \mathbf{b}_{\mathbb{i}} &= \mathbf{b}[\mathbf{i} \dots \mathbf{i}] + iD[\mathbf{p} - \mathbf{i}] + jD[\mathbf{q} - \mathbf{i}] + kD[\mathbf{r} - \mathbf{i}] + O(h^2) \\ &= \mathbf{b}(\mathbf{i}) + O(h^2) \ , \end{aligned}$$

which concludes the proof. $\diamond$

An application of this approximation property is discussed in the following section.

11.5 Surface generation

As a consequence of section 11.4, repeated subdivision of a Bézier net produces arbitrarily good approximations of the underlying surface. We discuss three subdivision strategies.

(1) Subdividing triangles at their centers, as illustrated in Figure 11.5, leaves

the maximum diameter of the reference triangles unchanged. Hence, the Bézier nets over these triangulations do **not** converge to the surface.

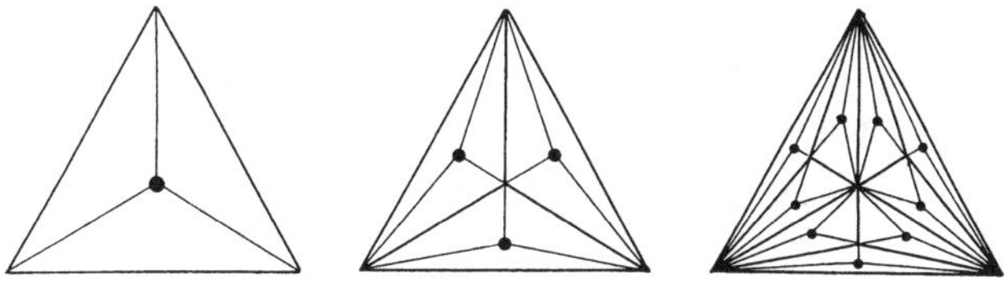

Figure 11.5: Subdivision at centers.

(2) Subdividing every triangle uniformly, as illustrated in Figures 11.6 and 11.7, generates a sequence of Bézier nets which converges to the surface.

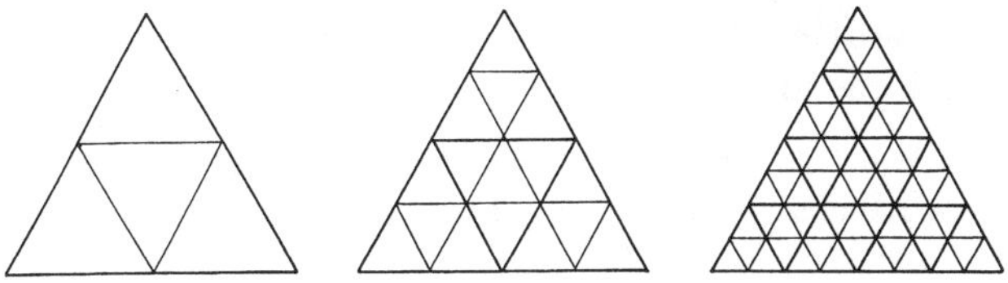

Figure 11.6: Uniform subdivision.

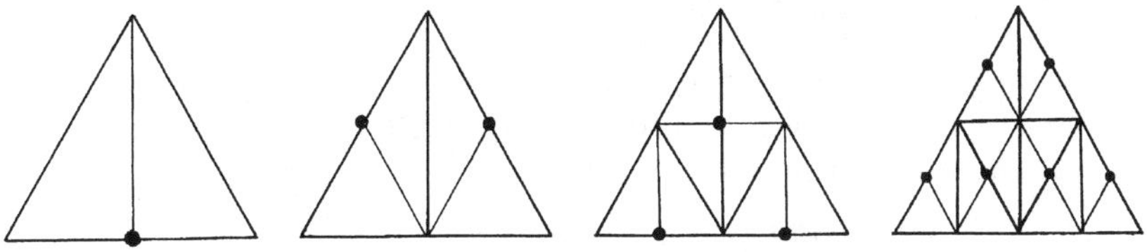

Figure 11.7: Repeated bisection.

(3) The uniform subdivision scheme shown in Figure 11.6 is either expensive to compute or uses non-convex combinations, see 11.3. Thus, for the purpose of surface generations, it is best to use the strategy illustrated in Figure 11.7. This refinement is inexpensive to compute, and one needs to evaluate only convex combinations.

Comparisons with other surface generation methods reveal that subdivision according to Figure 11.7 provides the fastest method known [Peters '94].

11.6 The symmetric polynomial of the derivative

The directional derivative $D_{\mathbb{v}}\mathbf{b}(\mathbb{u})$ of a polynomial surface with respect to a direction $\mathbb{v} = [v_0\, v_1\, v_2]^t$, $|\mathbb{v}| = v_0 + v_1 + v_2 = 0$, can also be written in terms of the symmetric polynomial $\mathbf{b}[\mathbb{u}_1 \ldots \mathbb{u}_n]$.

From 10.5, or simply by differentiating the symmetric polynomial, it follows that

$$\begin{aligned} D_{\mathbb{v}}\mathbf{b}(\mathbb{u}) &= n(v_0\mathbf{b}[\mathbb{e}_0\mathbb{u} \ldots \mathbb{u}] + v_1\mathbf{b}[\mathbb{e}_1\mathbb{u} \ldots \mathbb{u}] + v_2\mathbf{b}[\mathbb{e}_2\mathbb{u} \ldots \mathbb{u}]) \\ &= n\mathbf{b}[\mathbb{v}\,\mathbb{u} \ldots \mathbb{u}]\ . \end{aligned}$$

Obviously, $n\mathbf{b}[\mathbb{v}\mathbb{u}_2 \ldots \mathbb{u}_n]$ represents the $(n-1)$-affine symmetric polynomial of $D_{\mathbb{v}}\mathbf{b}(\mathbb{u})$.

The $\mathbb{u}_i$ represent points while $\mathbb{v}$ represents a vector with respect to the reference triangle. Further, $\mathbf{b}[\mathbb{v}\;\mathbb{u}_2 \ldots \mathbb{u}_n]$ represents a vector which is affine in $\mathbb{u}_2, \ldots, \mathbb{u}_n$ and linear in $\mathbb{v}$.

Repeating this process, we obtain the symmetric polynomial of any mixed directional derivative $\mathbf{c}(\mathbb{u}) = D_{\mathbb{v}_r} \ldots D_{\mathbb{v}_1}\mathbf{b}(\mathbb{u})$ with respect to r vectors $\mathbb{v}_1, \ldots, \mathbb{v}_r$,

$$\mathbf{c}[\mathbb{u}_{r+1} \ldots \mathbb{u}_n] = \frac{n!}{(n-r)!}\mathbf{b}[\mathbb{v}_1 \ldots \mathbb{v}_r\, \mathbb{u}_{r+1} \ldots \mathbb{u}_n]\ .$$

11.7 Simple C^r joints

Subdivision provides a convenient tool to describe certain differentiability conditions of two polynomial surfaces $\mathbf{b}(\mathbf{x})$ and $\mathbf{c}(\mathbf{x})$ given by their Bézier points $\mathbf{b}_{\mathbb{i}}$ and $\mathbf{c}_{\mathbb{i}}$ over $\mathbf{pq}$ and $\mathbf{sqr}$, respectively, see Figure 11.8.

From 10.5, it follows that the derivatives up to order r over the line $\mathbf{qr}$ determine, and are determined, by the Bézier points $\mathbf{b}_{\mathbb{i}}$ and $\mathbf{c}_{\mathbb{i}}$ for $i = 0, \ldots, r$. This leads to Farin's version of Stärk's theorem, see [Farin '86, p. 98] and [Sabin '77, p. 85].

> *The derivatives of* $\mathbf{b}$ *and* $\mathbf{c}$ *up to order* r *agree over* $\mathbf{qr}$ *if and only if the first* $r+1$ *rows of Bézier points of* $\mathbf{b}$ *and* $\mathbf{c}$ *over* $\mathbf{sqr}$ *agree, i.e.,* $\mathbf{b}[\mathbf{s} \overset{i}{\ldots} \mathbf{s}\; \mathbf{q} \overset{j}{\ldots} \mathbf{q}\; \mathbf{r} \overset{k}{\ldots} \mathbf{r}] = \mathbf{c}_{\mathbb{i}}$, $i = 0, \ldots, r$.

Over $\mathbf{pqr}$ and $\mathbf{sqr}$, the polynomial $\mathbf{b}[\mathbf{x} \overset{r}{\ldots} \mathbf{x}\mathbf{q} \overset{l}{\ldots} \mathbf{q}\mathbf{r} \overset{n-r-l}{\ldots} \mathbf{r}]$ has the Bézier points $\mathbf{b}_{\mathbb{i}}$ and $\mathbf{c}_{\mathbb{i}}$, respectively, where $i \le r$, $j \le l$ and $k \ge n-r-l$. The $\mathbf{c}_{\mathbb{i}}$ can be computed from the $\mathbf{b}_{\mathbb{i}}$ using de Casteljau's algorithm, see 11.3 and Figure 11.8 and 11.9.

Using the main theorem 11.2, this can be rephrased in the following way.

The derivatives of $\mathbf{b}$ *and* $\mathbf{c}$ *up to order* r *agree over* $\mathbf{qr}$ *if and only if for all* $l = 0, \ldots, n-r$ *the two polynomials* $\mathbf{b}[\mathbf{x} \overset{r}{\ldots} \mathbf{x}\,\mathbf{q} \overset{l}{\ldots} \mathbf{q}\,\mathbf{r} \overset{n-r-l}{\ldots} \mathbf{r}]$ *and* $\mathbf{c}[\mathbf{x} \overset{r}{\ldots} \mathbf{x}\,\mathbf{q} \overset{l}{\ldots} \mathbf{q}\,\mathbf{r} \overset{n-r-l}{\ldots} \mathbf{r}]$ *are equal.*

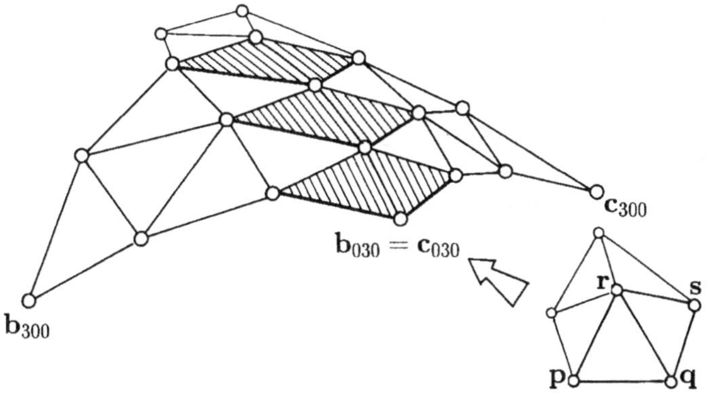

Figure 11.8: Sabin's simple C^1 joint.

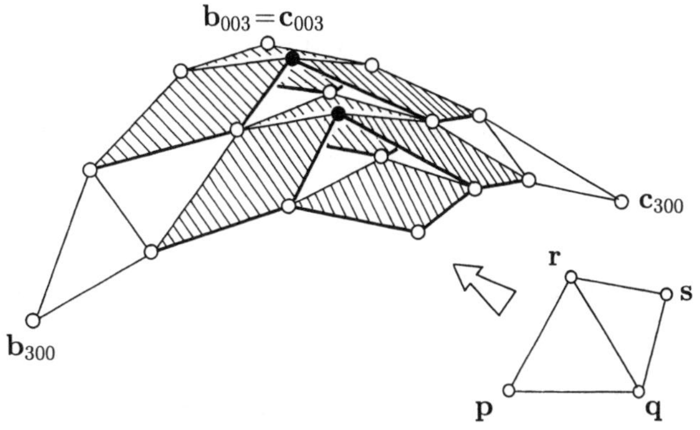

Figure 11.9: Farin's simple C^2 joint.

Remark 6: The shaded quadrilaterals in Figure 11.8 and 11.9 are different affine images of the quadrilateral **pqrs**. Consequently, any k triangular patches $\mathbf{b}^i(\mathbf{x})$, $i = 1, \ldots, m$, enclosing a common vertex have simple C^1 joints at this vertex if and only if their parameter triangles form an m-gon that is an affine image of the m-gon formed by the respective corner triangles of the associated Bézier nets.

Remark 7: Since two polynomials are equal if and only if their polar forms are equal, $\mathbf{b}(\mathbf{x})$ and $\mathbf{c}(\mathbf{x})$ have identical derivatives up to order r over the line $\mathbf{qr}$ if and only if their polar forms satisfy the equation

$$\mathbf{b}[\mathbf{x}_1 \ldots \mathbf{x}_r \, \mathbf{p} \overset{j}{\ldots} \mathbf{p} \, \mathbf{q} \overset{k}{\ldots} \mathbf{q}] = \mathbf{c}[\mathbf{x}_1 \ldots \mathbf{x}_r \, \mathbf{p} \overset{j}{\ldots} \mathbf{p} \, \mathbf{q} \overset{k}{\ldots} \mathbf{q}]$$

for arbitrary variables $\mathbf{x}_1, \ldots, \mathbf{x}_r$ and for all j and k with $r + j + k = n$. This condition is used in [Lai '91] to characterize multivariate C^r splines over arbitrary triangulations.

11.8 Degree elevation

A polynomial surface of degree n also has a Bézier representation of any degree m higher than n. As in the case of curves, a conversion to a higher degree representation is called degree **elevation**.

Given an nth degree Bézier representation,

$$\mathbf{b}(\mathbf{x}) = \sum \mathbf{b}_{\mathbb{i}} B_{\mathbb{i}}^n(\mathbb{u}) \ , \quad \mathbf{x} = [\mathbf{pqr}]\mathbb{u} \ ,$$

of some polynomial surface $\mathbf{b}(\mathbb{u})$ over a triangle $\mathbf{pqr}$, we show how to obtain its Bézier representation of degree $n+1$. In analogy to the derivation for curves in 3.11, we use the symmetric polynomial $\mathbf{b}[\mathbf{x}_1 \ldots \mathbf{x}_n]$ of $\mathbf{b}(\mathbf{x})$. The polynomial

$$\mathbf{c}[\mathbf{x}_0 \ldots \mathbf{x}_n] = \frac{1}{n+1} \sum_{l=0}^{n} \mathbf{b}[\mathbf{x}_0 \ldots \mathbf{x}_l^* \ldots \mathbf{x}_n]$$

is multiaffine, symmetric and agrees with $\mathbf{b}(\mathbf{x})$ on its diagonal. Hence, due to the main theorem in 11.2, it follows that the points

$$\mathbf{b}_{\mathbb{j}} = \mathbf{c}[\mathbf{p} \overset{j_0}{\ldots} \mathbf{p} \, \mathbf{q} \overset{j_1}{\ldots} \mathbf{q} \, \mathbf{r} \overset{j_2}{\ldots} \mathbf{r}]$$

are the Bézier points of $\mathbf{b}(\mathbf{x})$ over $\mathbf{pqr}$ in its representation for degree $n+1$. Consequently,

$$\begin{aligned} \mathbf{b}_{\mathbb{j}} = \quad & \frac{j_0}{n+1} \mathbf{b}[\mathbf{p} \overset{j_0-1}{\ldots} \mathbf{p} \, \mathbf{q} \overset{j_1}{\ldots} \mathbf{q} \, \mathbf{r} \overset{j_2}{\ldots} \mathbf{r}] \\ + \quad & \frac{j_1}{n+1} \mathbf{b}[\mathbf{p} \overset{j_0}{\ldots} \mathbf{p} \, \mathbf{q} \overset{j_1-1}{\ldots} \mathbf{q} \, \mathbf{r} \overset{j_2}{\ldots} \mathbf{r}] \\ + \quad & \frac{j_2}{n+1} \mathbf{b}[\mathbf{p} \overset{j_0}{\ldots} \mathbf{p} \, \mathbf{q} \overset{j_1}{\ldots} \mathbf{q} \, \mathbf{r} \overset{j_2-1}{\ldots} \mathbf{r}] \ . \end{aligned}$$

Figure 11.10 illustrates the associated construction for $n = 2$.

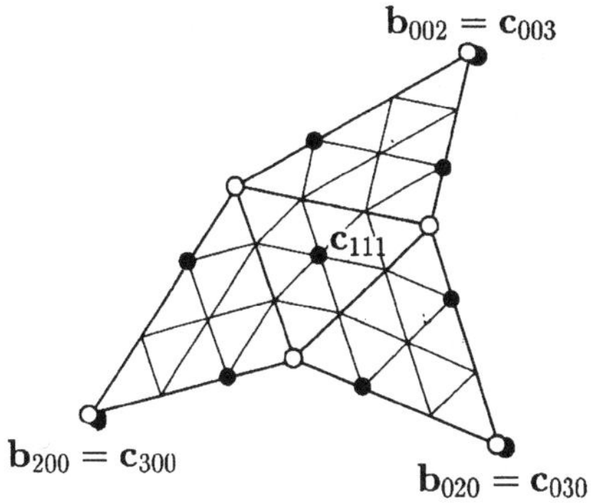

Figure 11.10: Degree elevation by one.

11.9 Convergence under degree elevation

Repeating the degree elevation process, we obtain any higher degree representation

$$\mathbf{b}(\mathbf{x}) = \sum \mathbf{d}_{\mathbb{k}} B_{\mathbb{k}}^{m}(\mathbb{u}) \ , \quad m > n \ .$$

The new Bézier points $\mathbf{d}_{\mathbb{k}}$ can be expressed in terms of the points $\mathbf{b}_{\mathbb{i}}$. With $r = m - n$, we write

$$\mathbf{b}(\mathbf{x}) = \sum_{|\mathbb{i}|=n} \mathbf{b}_{\mathbb{i}} B_{\mathbb{i}}^{n} \cdot 1$$

as

$$\begin{aligned} \mathbf{b}(\mathbf{x}) &= \sum_{|\mathbb{i}|=n} \mathbf{b}_{\mathbb{i}} B_{\mathbb{i}}^{n} \cdot \sum_{|\mathbb{j}|=r} B_{\mathbb{j}}^{r} \\ &= \sum_{|\mathbb{k}|=m} \left(\sum_{|\mathbb{i}|=n} \mathbf{b}_{\mathbb{i}} \beta_{\mathbb{i}\mathbb{k}} \right) B_{\mathbb{k}}^{m} \ , \end{aligned}$$

where

$$\beta_{\mathbb{i}\mathbb{k}} = \frac{B_{\mathbb{i}}^{n} B_{\mathbb{k}-\mathbb{i}}^{r}}{B_{\mathbb{k}}^{m}} = \frac{\binom{n}{\mathbb{i}}\binom{r}{\mathbb{k}-\mathbb{i}}}{\binom{m}{\mathbb{k}}} \ .$$

Thus, we obtain Zhou's formula

$$\mathbf{d}_{\mathbb{k}} = \sum_{|\mathbb{i}|=n} \mathbf{b}_{\mathbb{i}} \beta_{\mathbb{i}\mathbb{k}} \ ,$$

see [Farin '86, de Boor '87]. Let $\mathbb{i} = (i_1 i_2 i_3)$ and $\mathbb{k} = (k_1 k_2 k_3)$. Then, $\beta_{\mathbb{i}\mathbb{k}}$ can be written as

$$\binom{n}{\mathbb{i}}\left(\frac{k_1}{m}\cdots\frac{k_1-i_1+1}{m-i_1+1}\right)\left(\frac{k_2}{m-i_1}\cdots\frac{k_2-i_2+1}{m-i_2-i_2+1}\right)\left(\frac{k_3}{m-i_1-i_2}\cdots\frac{k_3-i_3+1}{m-n+1}\right),$$

from which we conclude that

$$\beta_{\mathbb{i}\mathbb{k}} \le \binom{n}{\mathbb{i}}\left(\frac{k_1}{m}+\frac{n}{m-n}\right)^{i_1}\left(\frac{k_2}{m}+\frac{n}{m-n}\right)^{i_2}\left(\frac{k_3}{m}+\frac{n}{m-n}\right)^{i_3}$$

and

$$\beta_{\mathbb{i}\mathbb{k}} \ge \binom{n}{\mathbb{i}}\left(\frac{k_1}{m}-\frac{n}{m}\right)^{i_1}\left(\frac{k_2}{m}-\frac{n}{m}\right)^{i_2}\left(\frac{k_3}{m}-\frac{n}{m}\right)^{i_3} .$$

Thus, we find

$$\beta_{\mathbb{i}\mathbb{k}} = B^n_{\mathbb{i}}(\mathbb{k}/m) + O(1/m)$$

and, therefore,

$$\mathbf{d}_{\mathbb{k}} = \sum \mathbf{b}_{\mathbb{i}} B^n_{\mathbb{i}}(\mathbb{k}/m) + O(1/m) \ .$$

Consequently, the mth degree Bézier nets of $\mathbf{b}(\mathbf{x})$ converge to $\mathbf{b}(\mathbf{x})$ linearly in $1/m$, see [Farin '79, Trump & Prautzsch '96].

11.10 Conversion to tensor product Bézier representation

Let $\mathbf{b}(\mathbf{x})$ be a bivariate polynomial with polar form $\mathbf{b}[\mathbf{x}_1 \dots \mathbf{x}_n]$, and let $\mathbf{x}_{st} = \mathbf{x}(s,t)$ be any biaffine map that maps the unit square $[0,1]^2$ onto a convex quadrilateral, see Figure 11.11. Then the reparametrized polynomial

$$\mathbf{c}(s,t) = \mathbf{b}(\mathbf{x}(s,t))$$

is a tensor product polynomial of degree (n,n) in (s,t). Its tensor product polar form is given by

$$\mathbf{c}[s_1 \dots s_n, t_1 \dots t_n] = \frac{1}{n!}\sum_{\tau} \mathbf{b}[\mathbf{x}(s_1,\tau_1)\dots\mathbf{x}(s_n,\tau_n)] \ ,$$

where the sum extends over all permutations $(\tau_1,\dots,\tau_n)$ of $(t_1,\dots,t_n)$. To verify this, one checks that $\mathbf{c}$ satisfies the three characterizing properties: the diagonal, symmetry and affinity property.

Knowing the tensor product polar form, we can apply the main theorem 9.2 to obtain the Bézier points of $\mathbf{c}(s,t)$ over $[0,1]^2$. These are the points

$$\mathbf{c}_{ij} = \mathbf{c}[0 \overset{n-i}{\dots} 0\, 1 \overset{i}{\dots} 1, 0 \overset{n-j}{\dots} 0\, 1 \overset{j}{\dots} 1] \ ,$$

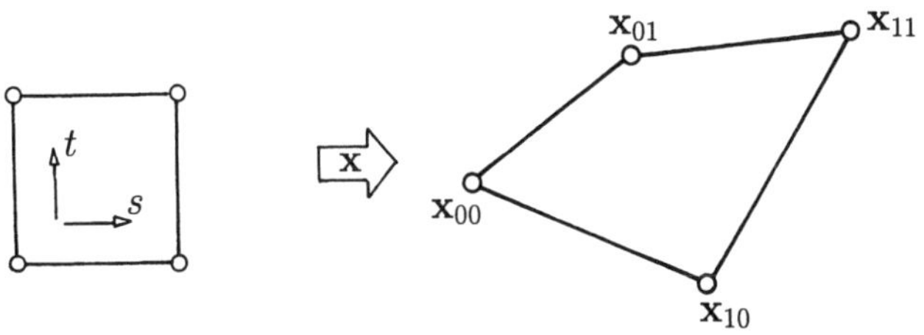

Figure 11.11: A biaffine reparametrization.

which can be written as

$$\mathbf{c}_{ij} = \sum_{k=0}^{j} \beta_{ijk} \mathbf{b}[\mathbf{x}_{00} \overset{n+k-i-j}{\ldots} \mathbf{x}_{00}\mathbf{x}_{01} \overset{j-k}{\ldots} \mathbf{x}_{01}\mathbf{x}_{10} \overset{i-k}{\ldots} \mathbf{x}_{10}\mathbf{x}_{11} \overset{k}{\ldots} \mathbf{x}_{11}] \ ,$$

with $n!\beta_{ijk}$ the number of permutations of $(0 \overset{n-j}{\ldots} 01 \overset{j}{\ldots} 1)$ such that exactly k ones sit in the last i positions. Hence,

$$\begin{aligned} \beta_{ijk} &= \frac{1}{n!} \cdot \binom{j}{k} i \ldots (i+1-k) \cdot (n-i) \ldots (n-i+1-j+k) \cdot (n-j)! \\ &= \frac{k!}{n!} \binom{i}{k} \binom{j}{k} \frac{(n-i)!(n-j)!}{(n+k-i-j)!} \ . \end{aligned}$$

From 11.3, we recall that the points

$$\mathbf{b}[\mathbf{x}_{00} \ldots \mathbf{x}_{00}\mathbf{x}_{01} \ldots \mathbf{x}_{01}\mathbf{x}_{10} \ldots \mathbf{x}_{10}\mathbf{x}_{11} \ldots \mathbf{x}_{11}]$$

arise when we subdivide the Bézier net of $\mathbf{b}(\mathbf{x})$ over $\mathbf{x}_{00}\mathbf{x}_{01}\mathbf{x}_{10}$ in $\mathbf{x}_{11}$, see also [DeRose et al. '93].

11.11 Conversion to triangular Bézier representation

A tensor product polynomial $\mathbf{b}(x, y)$ of degree (m, n) is of total degree $\leq m + n$. Thus, it has a Bézier representation of degree $l = m + n$ over any triangle **pqr**. To compute it, let $\mathbf{b}[x_1 \ldots x_m, y_1 \ldots y_n]$ be the tensor product polar form of $\mathbf{b}(x, y)$. Then, the (non-tensor product) polar form of $\mathbf{b}(\mathbf{x})$ is given by

$$\mathbf{c}[\mathbf{x}_1 \ldots \mathbf{x}_l] = \frac{1}{l!} \sum \mathbf{b}[x_{i_1} \ldots x_{i_m}, y_{i_{m+1}} \ldots y_{i_l}] \ ,$$

where the sum extends over all permutations $(i_1 \ldots i_l)$ of $(1 \ldots l)$ and $\mathbf{x}_i = (x_i, y_i)$. Namely, $\mathbf{c}$ obviously satisfies the three characterizing properties: the diagonal, symmetry and affinity property.

The Bézier points of $\mathbf{b}(\mathbf{x})$ over a triangle with vertices $\mathbf{p} = [p_1, p_2]^t$, $\mathbf{q} = [q_1, q_2]^t$ and $\mathbf{r} = [r_1, r_2]^t$ can be obtained by the main theorem 11.2.

They are the points

$$\begin{aligned}\mathbf{b}_{ijk} &= \mathbf{c}[\mathbf{p} \overset{i}{\ldots} \mathbf{pq} \overset{j}{\ldots} \mathbf{qr} \overset{k}{\ldots} \mathbf{r}] \\ &= \sum_{\substack{\alpha+\beta+\gamma=m \\ (\alpha,\beta,\gamma)\le(i,j,k)}} \delta^{ijk}_{\alpha\beta\gamma} \mathbf{b}[p_1 \overset{\alpha}{\ldots} p_1 q_1 \overset{\beta}{\ldots} q_1 r_1 \overset{\gamma}{\ldots} r_1, \\ &\qquad\qquad p_2 \overset{i-\alpha}{\ldots} p_2 q_2 \overset{j-\beta}{\ldots} q_2 r_2 \overset{k-\gamma}{\ldots} r_2] ,\end{aligned}$$

where

$$\delta^{ijk}_{\alpha\beta\gamma} = \binom{i}{\alpha}\binom{j}{\beta}\binom{k}{\gamma} m!n! .$$

Namely, there are $\binom{i}{\alpha}$ choices to pick α from i many p_1's etc., and there are $m!$ permutations of the first m variables $p_1, \overset{\alpha}{\ldots}, p_1, q_1, \overset{\beta}{\ldots}, q_1, r_1 \overset{\gamma}{\ldots}, r_1$ and $n!$ permutations of the last n variables. $\diamond$

From 9.5 we recall that the points

$$\mathbf{b}[p_1 \ldots p_1 q_1 \ldots q_1 r_1 \ldots r_1, p_2 \ldots p_2 q_2 \ldots q_2 r_2 \ldots r_2]$$

arise in the tensor product de Casteljau algorithm used to compute $\mathbf{b}(\mathbf{r})$ from the Bézier points over $[\mathbf{p}, \mathbf{q}]$.

11.12 Problems

1 Consider a functional polynomial with convex Bézier polyhedron. Show that degree elevation and uniform subdivision, as illustrated in Figure 11.6 preserve the convexity of the Bézier polyhedron.

2 There are parametric Bézier patches for which the statement in Problem 1 does not hold. Provide an example.

3 Generalize the intersection algorithm presented in Section 3.7 to triangular Bézier patches.

4 Let $\mathbf{v}_1 = \mathbf{q}-\mathbf{p}$ and $\mathbf{v}_2 = \mathbf{r}-\mathbf{q}$. Show that a Bézier net $\mathbf{b}_{\mathbf{i}}$ over $\mathbf{pq}$ is planar if all $\Delta_{\mathbf{v}_1}\Delta_{\mathbf{v}_1}\mathbf{b}_{\mathbf{i}}, \Delta_{\mathbf{v}_1}\Delta_{\mathbf{v}_2}\mathbf{b}_{\mathbf{i}}, \Delta_{\mathbf{v}_2}\Delta_{\mathbf{v}_2}\mathbf{b}_{\mathbf{i}}$ are zero. Thus, the maximum of these differences is a measure for the flatness of a triangular Bézier patch.

5 Storing triangular Bézier nets should be dealt with efficiently storage- and accesswise. An efficient way to deal with these Bézier nets is to store the points $\mathbf{b}_{ijk}$, $i + j + k = n$, in a linear array, say

$$\mathbf{a}[L] = \mathbf{b}_{ijk} \ ,$$

where $L = L(i, j)$ runs from 1 to $\binom{n+2}{2}$.
Fast access is guaranteed if the function L is stored in a matrix $L = [L_{ij}]$.
Provide an explicit formula for a function L used in Problem 5.

12 Interpolation

12.1 Hermite interpolation — 12.2 The Clough-Tocher interpolant — 12.3 The Powell-Sabin interpolant — 12.4 Surfaces of arbitrary topology — 12.5 Singular parametrization — 12.6 Quintic C^1 splines of arbitrary topology — 12.7 Problems

Given the values and derivatives of a bivariate function, it is quite easy to construct smooth piecewise polynomial interpolants using their Bézier representation. However, there is no straightforward extension to arbitrary parametric surfaces as, for example, spheres. General C^r joints or singular parametrizations are necessary to build such interpolants.

12.1 Triangular Hermite interpolation

Given a triangulation $\mathcal{T}$ of some polygonal domain in $\mathbb{R}^2$, one can construct a piecewise polynomial surface of degree $4r+1$ that interpolates any given derivatives up to order $2r$ at the vertices of $\mathcal{T}$ and is r-times differentiable. The Bézier representation is a very handy tool to describe this construction.

Figure 12.1 shows one triangle of a triangulation and the Bézier abscissae, see 10.3, of the interpolant over that triangle for $r = 2$.

There are three kinds of Bézier abscissae. The Bézier ordinates at the abscissae ● are defined by the prescribed derivatives, the Bézier ordinate at ○ can be chosen arbitrarily, and the Bézier ordinates at the abscissae ◓, ◕, ◔ depend on the corresponding ordinates over the adjacent triangles according to the C^r-conditions. One can choose the ordinates at ◓, ◕, ◔ in one triangle arbitrarily. Then, the corresponding Bézier ordinates in the adjacent triangles are determined by the C^r-conditions.

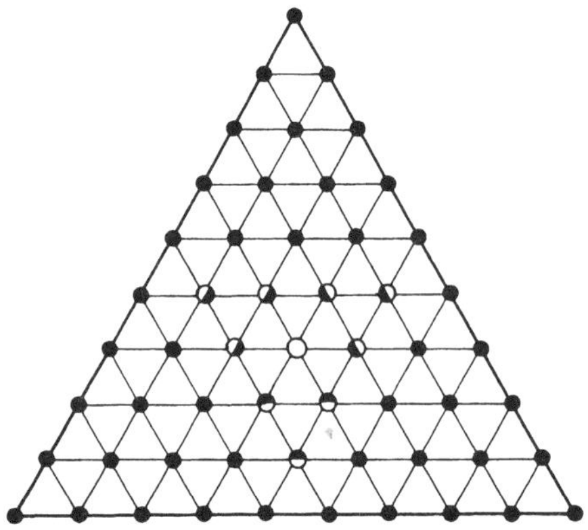

Figure 12.1: Bézier abscissae for $r = 2$.

12.2 The Clough-Tocher interpolant

Interpolants of lower degree than in 12.1 for arbitrary input data can only be constructed at the expense of more patches. The piecewise cubic C^1 interpolant described next is due to [Clough & Tocher '65].

Given a triangulation $\mathcal{T}$ of some polygonal domain in $\mathbb{R}^2$, we subdivide each triangle in $\mathcal{T}$ at some interior point into three micro triangles, as illustrated in Figure 12.2. To any set of positional values and first derivatives at the vertices of $\mathcal{T}$ we can then construct interpolating C^1-functions which are cubic over each micro triangle.

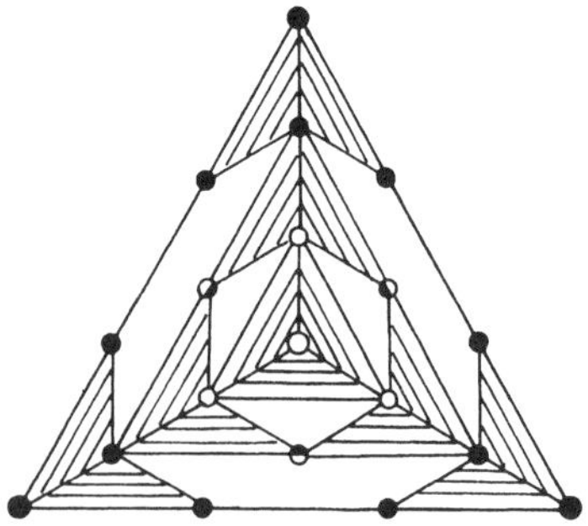

Figure 12.2: A macro triangle split into its three micro triangles.

Figure 12.2 shows the Bézier abscissae of such an interpolant over some triangle of $\mathcal{T}$. The Bézier ordinates at the abscissae $\bullet$ are given by the interpolation constraints, at the vertices. The ordinates at the abscissae ◒, ◑, ◐ are determined by the C^1 conditions along the edges in $\mathcal{T}$, as described in 12.1. The ordinates at the abscissae ∘ are determined by the C^1-conditions along the edges of the refined triangulation, which means that the four points corresponding to the vertices of every quadrilateral shaded in Figure 12.2 lie in a plane.

Remark 1: The interpolant above is even twice differentiable at the subdivision points. For a proof, consider the quadratic defined by the Bézier ordinates over the 6 interior abscissae •, ◒, ◑, ◐ shown in Figure 12.2. Since the 10 Bézier ordinates over all 10 inner abscissae •, ◒, ◑, ◐, ∘ in Figure 12.2 also represent the subdivided Bézier net of the quadratic the claimed differentiability follows from Stärk's theorem in 11.7.

12.3 The Powell-Sabin interpolant

The interpolation problem 12.2 can also be solved by piecewise quadratic C^1 interpolants at the expense of more patches [Powell & Sabin '77]. Here, each macro triangle is split into six micro triangles by some interior splitting point ∘ and points ⋄ on its edges, see Figure 12.3. The line connecting the interior splitting points of two adjacent triangles must cross the common edge in a point ⋄. If the splitting points are the incenters of the triangles this condition is satisfied, see also Problem 5.

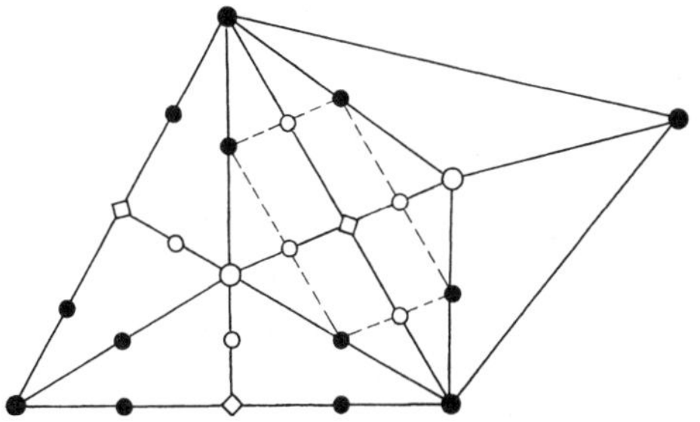

Figure 12.3: Bézier abscissae of a Powell-Sabin split.

The Bézier ordinates at the abscissae • are determined by the interpolation constraints, the Bézier ordinates at the abscissae ◇ and ∘ are determined by the C^1-conditions along the splitting edges.

The nine Bézier ordinates belonging to the parallelogram shown with dashed lines in Figure 12.3 lie on a bilinear surface. This implies C^1 continuity between adjacent macro triangles.

12.4 Surfaces of arbitrary topology

The simple C^k joint described in 11.7 is appropriate for the construction of surfaces with a regular pattern of patches, but it is not sufficient for the modelling of arbitrarily smooth surfaces.

In particular, it is impossible to construct closed surfaces with regular triangular patches using only the simple C^1 joint, except for surfaces of genus 1. Figure 12.4 shows closed surfaces of genus $0, 1, 2$ and 3.

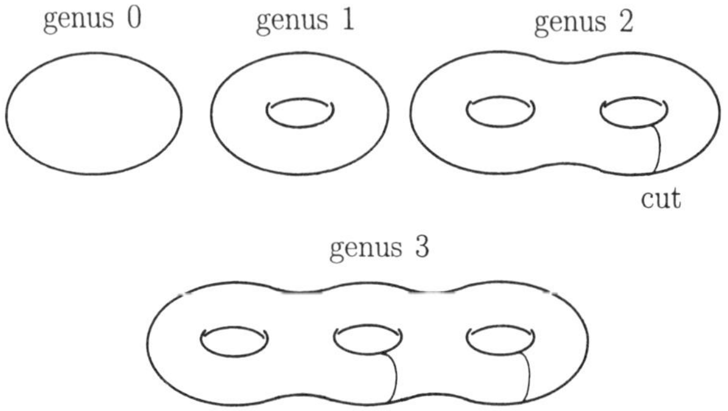

Figure 12.4: Closed surfaces of different genera.

For a proof, assume that there was a closed surface of genus 0 composed of regular triangular patches with simple C^1 joints. After splitting the surface into two components, as illustrated in Figure 12.5, it follows from 11.7, Remark 6, that each component can be parametrized by a regular C^1-map over a triangulated and simply connected domain. Further, there must be one affine map mapping all boundary triangles of the first domain onto triangles surrounding the second domain. Obviously this is impossible. $\diamond$

Contrary to this, a torus, or any other surface of genus 1, can be parametrized by a regular C^1-map over a rectangle, see Problem 1. Thus, if a surface of genus $n \geq 2$ was composed of regular triangular patches with the C^1 joint of 11.7, one could cut it, as illustrated in Figure 12.4 and parametrize it over a triangulated rectangular domain with $2(n-1)$ holes such that the boundaries of the hole correspond to the cut.

Again, there must be an affine map mapping the boundary triangles around one of the holes onto triangles adjacent to the boundary triangles of the corresponding one. Yet, this is impossible again. Therefore, it is impossible to construct such a surface.

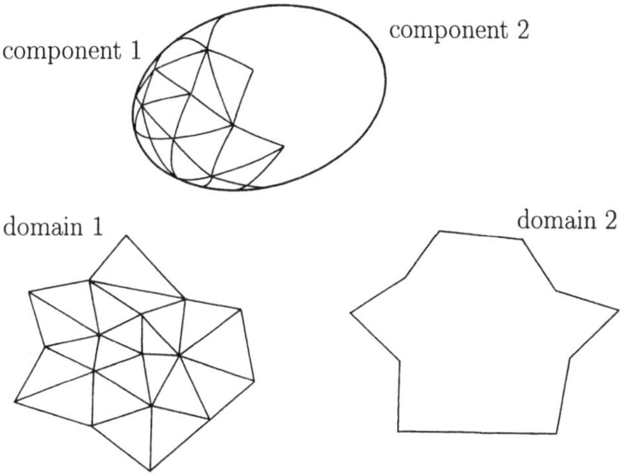

Figure 12.5: Splitting a surface of genus 0 into two components.

12.5 Singular parametrization

In order to model arbitrarily smooth surfaces, we need to use either general C^1 joints, as described in Chapter 14, or must admit singular parametrizations. For triangular patches, we could use the singularity discussed in 9.10 for tensor product patches. However, the weaker singularity illustrated in Figure 12.6 suffices.

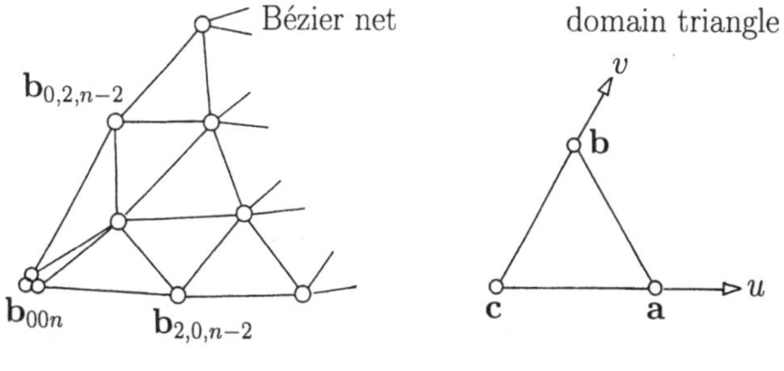

Figure 12.6: Bézier points of a singular parametrization at a vertex.

Figure 12.6 shows a part of a Bézier net of a triangular polynomial patch

$$\mathbf{b}(u,v) = \sum \mathbf{b}_{\mathbf{i}} B_{\mathbf{i}}^n(u,v,1-u-v) \ ,$$

where

$$\mathbf{b}_{00n} = \mathbf{b}_{1,0,n-1} = \mathbf{b}_{0,1,n-1}$$

and

$$\mathbf{b}_{1,1,n-2} = \alpha \mathbf{b}_{2,0,n-2} + \beta \mathbf{b}_{0,2,n-2} + \gamma \mathbf{b}_{00n}$$

with $\alpha, \beta > 0$, $\alpha + \beta + \gamma = 1$ and independent points $\mathbf{b}_{00n}, \mathbf{b}_{2,0,n-2}, \mathbf{b}_{0,2,n-2}$. The derivatives of $\mathbf{b}(u, v)$ vanish at $\mathbf{u} = (u, v) = (0, 0)$. However, there exists a reparametrization $\mathbf{u}(\mathbf{x})$ such that $\mathbf{c}(\mathbf{x}) = \mathbf{b}(\mathbf{u}(\mathbf{x}))$ is regular at $\mathbf{x} = \mathbf{u}^{-1}(\mathbf{o})$. After a suitable affine transformation, we obtain

$$\mathbf{b}_{00n} = \mathbf{0} \ , \quad \mathbf{b}_{2,0,n-2} = \frac{2}{n(n-1)} \begin{bmatrix} 1 \\ 0 \\ 0 \end{bmatrix} \ , \quad \mathbf{b}_{0,2,n-2} = \frac{2}{n(n-1)} \begin{bmatrix} 0 \\ 1 \\ 0 \end{bmatrix} .$$

Hence, the Taylor expansion of $\mathbf{b}(\mathbf{u})$ at $\mathbf{u} = (0, 0)$ is of the form

$$\mathbf{b}(\mathbf{u}) = \begin{bmatrix} \mathbf{x}(\mathbf{u}) \\ 0 \end{bmatrix} + \mathbf{d}(\mathbf{u}) \ ,$$

where

$$\mathbf{x}(\mathbf{u}) = \begin{bmatrix} u^2 + 2\alpha uv \\ v^2 + 2\beta uv \end{bmatrix} = O(\|\mathbf{u}\|^2)$$

and $\|\mathbf{d}(\mathbf{u})\| = O(\|\mathbf{u}\|^3)$. Obviously, x and y are strictly monotone in u and v for $u, v \geq 0$. Hence, $\mathbf{x}(\mathbf{u})$ is one-to-one for $u, v \geq 0$. Furthermore, $\mathbf{x}(\mathbf{u})$ is regular for $\mathbf{u} \neq \mathbf{0}$. Consequently, $\mathbf{c}(\mathbf{x}) = \mathbf{b}(\mathbf{u}(\mathbf{x}))$ is continuously differentiable if $\mathbf{c}(\mathbf{x})$ has continuous partial derivatives. These partials are given by

$$[\mathbf{c}_x \mathbf{c}_y] = [\mathbf{d}_u \mathbf{d}_v] \begin{bmatrix} x_u & x_v \\ y_u & y_v \end{bmatrix}^{-1} = \frac{1}{x_u y_v - x_v y_u} [\mathbf{d}_u \mathbf{d}_v] \begin{bmatrix} y_v & -x_v \\ -y_u & x_u \end{bmatrix} ,$$

which shows that

$$\|\mathbf{c}_x\| = O(\|\mathbf{u}\|) = O(\sqrt{\|\mathbf{x}\|}) \ .$$

The same argument can be made for $\mathbf{d}_y$. Hence, $\mathbf{c}(\mathbf{x})$ and, therefore, $\mathbf{b}(\mathbf{u}(\mathbf{x}))$ are continuously differentiable.

Remark 2: Even if $\alpha, \beta < 0$ and $4\alpha\beta > 1$, one can show that $\mathbf{b}(\mathbf{u})$ has a continuous tangent plane [Reif '95a].

12.6 Quintic C^1 splines of arbitrary topology

Singular parametrizations can be used to construct arbitrary C^1 surfaces composed of triangular patches with prescribed positions and tangent planes at their vertices. The following description of one such surface construction is meant to provide the general concept but not a complete solution for immediate practical use.

Consider a triangular net with specified normals at each vertex, as illustrated in Figure 12.7. For each face of the net, one constructs a triangular quintic Bézier patch, as illustrated schematically in Figure 12.8. The corner Bézier point •, together with the coalescing next Bézier points ○, are defined by the vertices of the triangular net.

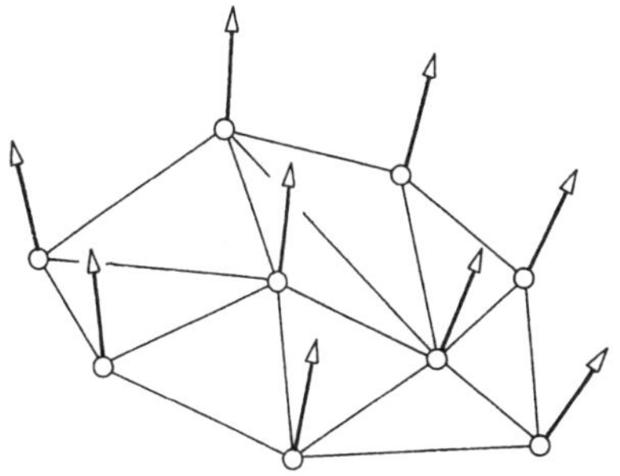

Figure 12.7: Triangular net with normals.

The next ring of Bézier points $*$ and $\diamond$ around such a set of coalescing Bézier points is chosen on the prescribed tangent plane at the corner and the points $\diamond$ such that the shaded quadrilaterals are parallelograms. The remaining Bézier points ◒, ◕, ◐ are chosen such that the remaining shaded quadrilaterals are also parallelograms.

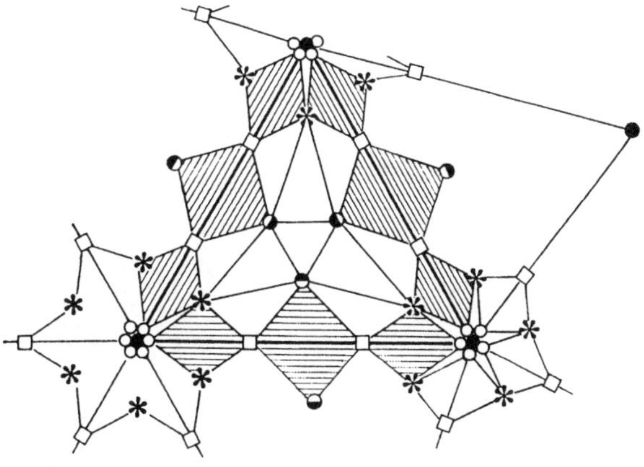

Figure 12.8: Bézier net of a piecewise quintic C^1 interpolant with singularities.

12.7 Problems

1 Construct a closed surface of genus 1, i.e., a torus-like surface, with regular patches having simple C^1 joints.

2 Show that one cannot model open C^1 surfaces whose boundaries have less than three corners with regular patches having simple C^1 joints. Figure 12.9 shows two such surfaces.

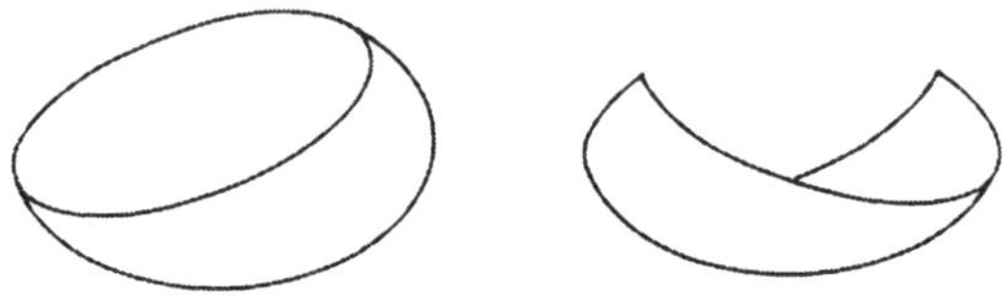

Figure 12.9: Open surfaces with no corner and two corners.

3 Show that one can model C^1 surfaces of arbitrary topology with singular triangular quadratic patches having simple C^1 joints.

4 Show that the surface from Problem 3, in general, contains flat patches.

5 Given a triangulation of $\mathbb{R}^2$, connect the centroids of all pairs of adjacent triangles. Show, by means of an example, that the connecting edges can pass through a third triangle.

6 Refute, by means of an example, that a Powell-Sabin element has a convex or concave Bézier polyhedron if the three tangent planes at the corners intersect above an interior point of the domain triangle; see also [Floater '97, Carnicer & Dahmen '92, Bangert & Prautzsch '99] .

7 Generalize the Powell-Sabin interpolation scheme for multivariate functions, see also [Bangert & Prautzsch '99].

13 Constructing smooth surfaces

13.1 The general C^1 joint — 13.2 The vertex enclosure problem — 13.3 The parity phenomenon — 13.4 Joining two triangular cubic patches — 13.5 A triangular G^1 interpolant — 13.6 Problems

The simple C^1 joint discussed in 11.7 is too restrictive for modelling smooth regular surfaces of arbitrary shape. Here, we present general C^1-conditions and a interpolation scheme that allows to design regular surfaces of arbitrary topology with triangular patches.

13.1 The general C^1 joint

Let $\mathbf{p}(x, y)$ and $\mathbf{q}(x, y)$ be two regular C^1 surface patches with a common boundary at $x = 0$, i.e.,

$$\mathbf{p}(0, y) = \mathbf{q}(0, y)$$

for all $y \in [0, 1]$. Figure 13.1 gives an illustration. The patches $\mathbf{p}$ and $\mathbf{q}$ neither need to be polynomial, nor three- or four-sided.

One says that $\mathbf{p}$ and $\mathbf{q}$ have a **general $\mathbf{C}^1$** or **geometric $\mathbf{C}^1$**- or, in short, **$\mathbf{G}^1$ joint** in $x = 0$ if they have equal normals along this parameter line, i.e., if

$$\frac{\mathbf{p}_x \times \mathbf{p}_y}{\|\mathbf{p}_x \times \mathbf{p}_y\|} = \frac{\mathbf{q}_x \times \mathbf{q}_y}{\|\mathbf{q}_x \times \mathbf{q}_y\|} \quad \text{for} \quad x = 0 \ .$$

Equivalently, one can characterize G^1-continuity by requiring that there are **connection functions** $\lambda(y), \mu(y)$ and $\nu(y)$ such that for $x = 0$ and all y

$$(1) \qquad \lambda \mathbf{p}_x = \mu \mathbf{q}_x + \nu \mathbf{q}_y \quad \text{and} \quad \lambda\mu < 0 \ ,$$

except for isolated zeros.

In particular, if $\mathbf{p}$ *and* $\mathbf{q}$ *have a* G^1 *joint and are polynomials,*

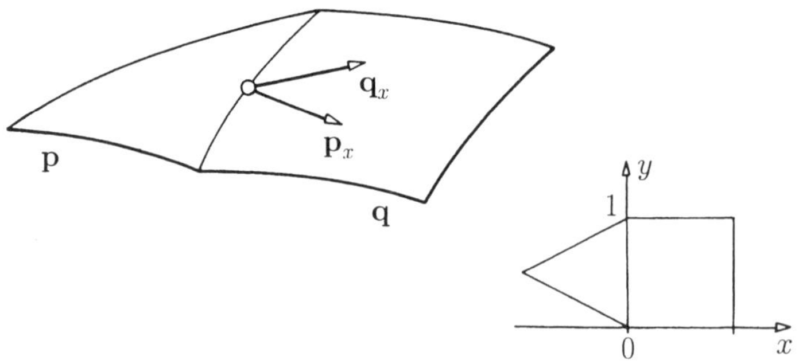

Figure 13.1: Two patches with a common boundary curve.

then the connection functions are also polynomials, and, up to a common factor, we have

$$\begin{aligned}\text{degree } \lambda &\leq \text{degree } \mathbf{q}_x(0,y) + \text{degree } \mathbf{q}_y(0,y) \ , \\ \text{degree } \mu &\leq \text{degree } \mathbf{p}_x(0,y) + \text{degree } \mathbf{q}_y(0,y) \ , \\ \text{degree } \nu &\leq \text{degree } \mathbf{p}_x(0,y) + \text{degree } \mathbf{q}_x(0,y) \ .\end{aligned}$$

For a proof, we compute the vector product of equation (1) with $\mathbf{q}_x$ and $\mathbf{q}_y$. This gives

$$\begin{aligned}\lambda \mathbf{p}_x \times \mathbf{q}_x &= \nu \mathbf{q}_y \times \mathbf{q}_x \ , \quad \text{and} \\ \lambda \mathbf{p}_x \times \mathbf{q}_y &= \mu \mathbf{q}_x \times \mathbf{q}_y \ .\end{aligned}$$

Recall that $\mathbf{q}$ is regular. Hence, at least one coordinate, say the first of $[\mathbf{q}_x \times \mathbf{q}_y]$, denoted by $[\mathbf{q}_x \times \mathbf{q}_y]_1$, is non-zero. Since equation (1) can be multiplied by a factor, we may assume that

$$\lambda = [\mathbf{q}_x \times \mathbf{q}_y]_1 \ .$$

This implies

$$\mu = [\mathbf{p}_x \times \mathbf{q}_y]_1 \quad \text{and} \quad \nu = -[\mathbf{p}_x \times \mathbf{q}_x]_1 \ ,$$

which proves the assertion. ◇

Remark 1: Often, one sets $\lambda = 1$. Then μ and ν are rational, in general.

Remark 2: The proof given above also holds for rational polynomials $\mathbf{p}$ and $\mathbf{q}$. Then, the functions λ, μ and ν are rational up to a common factor with the same degree estimates as above.

Remark 3: Any G^1 joint is a simple C^1 joint after a suitable parameter transformation. Namely, if $\mathbf{p}$ and $\mathbf{q}$ satisfy the G^1-condition (1), then $\mathbf{a}(x,y) = \mathbf{p}(\lambda x, y)$ and $\mathbf{b}(x,y) = \mathbf{q}(\mu x, \nu x + y)$ have a simple C^1 joint, see

9.7 and 11.7.

Remark 4: Whether two patches have a G^1 joint does not depend on their parametrization. However, the connection functions do depend on the parametrization. The maximum degree of the connection function is invariant under affine reparametrization.

13.2 Joining two triangular cubic patches

Consider two triangular cubic patches

$$\mathbf{p}(\mathbb{u}) = \sum \mathbf{p}_{\mathbb{i}} B^3_{\mathbb{i}}(\mathbb{u}) \quad \text{and} \quad \mathbf{q}(\mathbb{u}) = \sum \mathbf{q}_{\mathbb{i}} B^3_{\mathbb{i}}(\mathbb{u}) \ ,$$

where $\mathbb{0} \leq \mathbb{i} = (i, j, k)$ and $|\mathbb{i}| = i + j + k = 3$ and $\mathbf{p}_{\mathbb{i}} = \mathbf{q}_{\mathbb{i}}$ for $i = 0$ such that $\mathbf{p}$ and $\mathbf{q}$ join continuously at $n = 0$ and have common tangent planes at $\mathbb{e}_1 = (0, 1, 0)$ and $\mathbb{e}_2 = (0, 0, 1)$. This configuration is illustrated in Figure 13.2. The shaded quadrilaterals are planar but not necessarily affine.

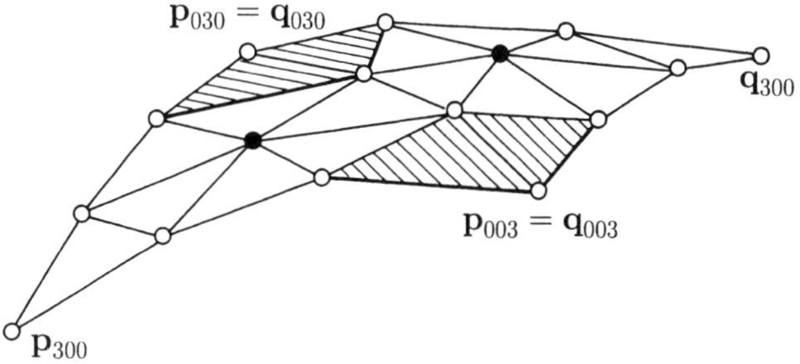

Figure 13.2: Moving interior Bézier points so as to achieve a G^1 joint.

In general, we can move both interior points $\mathbf{p}_{111}$ *and* $\mathbf{q}_{111}$ *so that* $\mathbf{p}$ *and* $\mathbf{q}$ *join* G^1*-continuously along* $u = 0$.

In particular, we show how to obtain such a smooth joint with linear connection functions $\lambda(v)$, $\mu(v)$ and $\nu(v)$. Then, the G^1-condition for $\mathbf{p}$ and $\mathbf{q}$ along $u = 0$ becomes a cubic equation in $w = 1 - v$. Denoting the partial derivatives with respect to the directions $\mathbb{e}_0 - \mathbb{e}_2$ and $\mathbb{e}_2 - \mathbb{e}_1$ by subindices 0 and 1, respectively, this cubic equation is

$$\lambda \mathbf{p}_0 + \mu \mathbf{q}_0 = \nu \mathbf{q}_1 \ , \quad \lambda\mu > 0 \ .$$

At $v = 0$, we know the derivatives $\mathbf{p}_0, \mathbf{q}_0$ and $\mathbf{q}_1$. Hence, this equation establishes a linear system for $\lambda_0 = \lambda(0)$, $\mu_0 = \mu(0)$ and $\nu_0 = \nu(0)$, with a one parameter family of solutions. Similarly, there is a one parameter family of solutions λ_1, μ_1 and ν_1 at $v = 1$. We choose arbitrary solutions at $v = 0$ and $v = 1$, which determine the linear functions λ, μ and ν. Since a cubic is determined by its values and derivatives at two points, here $v = 0$ and $v = 1$, we are also interested in the derivative and, therefore, differentiate the G^1-condition along $u = 0$. Thus, we obtain

$$\lambda \mathbf{p}_{01} + \mu \mathbf{q}_{01} = \nu \mathbf{q}_{11} + \nu' \mathbf{q}_1 - \lambda' \mathbf{p}_0 + \mu' \mathbf{q}_0 \ .$$

Expressing $\mathbf{p}_{01}$, $\mathbf{q}_{01}$ etc. in terms of the Bézier points, we obtain at $\mathbb{e}_1$ the equations

$$\begin{aligned} \mathbf{p}_{01} &= 6(\mathbf{p}_{003} + \mathbf{p}_{111} - \mathbf{p}_{012} - \mathbf{p}_{102}) \ , \\ \mathbf{q}_{01} &= 6(\mathbf{q}_{021} + \mathbf{q}_{111} - \mathbf{q}_{012} - \mathbf{q}_{120}) \\ &\text{etc.} \end{aligned}$$

and similar expressions for $\mathbf{p}_{01}, \mathbf{q}_{01}$, etc. at $\mathbb{e}_2$. The points $\mathbf{q}_{11}, \mathbf{q}_1, \mathbf{q}_0$, and $\mathbf{p}_0$ do not depend on $\mathbf{p}_{111}$ and $\mathbf{q}_{111}$ for $v = 0$ and $v = 1$. Substituting these expressions into the differentiated G^1-condition leads to a linear system for $\mathbf{p}_{111}$ and $\mathbf{q}_{111}$ given by

$$[\mathbf{p}_{111} \mathbf{q}_{111}] \begin{bmatrix} \lambda_0 & \lambda_1 \\ \mu_0 & \mu_1 \end{bmatrix} = [\mathbf{w}_0 \mathbf{w}_1] \ ,$$

where $\mathbf{w}_0$ and $\mathbf{w}_1$ are combinations of known Bézier points $\mathbf{p}_{\mathbb{i}}$ and $\mathbf{q}_{\mathbb{i}}$, except $\mathbf{p}_{111}$ and $\mathbf{q}_{111}$. This system has a solution if the matrix

$$\begin{bmatrix} \lambda_0 & \lambda_1 \\ \mu_0 & \mu_1 \end{bmatrix}$$

is invertible. Hence, a solution exists, unless $\lambda(y) : \mu(y) = \textit{constant}$. Only for the configuration illustrated in Figure 13.3, a solution might not exist. ◇

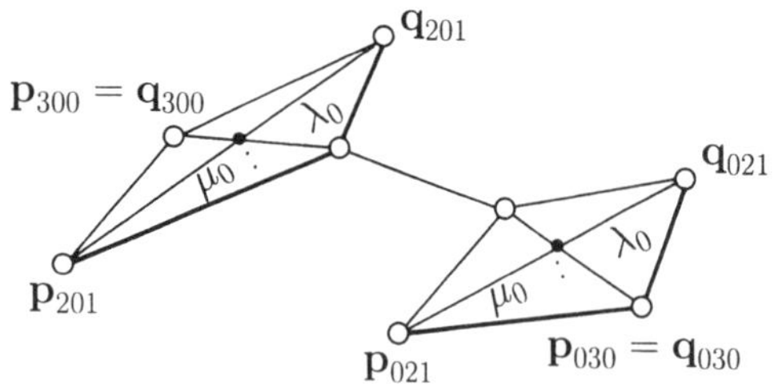

Figure 13.3: Critical configuration.

In fact, there is no solution if $\lambda(y) : \mu(y)$ is constant, if both quadrilaterals are not affine and if the common boundary $\mathbf{p}(0, y) = \mathbf{q}(0, y)$ is a regular cubic, i.e., if $\mathbf{q}_y(0, y)$ is a quadratic not passing through the origin.

Namely, rewriting the G^1-condition as

$$\mathbf{p}_x - \frac{\mu}{\lambda}\mathbf{q}_x = \frac{\nu}{\lambda}\mathbf{q}_y$$

results in a quadratic on the left. Since $\mathbf{q}_y$ is also quadratic without real root, it follows that ν/λ must be constant. This, finally, contradicts the assumption that the two quadrilaterals shown in Figure 13.3 are not affine.

If $\mathbf{q}(0, y)$ is quadratic or non-regular, a solution exists with linear functions λ, μ and ν, see Problem 3.

13.3 A triangular G¹ interpolant

In 1985, Bruce Piper [Piper '87] presented a scheme to construct a piecewise quartic G^1 surface interpolating a triangular network of cubic curves, as illustrated in Figure 13.4. We review the basic construction, but rule out critical situations so that cubic patches suffice.

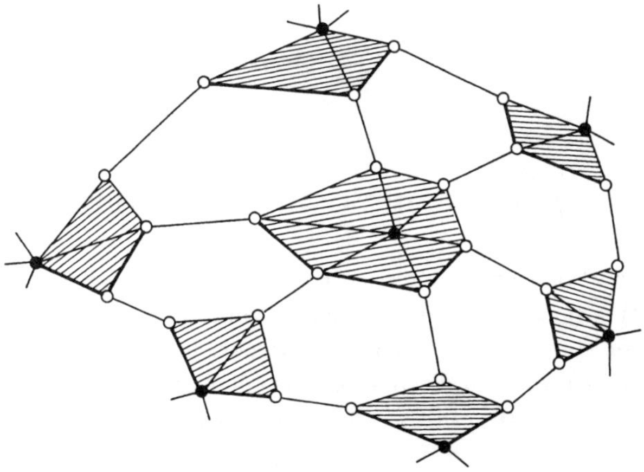

Figure 13.4: A triangular G^1-net of cubic curves.

Adjacent "triangles" of a cubic net exhibit the configurations discussed in 13.2. For simplicity, we assume that there are no critical configurations as in Figure 13.3. Then, any "triangle" can be interpolated by a macro patch consisting of three cubic patches, as described below. Figure 13.5 shows the Bézier points of such a macro patch schematically.

The Bézier points ○ on the boundary are given by the cubic net. The Bézier points ● are the centroids of their three neighbors ○ with which they form a planar quadrilateral which is shaded in the Figure.

The Bézier points ◒ are computed as in 13.2 such that adjacent macro patches have a G^1 joint.

The Bézier points □ are the centroids of the three neighbors ◒ ● ◒ with which they form a planar quadrilateral (shaded in the Figure).

The Bézier point ■ is the centroid of the shaded triangle □ □ □.

Hence, adjacent patches of the macro patch have simple C^1 joints in analogy to the Clough-Tocher element, see 12.2.

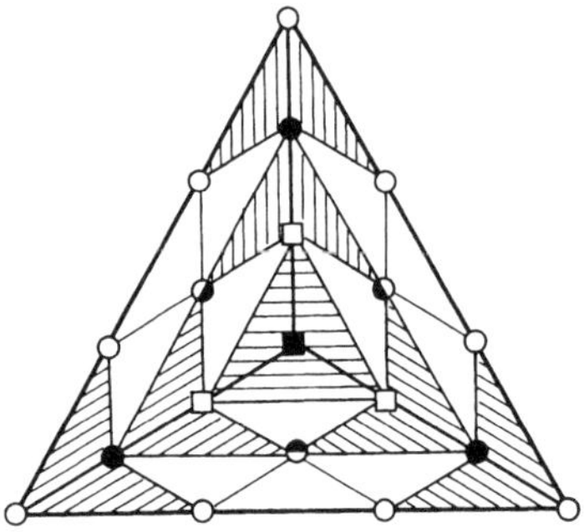

Figure 13.5: Macro patch.

13.4 The vertex enclosure problem

Piper implicitly solves with his construction a special G^1-problem. The G^1-conditions for adjacent patches sharing a common vertex form a cyclic system. Whether this problem is solvable or not, is referred to as the **vertex enclosure problem**. To discuss this problem, we consider n trilateral or quadrilateral patches $\mathbf{p}^i(x, y)$, $i = 1, \ldots, n$, such that

$$\mathbf{p}^i(0, z) = \mathbf{p}^{i+1}(z, 0) \ ,$$

where $\mathbf{p}^{n+1} = \mathbf{p}^1$, as illustrated in Figure 13.6, and

$$\text{(2)} \qquad \lambda_i \mathbf{p}_x^i(0, z) = \mu_i \mathbf{p}_y^{i+1}(z, 0) + \nu_i \mathbf{p}_x^{i+1}(z, 0) \ ,$$

with any $3n$ connection functions $\lambda_i(z), \mu_i(z)$ and $\nu_i(z)$.

For $z = 0$, these equations form constraints on the derivatives $\mathbf{p}_x^i$ and the connection functions, which can easily be satisfied. More difficult are the

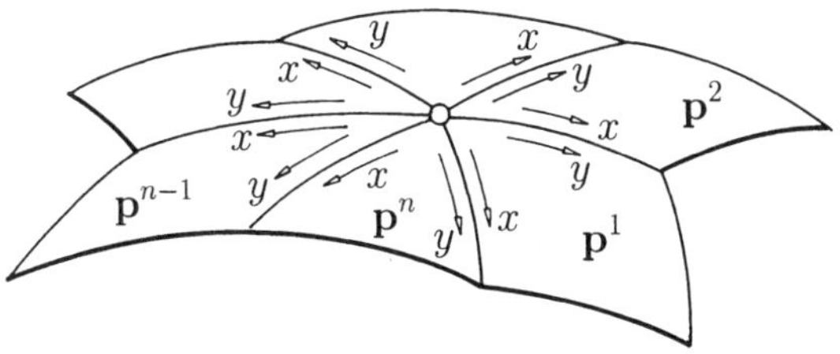

Figure 13.6: A vertex enclosed by n patches, where $n = 6$.

twist constraints, which we obtain by differentiating the G^1-conditions (2),

$$\lambda_i' \mathbf{p}_x^i + \lambda_i \mathbf{p}_{xy}^i = \mu_i' \mathbf{p}_y^{i+1} + \mu_i \mathbf{p}_{xy}^{i+1} + \nu_i' \mathbf{p}_x^{i+1} + \nu_i \mathbf{p}_{xx}^{i+1} .$$

For $z = 0$, these equations form the cyclic linear system

$$[\mathbf{p}_{xy}^1 \cdots \mathbf{p}_{xy}^n] \begin{bmatrix} \lambda_1 & & & -\mu_1 \\ -\mu_2 & \lambda_2 & & \\ & & \ddots & \\ & & -\mu_n & \lambda_n \end{bmatrix} = [\mathbf{r}_1 \ldots \mathbf{r}_n] ,$$

where

$$\mathbf{r}_i = -\lambda_i' \mathbf{p}_x^i + \mu_i' \mathbf{p}_y^i + \nu_i' \mathbf{p}_x^{i+1} + \nu_i \mathbf{p}_{xx}^{i+1} .$$

We abbreviate this system by $TA = R$.

13.5 The parity phenomenon

The cyclic matrix A of the twist constraints exhibits the following phenomenon. The rank of A is n if n is odd, and it is $n - 1$ if n is even. Thus, A is non-singular only for odd n. Consequently, the twist constraints are solvable if the number n of patches is odd, while, in general, there is no solution if the number is even. In order to verify this surprising fact, we observe that

$$\begin{aligned} \det A &= \lambda_1 \ldots \lambda_n - \mu_1 \ldots \mu_n \\ &= \Pi\lambda_i - \Pi\mu_i . \end{aligned}$$

Computing, the vector product of

$$\lambda_i \mathbf{p}_x^i = \mu_i \mathbf{p}_y^{i+1} + \nu_i \mathbf{p}_x^{i+1} ,$$

with $\mathbf{p}_x^{i+1} = \mathbf{p}_y^i$ gives

$$\lambda_i : \mu_i = [\mathbf{p}_x^{i+2} \times \mathbf{p}_x^{i+1}] : [\mathbf{p}_x^{i+1} \times \mathbf{p}_x^i] \ ,$$

which implies that

$$\Pi\lambda_i : \Pi\mu_i = (-1)^n$$

and, since all $\lambda_i \neq 0$,

$$\det A \begin{cases} = 0 & \text{if } n \text{ is even} \\ \neq 0 & \text{if } n \text{ is odd} \end{cases} \ .$$

Since the submatrix of A obtained by deleting the first row and column has full rank, A has at least rank $n-1$, which concludes the proof.

Remark 5: The twist constraints $AT = R$ are solvable if the data originates from patches $\mathbf{p}^1, \ldots, \mathbf{p}^n$ forming a G^1 surface. In particular, this is the case when the $\mathbf{p}^i$ form a piecewise polynomial reparametrization of a polynomial patch.

Remark 6: If $n = 4$ and $\lambda_i'(0) = \mu_i'(0) = \nu_i'(0) = \nu_i(0) = 0$, for $i = 1, 2, 3, 4$, as illustrated in Figure 13.7, then the twist constraints are solvable, see 9.7.

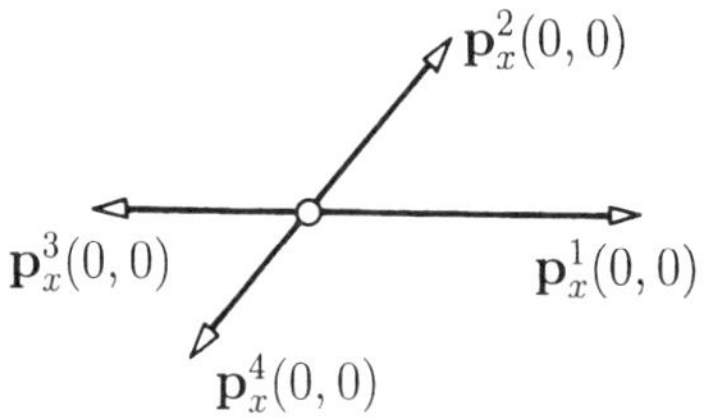

Figure 13.7: Equal opposite tangents.

Remark 7: One obtains always solvable twist constraints by splitting each patch $\mathbf{p}^i$ as in Piper's construction described in 13.3, see [Peters '91].

13.6 Problems

1 Show that the problem from Section 13.2 can always be solved if $\mathbf{p}$ and $\mathbf{q}$ are quartic patches.

2 Solve the problem from Section 13.2, where

$$\begin{array}{ccccccc} & \mathbf{p}_{201} & & \mathbf{p}_{111} & & \mathbf{p}_{012} & \\ \mathbf{p}_{300} & & \mathbf{p}_{210} & & \mathbf{p}_{120} & & \mathbf{p}_{030} \\ \| & & \| & & \| & & \| \\ \mathbf{q}_{300} & & \mathbf{q}_{210} & & \mathbf{q}_{120} & & \mathbf{q}_{030} \\ & \mathbf{q}_{201} & & \mathbf{q}_{111} & & \mathbf{q}_{012} & \end{array} \quad = \quad \begin{array}{ccccccc} & \begin{bmatrix} 1 \\ 1 \\ 0 \end{bmatrix} & & \begin{bmatrix} \mathbf{p}_{111} \end{bmatrix} & & \begin{bmatrix} 6 \\ 1 \\ 1 \end{bmatrix} & \\ \begin{bmatrix} 0 \\ 0 \\ 0 \end{bmatrix} & & \begin{bmatrix} 2 \\ 0 \\ 0 \end{bmatrix} & & \begin{bmatrix} 4 \\ 0 \\ 1 \end{bmatrix} & & \begin{bmatrix} 6 \\ 0 \\ 1 \end{bmatrix} \\ & \begin{bmatrix} 1 \\ -1 \\ 0 \end{bmatrix} & & \begin{bmatrix} \mathbf{q}_{111} \end{bmatrix} & & \begin{bmatrix} 5 \\ -1/2 \\ 1 \end{bmatrix} & \end{array} .$$

3 Show that the problem stated in 13.2, see Figure 13.3, has a solution with linear functions $\lambda = \mu$ and ν if $\mathbf{q}(0, v, 1-v)$ is quadratic or non-regular. Hint: If $\mathbf{q}(0, v, 1-v)$ is quadratic, choose $\lambda = \mu = 1$. If $\mathbf{q}_1(0, v, 1-v) = \mathbf{o}$ for $v = v_0$, set $\lambda = \mu = (v - v_0)c_1$ and $\nu = vc_2$, where the constants c_1 and c_2 are chosen so that the G^1-condition is satisfied at $v = 1$.

4 Show that the twist constraints in 13.4 have a solution if and only if

$$\mathbf{s} = \mathbf{t}_1 \left(1 - \frac{\mu_1 \cdots \mu_n}{\lambda_1 \cdots \lambda_n}\right)$$

can be solved for $\mathbf{t}_1$, where

$$\mathbf{s} = \frac{\mu_1 \cdots \mu_{n-1}}{\lambda_1 \cdots \lambda_n} \mathbf{r}_n + \cdots + \frac{\mu_1}{\lambda_1 \lambda_2} \mathbf{r}_2 + \frac{1}{\lambda_1} \mathbf{r}_1 .$$

5 The equation $\mathbf{s} = \mathbf{o}$ can be viewed as a linear system for $\nu_1'(0), \nu_2'(0)$ and $\nu_1(0)$. It has a (unique) solution unless $\mathbf{p}_{xx}^2 = \mathbf{o}$.

6 Consider two biquartic patches

$$\begin{aligned} \mathbf{p}(x, y) &= \sum \mathbf{p}_{ij} B_{ij}^{44}(x, y) \quad \text{and} \\ \mathbf{q}(x, y) &= \sum \mathbf{q}_{ij} B_{ij}^{44}(x, y) \end{aligned}$$

such that the boundary curves $\mathbf{p}(x, 0)$ and $\mathbf{q}(x, 0)$ agree and are cubic, see Figure 13.8.

Let λ, μ and ν be any connection functions of degree $1, 1$ and 3, respectively, such that the G^1-condition

$$\lambda \mathbf{p}_y(x, 0) = \mu \mathbf{q}_y(x, 0) + \nu \mathbf{q}_x(x, 0)$$

and its derivative with respect to x hold at $x = 0$ and $x = 1$. Show that

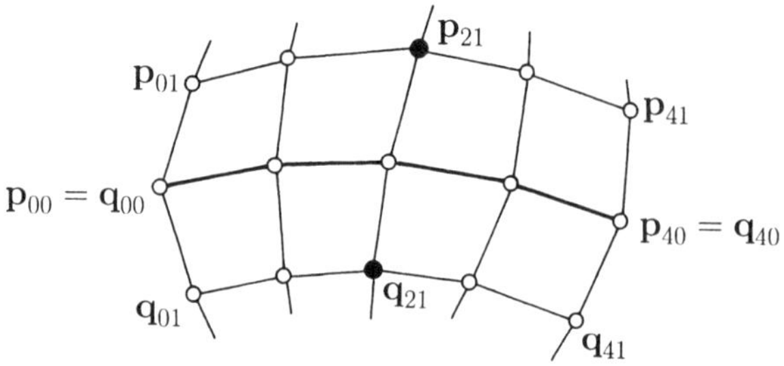

Figure 13.8: Moving two Bézier points so as to obtain a G^1 joint.

one can change $\mathbf{p}_{21}$ and $\mathbf{q}_{21}$, in general, so that the G^1-condition holds for all x.

14 G^k-constructions

14.1 The general C^k joint — 14.2 G^k joints by cross curves — 14.3 G^k joints by the chain rule — 14.4 G^k surfaces of arbitrary topology — 14.5 Smooth n-sided patches — 14.6 Multi-sided patches in the plane — 14.7 Problems

Two patches join smoothly if they can be (re-)parametrized so that their derivatives up to some order are identical along a common boundary curve. For any fixed reparametrization, this smoothness condition means that the derivatives of both patches at any common point are related by a linear transformation. This is analogous to the curve case.

In this chapter, we discuss these smoothness conditions and use them to build surfaces of arbitrary topological form and arbitrarily high smoothness order.

14.1 The general C^k joint

Two regular patches $\mathbf{p}$ and $\mathbf{q}$ with a common boundary curve $\mathbf{b}$ are said to have a **general C^k joint** along $\mathbf{b}$ if they have a simple C^k joint locally for each point $\mathbf{b}_0$ on $\mathbf{b}$ after some regular reparametrization. This means that, locally, there are regular reparametrizations $\mathbf{u}$ and $\mathbf{v}$ such that $\mathbf{p} \circ \mathbf{u}$ and $\mathbf{q} \circ \mathbf{v}$ have identical derivatives up to order k along $\mathbf{b}$, see Figure 14.2. It suffices to reparametrize only one patch, for example $\mathbf{q}$ by $\mathbf{v} \circ \mathbf{u}^{-1}$, see Figure 14.2. A general C^k joint is also referred to as a **G^k joint**.

If $\mathbf{p}$ and $\mathbf{q}$ have a G^k joint at some point $\mathbf{b}_0$, we obtain a local C^k parametrization $\mathbf{r}(x, y)$ of the union of both patches simply by a non-tangential projection π into some plane P, as illustrated in Figure 14.3.

The regular C^k-maps $\phi = \pi \circ \mathbf{p} \circ \mathbf{u}$ and $\psi = \pi \circ \mathbf{q} \circ \mathbf{v}$, have a C^k joint along $\pi(\mathbf{b})$ in a neighborhood of $\pi(\mathbf{b}_0)$. Therefore,

$$\mathbf{r}(x, y) = \begin{cases} \mathbf{p} \circ \mathbf{u} \circ \phi^{-1}(x, y) & \text{if } (x, y) \text{ lies in } \pi(\mathbf{p}) \\ \mathbf{q} \circ \mathbf{v} \circ \psi^{-1}(x, y) & \text{if } (x, y) \text{ lies in } \pi(\mathbf{q}) \end{cases}$$

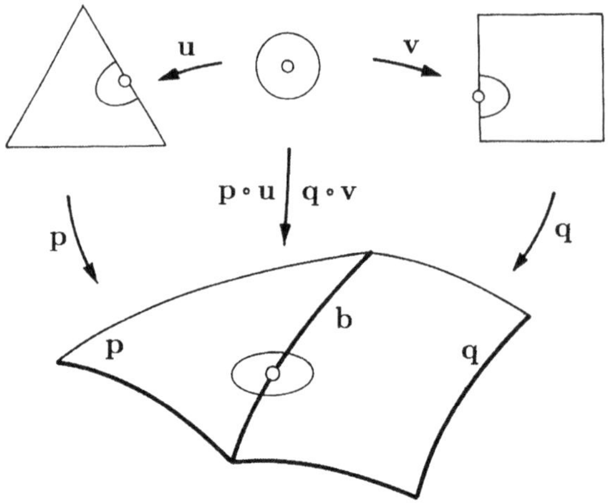

Figure 14.1: A general C^k joint.

is, indeed, a C^k-parametrization.

14.2 $\mathbf{G}^{\mathbf{k}}$ joints by cross curves

One way to check whether two patches $\mathbf{p}$ and $\mathbf{q}$ have a G^k joint, with $k \geq 1$, along a common boundary curve $\mathbf{b}(t)$ is provided by the following theorem.

> *If* $\mathbf{b}(t)$ *is differentiable, and if through every point* $\mathbf{b}_0$ *on* $\mathbf{b}(t)$ *there is a regular* k *times differentiable curve* $\mathbf{c}(s)$ *which lies on the union of* $\mathbf{p}$ *and* $\mathbf{q}$ *and crosses* $\mathbf{b}$ *non-tangentially, then* $\mathbf{p}$ *and* $\mathbf{q}$ *have a* G^k *joint.*

We prove this fact by induction over k. According to our set-up, $\mathbf{p}$ and $\mathbf{q}$ join continuously. Thus it suffices to prove the hypothesis under the assumption that $\mathbf{p}$ and $\mathbf{q}$ have a G^{k-1} joint along $\mathbf{b}$ for $k > 0$.

Let $\mathbf{p}$ and $\mathbf{q}$ be parametrized locally around a point $\mathbf{b}_0 = \mathbf{b}(t_0) = \mathbf{c}(s_0)$ by a projection π into some plane P as illustrated in Figure 14.2. We use the affine coordinates x and y with respect to the coordinate system formed by the projections of $\mathbf{b}_0, \mathbf{b}'(t_0), \mathbf{c}'(s_0)$.

Further, let

$$\bar{\mathbf{p}} = \mathbf{p}_{x \overset{i}{\cdots} x\, y \overset{j}{\cdots} y} \quad \text{and} \quad \bar{\mathbf{q}} = \mathbf{q}_{x \overset{i}{\cdots} x\, y \overset{j}{\cdots} y}$$

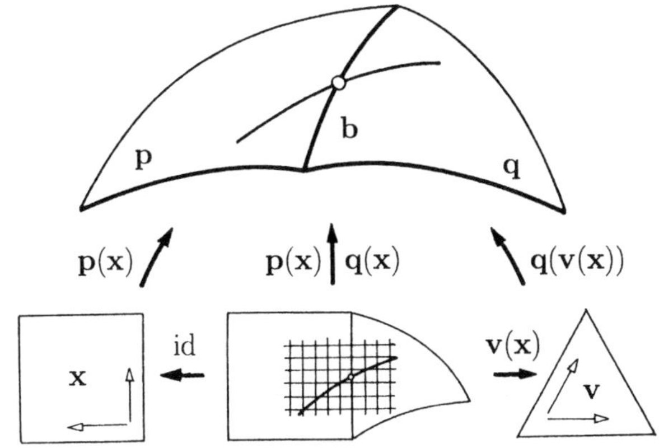

Figure 14.2: A general C^k joint brought into a simple form.

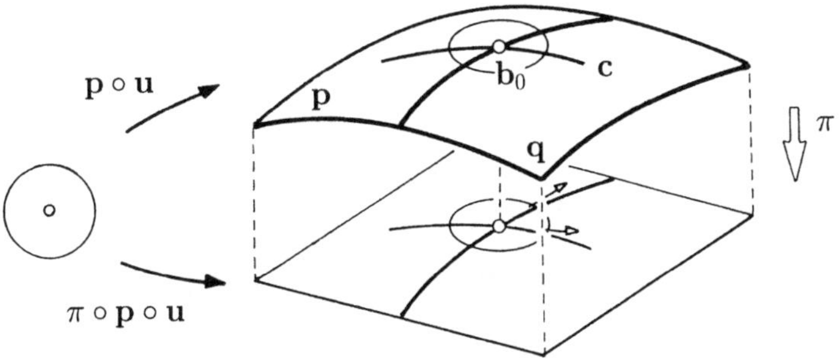

Figure 14.3: Parametrization by projection.

be any partial derivatives of $\mathbf{p}$ and $\mathbf{q}$ of order $i + j = k - 1$. Then, the induction assumption implies that

$$\overline{\mathbf{p}}(\pi \mathbf{b}(t)) = \overline{\mathbf{q}}(\pi \mathbf{b}(t)) \ .$$

Differentiating this equation with respect to t, we obtain for $t = t_0$

$$\overline{\mathbf{p}}_x(\mathbf{o}) = \overline{\mathbf{q}}_x(\mathbf{o}) \ .$$

Hence, almost all k-th partial derivatives of $\mathbf{p}$ and $\mathbf{q}$ are, equal and we need to show that

$$\frac{\partial^k}{\partial y^k}[\mathbf{p}(\mathbf{o}) - \mathbf{q}(\mathbf{o})] = \mathbf{o} \ .$$

Since the curve $\mathbf{c}(s)$ is k times differentiable, we get

$$\begin{aligned} \mathbf{o} &= \frac{\partial^k}{\partial s^k}[\mathbf{c}(s) - \mathbf{c}(s)]_{s=s_0} \\ &= \frac{\partial^k}{\partial s^k}[\mathbf{p}(\pi\mathbf{c}(s)) - \mathbf{q}(\pi\mathbf{c}(s))]_{s=s_0} \ . \end{aligned}$$

Using the chain rule, we can express this derivative in terms of the derivatives of $\pi\mathbf{c}$ and the partial derivatives of $\mathbf{p}$ and $\mathbf{q}$. Since almost all partial derivatives of $\mathbf{p}$ and $\mathbf{q}$ are equal and since $\pi\mathbf{c}'(s_0) = (0,1)$, one obtains

$$\mathbf{o} = \frac{\partial^k}{\partial y^k}[\mathbf{p}(\mathbf{x}) - \mathbf{q}(\mathbf{x})]_{x=0} \ ,$$

which concludes the proof.◇

Remark 1: If the boundary curve $\mathbf{b}(t)$ is not differentiable at $t = t_0$, but the quotient $[\mathbf{b}(t_n) - \mathbf{b}(t_0)]/[t_n - t_0]$ converges to different directions for at least two different sequences $t_n \to t_0$, then $\mathbf{p}$ and $\mathbf{q}$ have a G^k joint at $\mathbf{b}(t_0)$ if they have a G^{k-1} joint along $\mathbf{b}(t)$.

14.3 $\mathrm{G^k}$ joints by the chain rule

Let $\mathbf{p}(x,y)$ and $\mathbf{q}(u,v)$ be two regular surfaces such that $\mathbf{p}(0,y) = \mathbf{q}(0,y)$. As discussed in 14.1, they have a G^k joint along the parameter line $x = u = 0$ if, for every y, there is a local reparametrization $\mathbf{u}(x,y)$ for $\mathbf{q}$ such that all derivatives up to order k of $\mathbf{p}(x,y)$ and $\mathbf{q}(\mathbf{u}(x,y))$ are locally equal.

From 14.2, it follows that all mixed derivatives are equal if the partial derivatives of $\mathbf{p}$ and $\mathbf{q} \circ \mathbf{u}$ with respect to the first parameter x are identical. Hence, $\mathbf{p}$ and $\mathbf{q}$ have a G^k joint along the parameter line $x = u = 0$ if and only if there is a regular C^k-map

$$\mathbf{u}(x,y) = (u(x,y), v(x,y))$$

such that for all arguments $(0,y)$

$$\begin{aligned} \mathbf{p} &= \mathbf{q} \ , \\ \mathbf{p}_x &= \mathbf{q}_u \ u_x + \ \mathbf{q}_v \ v_x \ , \\ \mathbf{p}_{xx} &= \mathbf{q}_{uu} u_x^2 + 2\mathbf{q}_{uv} u_x v_x + \mathbf{q}_{vv} v_x^2 + \mathbf{q}_u u_{xx} + \mathbf{q}_v v_{xx} \ , \\ &\text{etc.} \ , \end{aligned}$$

where $u_x > 0$. Note that $\mathbf{u}(0,y) = (0,y)$.

In particular, if $\mathbf{p}$ and $\mathbf{q}$ are (rational) polynomials, it follows as in 13.1 from the G^k-conditions that $u_x(0,y), v_x(0,y), u_{xx}(0,y), \ldots$ are (rational) polynomials of certain degrees.

The G^k-conditions are simpler if $\mathbf{u}$ is linear in x. In this case, $\mathbf{u}_{x.\overset{s}{.}.x}$ is zero for $s = 2, \ldots, k$, and the G^k-conditions reduce to

$$
\begin{aligned}
\mathbf{p} &= \mathbf{q} \ , \\
\mathbf{p}_x &= \mathbf{q}_u \ \alpha \ + \ \mathbf{q}_v \ \beta \ , \\
\mathbf{p}_{xx} &= \mathbf{q}_{uu}\alpha^2 + 2\mathbf{q}_{uv}\alpha\beta + \mathbf{q}_{vv}\beta^2 \ , \\
&\ \vdots \\
\mathbf{p}_{x.\overset{k}{.}.x} &= \sum_{i+j=k} \binom{k}{i} \mathbf{q}_{u.\overset{i}{.}.uv.\overset{j}{.}.v}\alpha^i\beta^i \ ,
\end{aligned} \tag{1}
$$

where $\alpha = \alpha(y) = u_x(0, y)$ and $\beta = \beta(y) = v_x(0, y)$.

Note that the right hand sides also represent the partial derivatives of $\mathbf{q}$ with respect to the direction $[\alpha\ \beta]^t = \mathbf{u}_x$.

Remark 2: A mixed partial derivative $\mathbf{p}_{x.\overset{i}{.}.xy.\overset{j}{.}.y}$ can be expressed in terms of the mixed partial derivatives of $\mathbf{q}$ up to total order $i + j$ and the mixed partial derivatives of $\mathbf{u}(x, y)$ up to order (i, j). For example,

$$\mathbf{p}_{xy} = \mathbf{q}_{uu}u_x u_y + \mathbf{q}_{uv}(u_x v_y + u_y v_x) + \mathbf{q}_{vv}v_x v_y + \mathbf{q}_u u_{xy} + \mathbf{q}_v v_{xy} \ .$$

Remark 3: In particular, if $\mathbf{u}(x, y)$ is a dilation in x and y, i.e.,

$$\mathbf{u}(x, y) = [c_1 x\ c_2 y]^t \ ,$$

then all mixed partial derivatives $\mathbf{p}_{x.\overset{i}{.}.xy.\overset{j}{.}.y}$ can be expressed in terms of u_x, v_y and the mixed partial derivatives of $\mathbf{q}$ up to order (i, j).

14.4 G^k surfaces of arbitrary topology

In this section, we present a construction of smooth free-form surfaces. These surfaces interpolate the **vertices** $\mathbf{c}_i$ of a given quadrilateral net and smoothly contact prescribed polynomial surfaces $\mathbf{s}_i$ at these vertices.

The resulting surface is G^k-continuous and consists of tensor product patches parametrized over $[0, 1]^2$, where each patch corresponds uniquely to a quadrilateral of the given net and vice versa.

For notational simplicity, we assume that the given net is **orientable** and has no boundary.

We call a vertex **regular**, if it has exactly four neighbors and **irregular** otherwise. To simplify the complex construction, we assume further that irregular vertices are isolated, i.e., that they have only regular neighbors as, for example, in Figure 14.4.

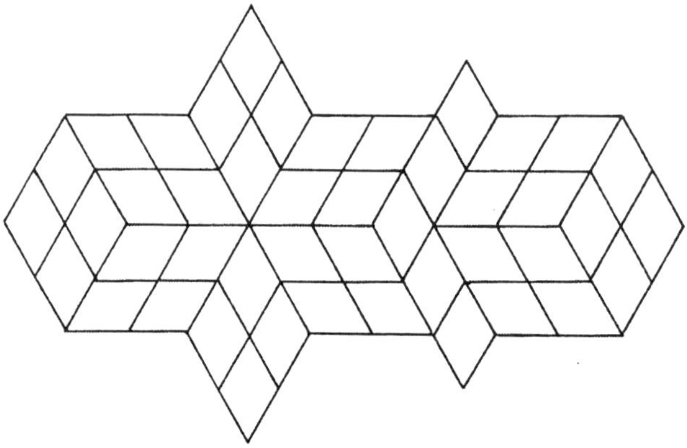

Figure 14.4: A quadrilateral net with isolated irregular vertices.

Before we come to the details of the construction, we briefly outline its three steps:

1. We reparametrize the surfaces $\mathbf{s}_i$ by rotations of the domain system.
2. For any two neighboring surfaces $\mathbf{s}_i$ and $\mathbf{s}_j$, we compute a Hermite interpolant $\mathbf{b}_{ij}$.
3. For each quadrilateral $\mathbf{c}_i\mathbf{c}_j\mathbf{c}_k\mathbf{c}_l$ of the net we compute a patch $\mathbf{p}$ that has G^k contact with $\mathbf{b}_{ij}, \mathbf{b}_{jk}, \mathbf{b}_{kl}$ and $\mathbf{b}_{li}$ along its four boundaries. If all vertices are regular, $\mathbf{p}$ has bidegree $2k+1$ and, otherwise, it has bidegree $2k^2+2k+1$.

Next we give a detailed description of these steps.

1. For each i, we assume that $\mathbf{s}_i(0,0) = \mathbf{c}_i$. Denoting the valence, i.e., the number of incoming edges, of $\mathbf{c}_i$ by ν, we rotate the domain system repeatedly by $\varphi_i = 360°/\nu$. This gives the reparametrized polynomials

$$\mathbf{s}_i^k(\mathbf{x}) = \mathbf{s}_i(R^k\mathbf{x}) ,$$

where

$$R = \begin{bmatrix} \cos\varphi_i & -\sin\varphi_i \\ \sin\varphi_i & \cos\varphi_i \end{bmatrix} .$$

We count the neighbors of $\mathbf{c}_i$ counterclockwise with respect to the normal. If

$\mathbf{c}_j$ is the k-th neighbor, we associate the scaled polynomial

$$\mathbf{s}_{ij}(\mathbf{x}) := \mathbf{s}_i^k\left(x, \frac{y}{\tan(\varphi_j/2)}\right)$$

with the directed edge $\mathbf{c}_i\mathbf{c}_j$. This is illustrated in Figure 14.5.

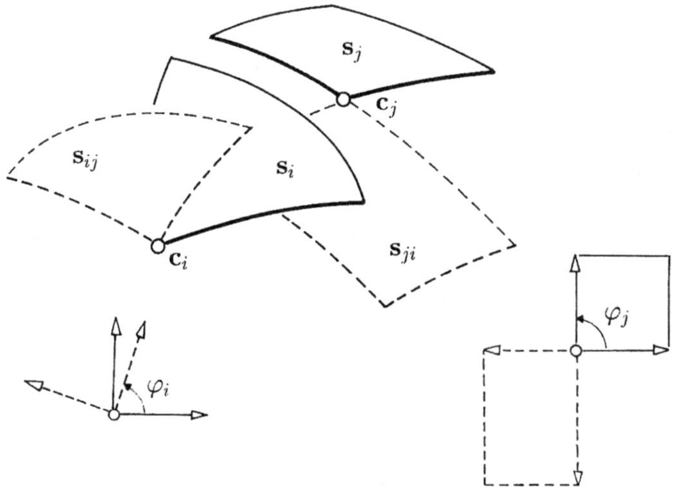

Figure 14.5: Reparametrization by domain rotation.

2. For each directed edge $\mathbf{c}_i\mathbf{c}_j$ we determine a polynomial $\mathbf{b}_{ij}(\mathbf{x})$ by Hermite interpolation of $\mathbf{s}_{ij}(\mathbf{x})$ and $\mathbf{s}_{ji}(\mathbf{x})$. If $\mathbf{c}_i$ is regular, then $\mathbf{b}_{ij}$ is constructed such that it has identical derivatives up to order (k,k) with $\mathbf{s}_{ij}$ at $\mathbf{x} = (0,0)$. We abbreviate this $C^{k,k}$ contact by

$$\mathbf{b}_{ij}(\mathbf{x}) \underset{\mathbf{x}=\mathbf{0}}{\overset{k,k}{\equiv}} \mathbf{s}_{ij}(\mathbf{x}) \ .$$

If $\mathbf{c}_i$ is irregular, then we require a C^{2k} contact,

$$\mathbf{b}_{ij}(\mathbf{x}) \underset{\mathbf{x}=\mathbf{0}}{\overset{2k}{\equiv}} \mathbf{s}_{ij}(\mathbf{x})$$

and, analogously, at $\mathbf{x} = (1,0)$ we require a $C^{k,k}$ contact, or a C^{2k} contact of $\mathbf{b}_{ij}(x,y)$ and $\mathbf{s}_{ji}(1-x,-y)$, respectively. This is illustrated in Figure 14.6.

We determine $\mathbf{b}_{ij}$ such that its cross boundary derivatives

$$\frac{\partial^r}{\partial y^r}\mathbf{b}_{ij}(x,0) \ , \ r = 0, \ldots, k \ ,$$

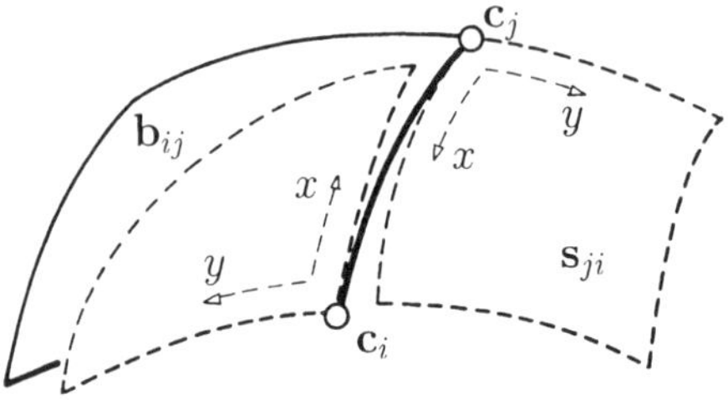

Figure 14.6: An edge polynomial.

are of minimal degree. Therefore, if $\mathbf{c}_i$ and $\mathbf{c}_j$ are regular, then $\mathbf{b}_{ij}$ has degree $2k + 1$ and, otherwise, if $\mathbf{c}_i$ or $\mathbf{c}_j$ is irregular, then $\frac{\partial^r}{\partial y^r}\mathbf{b}_{ij}$ has the degree $3k+1-r$. The Bézier points of $\mathbf{b}_{ij}$ are shown schematically in Figure 14.7 for $k = 2$, where $\mathbf{c}_i$ is irregular. The points determined by the C^{2k} contact at $\mathbf{c}_i$ are marked by triangles $\triangle$, the points determined by the $C^{k,k}$ contact at $\mathbf{c}_j$ by squares $\square$ and the points determined by the minimal degree constraints by circles $\circ$. The points marked by dots $\cdot$ are non-interesting.

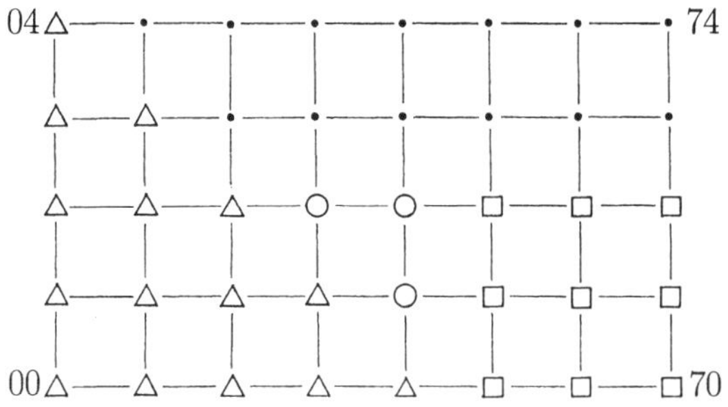

Figure 14.7: Schematic view of the Bézier points of an edge polynomial.

Analogously, we get a second polynomial $\mathbf{b}_{ji}$ for the oppositely directed edge $\mathbf{c}_j\mathbf{c}_i$. Due to our construction, both polynomials have a simple C^k contact along the parameter line $y = 0$,

$$\mathbf{b}_{ij}(x, y) \underset{\mathrm{y}=0}{\overset{k}{\equiv}} \mathbf{b}_{ji}(1 - x, -y) \ .$$

Further, due to our construction, any two polynomials $\mathbf{b}_{ij}$ and $\mathbf{b}_{ik}$ belonging to a vertex $\mathbf{c}_i$ have $G^{k,k}$- or G^{2k} contact at $\mathbf{c}_i$.

3. For each quadrilateral of the given net, we construct a patch $\mathbf{p}(u, v)$ of the final G^k surface from the four associated edge polynomials. For each edge

$\mathbf{c}_i\mathbf{c}_j$, we use a reparametrization

$$\mathbf{x}_{ij}(u,v) = \sum \mathbf{b}_{rs} B_{rs}^{2k,2k}(u,v)$$

of bidegree $2k$. The Bézier net of this reparametrization $\mathbf{x}_{ij}(u,v)$ is shown in Figure 14.8 for $k=2$ and $\varphi = \varphi_i = 360°/5$.

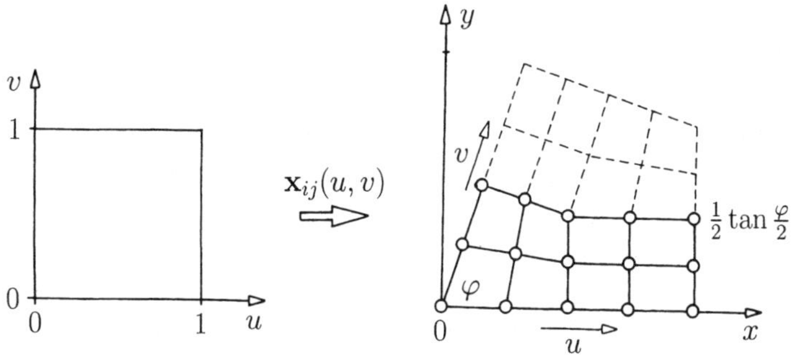

Figure 14.8: Reparametrization map $\mathbf{x}_{ij}$ for an edge polynomial for $k=2$ and $\mathbf{c}_i$ having valence 5.

In general, if $\mathbf{c}_j$ is regular, then the relevant Bézier points of $\mathbf{x}_{ij}$ are given by

$$\begin{aligned} \mathbf{b}_{rs} = & \begin{bmatrix} \cos\varphi_i \\ \sin\varphi_i \end{bmatrix} B_{01}^{11}(u,v) + \begin{bmatrix} 1 \\ \tan\frac{\varphi_i}{2} \end{bmatrix} B_{11}^{11}(u,v) \\ & + \begin{bmatrix} 0 \\ 0 \end{bmatrix} B_{00}^{11}(u,v) + \begin{bmatrix} 1 \\ 0 \end{bmatrix} B_{10}^{11}(u,v) \end{aligned}$$

for $(u,v) = \frac{1}{2k}(r,s)$ if $r,s \le k$. Otherwise these Bézier points are given by

$$\mathbf{b}_{rs} = \frac{1}{2k}\begin{bmatrix} r \\ s\sin\frac{\varphi_i}{2} \end{bmatrix} \qquad \text{for} \quad s \le k \le r\ .$$

The other Bézier points $\mathbf{b}_{rs}$ for $r > k$ can be chosen arbitrarily. This is indicated by the dashed lines in Figure 14.8. If $\mathbf{c}_j$ is irregular, $\mathbf{x}_{ij}$ is obtained from the corresponding map $\mathbf{x}_{ji}$ by the transformation

$$\mathbf{x}_{ij}(u,v) = \begin{bmatrix} 1 \\ 0 \end{bmatrix} + \begin{bmatrix} -1 & 0 \\ 0 & 1 \end{bmatrix} \mathbf{x}_{ji}(1-u,v)\ .$$

The map $\mathbf{x}_{ij}$ is the identity if $\mathbf{c}_i$ and $\mathbf{c}_j$ are regular.

Let $\mathbf{c}_1, \ldots, \mathbf{c}_4$ be the vertices of some quadrilateral in counterclockwise order. Then, the corresponding patch $\mathbf{p}(u,v)$ of the final G^k surface is constructed such that

$$\text{(2)} \qquad \mathbf{p}(u,v) \begin{cases} \overset{k}{\underset{v=0}{=}} \mathbf{b}_{12} \circ \mathbf{x}_{12}(u,v) \\ \overset{k}{\underset{u=1}{=}} \mathbf{b}_{23} \circ \mathbf{x}_{23}(v, 1-u) \\ \overset{k}{\underset{v=1}{=}} \mathbf{b}_{34} \circ \mathbf{x}_{34}(1-u, 1-v) \\ \overset{k}{\underset{u=0}{=}} \mathbf{b}_{41} \circ \mathbf{x}_{41}(1-v, u) \end{cases} .$$

By construction, the mixed partial derivatives of $\mathbf{p}$ up to order (k,k) at the vertices are well-defined by these conditions. Recall that the rth cross boundary derivative of an edge polynomial $\mathbf{b}_{ij}$ is of degree $\leq 3k+1-r$. Hence, it follows from (1) in 14.3 that an rth cross boundary derivative of $\mathbf{p}(u,v)$ is of degree $\leq 2kr + 3k + 1 - r \leq 2k^2 + 2k + 1$. Along an edge between two regular vertices the degree is lower, namely $\leq 2k+1$.

If all vertices are regular, then the conditions (2) define a unique patch $\mathbf{p}$ of bidegree $2k+1$. If one or two vertices are irregular, then $\mathbf{p}$, in general, must be of bidegree $2k^2 + 2k + 1$ to satisfy all boundary conditions (2). However, in this case, $\mathbf{p}$ is not completely defined by (2). Its inner $2k^2 \times 2k^2$ Bézier points can be chosen arbitrarily.

14.5 Smooth n-sided patches

The construction described in Section 14.4 leads to G^k surfaces of bidegree $2k^2 + 2k + 1$. This degree is rather high, and a G^2 surface of degree 13 may not be desirable in practical applications. Namely, storage costs are high and evaluation takes long and is prone to rounding errors.

To avoid this, a general method for constructing regular G^k surfaces of bidegree $2(k+1)$ is developed in [Prautzsch '97]. In the following, we present this method.

We recall from 14.4 and 13.4 that the major difficulty in building a smooth surface with quadrilateral (or with triangular) patches is to join three or more than four patches smoothly around some common vertex. Hence, we are interested in smooth n-sided surfaces that can be fit into n-sided holes with simple C^k joints. To construct such surfaces, we

first determine a reparametrization of the x,y-parameter plane by planar patches $\mathbf{x}_i(\mathbf{u}), \mathbf{u} \in [0,1]^2, i = 1, \ldots, 4n$, of bidegree $k+1$, forming an n-gon, as illustrated in Figure 14.9 for $n = 5$. The inner patches $\mathbf{x}_1, \ldots, \mathbf{x}_n = \mathbf{x}_1$ have C^0 joints, i.e.,

$$\mathbf{x}_i(0,u) = \mathbf{x}_{i+1}(u,0) \;,$$

while all other joints are simple C^k joints, i.e., for all u, v and $i = 1, \dots, n$ the following equations hold:

$$\begin{aligned}
\mathbf{x}_i(u,1) &\overset{k}{=} \mathbf{x}_{i+3n}(u,0) \ , \\
\mathbf{x}_i(1,v) &\overset{k}{=} \mathbf{x}_{i+2n}(0,v) \ , \\
\mathbf{x}_{i+3n}(1,v) &\overset{k}{=} \mathbf{x}_{i+n}(0,v) \ , \\
\mathbf{x}_{i+n}(u,0) &\overset{k}{=} \mathbf{x}_{i+2n}(u,1) \ , \\
\mathbf{x}_{1+2n}(u,0) &\overset{k}{=} \mathbf{x}_{4n}(0,u)
\end{aligned}$$

and for $i = 2, \dots, n$

$$\mathbf{x}_{i+2n}(u,0) \overset{k}{=} \mathbf{x}_{i+3n}(0,u) \ .$$

All joints are even G^∞ joints since the surface lies in a plane. In 14.6 we describe such patches $\mathbf{x}_i$, explicitly.

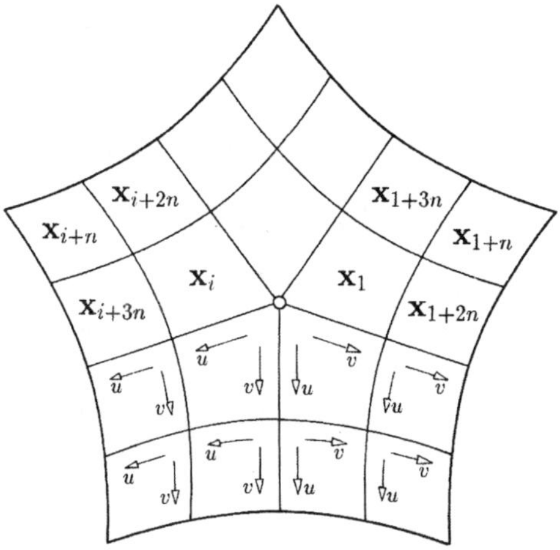

Figure 14.9: A planar n-sided macro patch for $n = 5$.

Second we pick any polynomial $\mathbf{p}(x,y)$ in $\mathbb{R}^3$ and reparametrized it by the maps $\mathbf{x}_i$. Thus, we obtain $4n$ patches

$$\mathbf{p}_i(\mathbf{u}) = \mathbf{p}(\mathbf{x}_i(\mathbf{u})) \ , \quad i = 1, \dots, 4n \ ,$$

with G^∞ joints. Moreover, two adjacent patches $\mathbf{p}_i$ and $\mathbf{p}_j$, with $i \geq n$ or $j \geq n$, have even a simple C^k joint.

We call the n-sided surface consisting of the patches $\mathbf{p}_i$ a **polynomial macro patch** or, in short, a **p-patch**. Such a patch has a large number of free parameters. We can easily modify the patches $\mathbf{p}_i$ and preserve the simple C^k- and genuine G^k joints.

Let $\mathbf{p}$ be a quadratic polynomial. Then, the p-patch is piecewise of bidegree $2k+2$. Its Bézier points are shown schematically in Figure 14.10 for $k=1$ and $n=5$. The Bézier points marked by solid dots determine the G^k joints among the inner n patches and are called **fixed**. The Bézier points marked by squares or without marks are **free**. We can change them arbitrarily. The remaining Bézier points are marked by circles and are **dependent**. We compute these so as to obtain simple C^k joints between any two adjacent patches $\mathbf{p}_i$ and $\mathbf{p}_j$, where $j > n$.

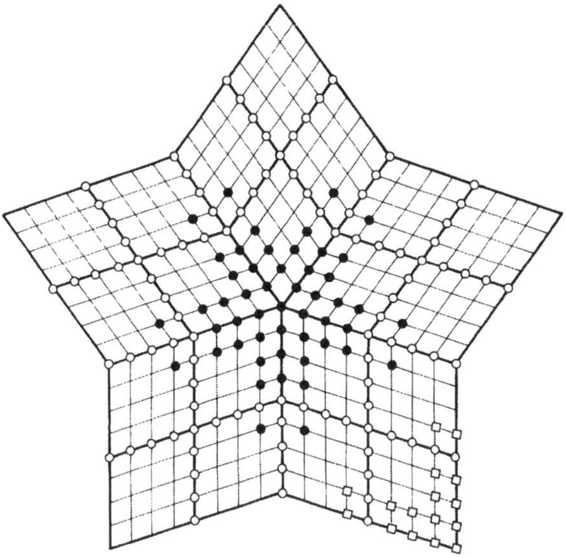

Figure 14.10: Bézier points of a five-sided macro, schematically.

Similarly, there are fixed, free and dependent Bézier points for any k and any n. In particular, we can choose the free Bézier points such that all n boundaries of the p-patch together with their cross boundary derivatives up to order k (or even $k+1$) are polynomial and not piecewise polynomial. Hence, given any polynomial patches $\mathbf{r}_1, \ldots, \mathbf{r}_{2n}$ of bidegree $\leq 2k+2$ with C^k joints, as illustrated schematically in Figure 14.11, there are n-sided p-patches of bidegree $2k+2$ fitting smoothly with C^k joints into the hole formed by $\mathbf{r}_1, \ldots, \mathbf{r}_{2n}$.

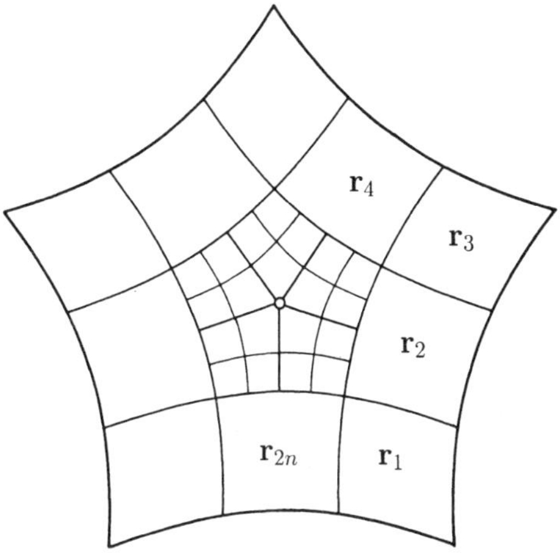

Figure 14.11: Filling a hole with a macro patch, schematically.

It is also possible to choose the free Bézier points of any two multi-sided p-patches such that they have four polynomial patches in common. For example, the Bézier points marked by empty squares, □, in Figure 14.10 can be chosen to be the fixed Bézier points of a second p-patch.

So it is possible to interpolate any quadrilateral net provided that each irregular vertex has only regular neighbors, by polynomial patches with simple and general C^k joints. For this interpolant, each face of the net corresponds to a single patch of bidegree $2k+1$ if all its vertices are regular or, otherwise, to four patches of bidegree $2k+2$. At regular vertices, the surface can interpolate any mixed partial derivatives up to order (k,k) and at extraordinary vertices all derivatives up to order 2.

14.6 Multi-sided patches in the plane

In 14.5 we have used a planar p-patch consisting of $4n$ patches $\mathbf{x}_1, \ldots, \mathbf{x}_{4n}$ of bidegree $k+1$ with simple C^0- and C^k joints, see Figure 14.9 . Here we present such planar p-patches explicitly. These patches are regular and injective without overlaps. We omit the lengthy technical proof.

Figure 14.12 shows the B-spline control points of the outer patches $\mathbf{x}_{n+1}, \ldots,$ $\mathbf{x}_{4n}$ for $k=4$ and $n=5$. They are marked by big circles. The origin is a multiple control point. The small circles indicate some of the control points for the alternative $k=3$ and $n=5$.

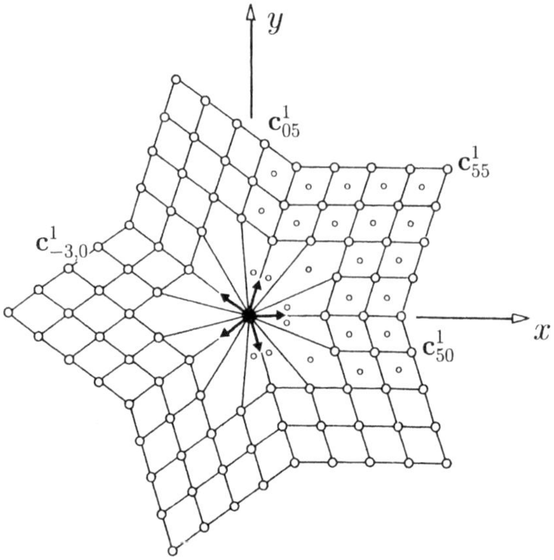

Figure 14.12: The entire spline control net for $\mathbf{x}_{n+1}, \ldots, \mathbf{x}_{4n}$.

Any four adjacent boundary patches $\mathbf{x}_{i+n}, \mathbf{x}_{i+3n}, \mathbf{x}_{i+1+2n}, \mathbf{x}_{i+1+n}$ form a tensor product B-spline surface,

$$\mathbf{s}_i(u,v) = \sum_{r=-3}^{k+1} \sum_{s=0}^{k+1} \mathbf{c}^i_{rs} N_r(u) N_s(v) \ , \quad (u,v) \in [k-2, k+2] \times [k+1, k+2] \ ,$$

where N_i denotes the uniform B-spline of degree $k+1$ with the knots $i, i+1, \ldots, i+k+2$. These surfaces $\mathbf{s}_i$ overlap in the corner patches $\mathbf{x}_{i+n}$.

The control points $\mathbf{c}^i_{00}, \ldots, \mathbf{c}^i_{k-2,k-2}$ are all zero, and for even k the other points $\mathbf{c}_{rs}$, where r and $s \geq \frac{k}{2} - 1$ are given by

$$\mathbf{c}^i_{rs} = (r - \frac{k}{2} + 1) \begin{bmatrix} \cos\varphi_{i-1} \\ \sin\varphi_{i-1} \end{bmatrix} + (s - \frac{k}{2} + 1) \begin{bmatrix} \cos\varphi_i \\ \sin\varphi_i \end{bmatrix} \ ,$$

with $\varphi_i = i \cdot 360°/n$. The remaining points are defined iteratively by the relation

$$\mathbf{c}^i_{rs} = \mathbf{c}^{i+1}_{s,k-2-r} \ , \quad -3 \leq r \leq \frac{k}{2} - 1 \quad \text{and} \quad \frac{k}{2} - 1 \leq s \leq k+1 \ .$$

The centers of the quadrilaterals in the control net form the control points $\mathbf{c}^i_{rs}$ for the $\mathbf{x}_i$ of odd bidegree k. The inner n patches $\mathbf{x}_1, \ldots, \mathbf{x}_n$ have C^k contact with the outer patches $\mathbf{x}_{n+1}, \ldots, \mathbf{x}_{4n}$. Hence, we know all their Bézier points except for the point $\mathbf{x}_{00}$ associated with the tensor product Bernstein polynomial $B_0^{k+1}(u) \cdot B_0^{k+1}(v)$. Because of symmetry reasons, we set $\mathbf{x}_{00} = 0$.

Remark 4: The above choice for the maps $\mathbf{x}_i$ is not the only one possible. For example, the characteristic map of the midpoint scheme, see 16.1 and 16.6, provides good candidates for the patches $\mathbf{x}_{n+1}, \ldots, \mathbf{x}_{4n}$ for $k = 1$ and $k = 2$.

Remark 5: The patches $\mathbf{x}_1, \ldots, \mathbf{x}_n$ have a singularity at the origin if $\mathbf{c}_{ij} = \mathbf{o}$ for all $i, j = 0, \ldots, k-1$. With these reparametrizations, the construction in 14.6 results in singularly parametrized C^k splines, see [Reif '98].

14.7 Problems

1 Compute the derivative

$$\frac{\partial^{i+j}}{\partial x^i \partial y^j} \mathbf{q}(\mathbf{u}(x, y))$$

in terms of the partial derivatives of $\mathbf{q}(u, v)$ and $\mathbf{u}(x, y)$.

2 In 14.4, it is assumed that each patch of the G^k surface has at least two regular vertices at opposite corners. Modify the construction so that one can drop this assumption. What is the degree of the resulting surface?

3 Describe a construction, similar to the one discussed in Section 14.4, of a regular surface consisting of triangular polynomial patches with G^k joints. What is the maximum degree of the patches?

4 Describe a construction, similar to the one discussed in Section 14.6, of a regular p-patch consisting of triangular polynomial patches with G^k joints. What is the maximum degree of the patches?

5 Describe a construction, similar to the one discussed in Section 14.6, of a multi-sided planar macro patch consisting of triangular patches with C^0- or C^k joints. Show that the minimal degree of such a macro patch is $3k/2 + 1$, see [Prautzsch & Reif '99].

15 Stationary subdivision for regular nets

15.1 Tensor product schemes — 15.2 General stationary subdivision and masks — 15.3 Convergence theorems — 15.4 Increasing averages — 15.5 Computing the difference schemes — 15.6 Computing the averaging schemes — 15.7 Subdivision for triangular nets — 15.8 Box splines over triangular grids — 15.9 Subdivision for hexagonal nets — 15.10 Half-box splines over triangular grids — 15.11 Problems

Under a stationary subdivision scheme, a regular control net is transformed into a regular control net whose vertices are affine combinations of the initial control points. The weights of these affine combinations can be given by masks or algebraically by a characteristic polynomial, as in the case of curve algorithms.

We discuss general and special stationary subdivision schemes for triangular and hexagonal nets and introduce box and half-box splines over a regular triangular net.

15.1 Tensor product schemes

Any two curve subdivision schemes define a tensor product subdivision scheme for regular quadrilateral nets.

As explained in 8.6 and 8.8, let $A = [\alpha_{j-2i}]$ and $B = [\beta_{j-2i}]$ be the subdivision matrices of two stationary curve schemes with characteristic polynomials $\alpha(x) = \sum \alpha_i x^i$ and $\beta(y) = \sum \beta_j y^j$, respectively. Let $\mathbf{c}_{ij}$, $i, j \in \mathbf{Z}$, be the vertices of a regular quadrilateral net. So, for simplicity we assume that the matrix $C = [\mathbf{c}_{ij}]$ has biinfinite rows and columns. This assumption also covers finite control nets since we can add zero control points.

We say that the sequence of control nets

$$C_m = (A^t)^m C B^m$$

obtained from C by subdividing m times all columns using A and all rows using B is obtained by the **tensor product scheme** given by A and B.

More precisely, any vertex $\mathbf{c}_{ij}^{m+1}$ of the net C_{m+1} is computed from the vertices $\mathbf{c}_{kl}^m$ of C_m by the refinement equation

$$\mathbf{c}_{ij}^{m+1} = \sum_k \sum_l \mathbf{c}_{kl}^m \, \alpha_{i-2k} \, \beta_{j-2l} \ .$$

Using multiindices $\mathbf{i}, \mathbf{j} \in \mathbb{Z}^2$ and the abbreviation $\gamma_{kl} = \alpha_k \beta_l$, the **refinement equation** takes on the same succinct form as the one given in 8.8 for curves:

$$\mathbf{c}_{\mathbf{i}}^{m+1} = \sum_{\mathbf{j}} \mathbf{c}_{\mathbf{j}}^m \gamma_{\mathbf{i}-2\mathbf{j}} \ .$$

If both A and B is the subdivision matrix $S_n = DM^n$ of the Lane-Riesenfeld algorithm for uniform splines of degree n, see 8.4, then the associated tensor product scheme can also be described by the following two operators.

The **doubling operator** $\mathcal{D}$ quadruples all control points of any control net C,

$$\mathcal{D}(C) = D^t C D = [\mathbf{c}_{\lfloor \mathbf{i}/2 \rfloor}] \ .$$

Let $\mathbf{e}_1 = [1\,0], \mathbf{e}_2 = [0\,1]$ and $\mathbf{e} = [1\,1]$. Then, the **averaging operator** $\mathcal{A}$ maps any net C to the net

$$\mathcal{A}(C) := M^t C M = \frac{1}{4}[\mathbf{c}_{\mathbf{i}} + \mathbf{c}_{\mathbf{i}-\mathbf{e}_1} + \mathbf{c}_{\mathbf{i}-\mathbf{e}_2} + \mathbf{c}_{\mathbf{i}-\mathbf{e}}]$$

that connects the centroids of any two meshes in C with a common edge.

Thus, the Lane-Riesenfeld algorithm for tensor product splines of bidegree n is given by the operator $\mathcal{M}_n := \mathcal{A}^n \mathcal{D}$. In particular, $\mathcal{M}_1 = \mathcal{A}\mathcal{D}$ represents the **refinement operator**, which maps any net C to the finer net $\mathcal{M}_1(C)$ that connects the midpoints of all edges in C with their endpoints and the centroids of both abutting (quadrilateral) meshes of C. Figure 15.1 gives an illustration. It shows a net C in light edges, the net $\mathcal{M}_1(C)$ in light and broken edges, and the net $\mathcal{M}_2(C)$ in bold edges.

Remark 1: From 8.2, it follows that any sequence of control nets

$$C^m = [\mathbf{c}_{ij}^m] = \mathcal{M}_n^m(C)$$

represents the same spline surface

$$\mathbf{s}(u, v) = \sum_{i,j} \mathbf{c}_{ij} N_i^n(u) N_j^n(v) \ ,$$

where N_i^n denotes the uniform B-spline of degree n over the knots $i, i+1, \ldots,$

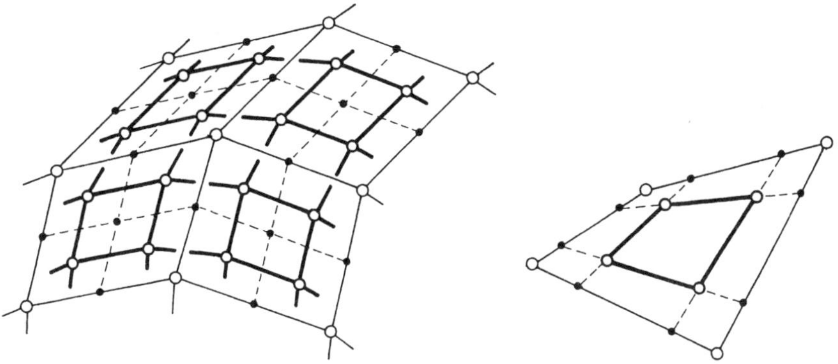

Figure 15.1: Refining and averaging a net.

$i+n+1$. Applying the convergence result in 6.3 twice, we obtain the estimate

$$\sup_{i,j} \|\mathbf{s}((i,j)/2^m) - \mathbf{c}_{ij}^m\| = O(1/4^m) \ ,$$

provided that the second derivatives of $\mathbf{s}$ are bounded over $\mathbb{R}^2$.

15.2 General stationary subdivision and masks

Any refinement equation

$$\mathbf{c}_{\mathbf{i}}^{m+1} = \sum_{\mathbf{k}} \mathbf{c}_{\mathbf{k}}^m \gamma_{\mathbf{i}-2\mathbf{k}}$$

with a finite number of non-zero and arbitrary coefficients $\gamma_{\mathbf{i}}$ represents a general **stationary subdivision scheme**. If the γ_{ij} are products of the form $\alpha_i\beta_j$, then this scheme is a tensor product scheme, as discussed in 15.1.

The refinement equation combines four different affine combinations: The indices $\mathbf{k}$ of the weights $\gamma_{\mathbf{k}}$ used to compute a point $\mathbf{c}_{\mathbf{i}}^{m+1}$ form the set $\mathbf{i}+2\mathbb{Z}^2$, which is either

$$\mathbb{Z}^2, \quad \mathbf{e}_1 + \mathbb{Z}^2, \quad \mathbf{e}_2 + \mathbb{Z}^2 \quad \text{or} \quad \mathbf{e} + \mathbb{Z}^2 \ .$$

The four (finite) matrices $[\gamma_{-2\mathbf{k}}], [\gamma_{\mathbf{e}_1-2\mathbf{k}}], [\gamma_{\mathbf{e}_2-2\mathbf{k}}]$ and $[\gamma_{\mathbf{e}-2\mathbf{k}}]$ are called **masks**. They, too, represent the subdivision scheme.

Remark 2: A necessary condition for the convergence of a stationary subdivision scheme is that each mask defines an affine combination, see 15.3. This means that the weights of any mask must sum to one. To avoid fractions, it is common, therefore, to represent a mask by some multiple of it. The proper mask is then obtained by dividing by the sum of all its weights. We will use this convention in the sequel.

Remark 3: The four masks of the refinement operator $\mathcal{M}_1$ defined in 15.1, are

$$\begin{bmatrix} 1 & 0 \\ 1 & 0 \end{bmatrix}, \begin{bmatrix} 1 & 1 \\ 1 & 1 \end{bmatrix}, \begin{bmatrix} 0 & 0 \\ 1 & 0 \end{bmatrix}, \begin{bmatrix} 0 & 0 \\ 1 & 1 \end{bmatrix}.$$

They are presented graphically on the left side of Figure 15.2. The right side of Figure 15.2 shows the four masks

$$\begin{bmatrix} 9 & 3 \\ 3 & 1 \end{bmatrix}, \begin{bmatrix} 3 & 9 \\ 1 & 3 \end{bmatrix}, \begin{bmatrix} 3 & 1 \\ 9 & 3 \end{bmatrix}, \begin{bmatrix} 1 & 3 \\ 3 & 9 \end{bmatrix}$$

of the operator $\mathcal{M}_2 = \mathcal{A}\mathcal{M}_\infty$ for biquadratic splines.

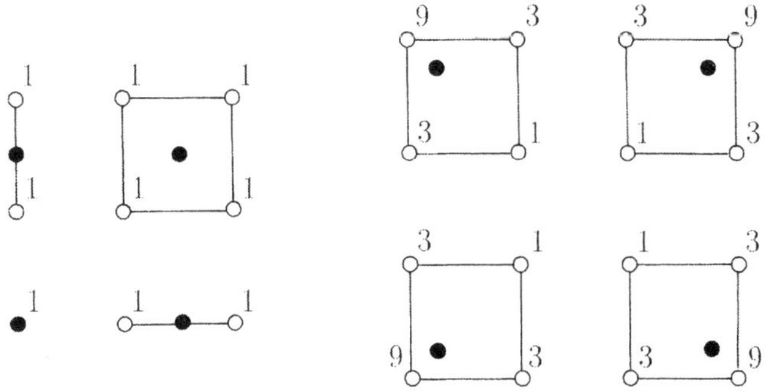

Figure 15.2: The four masks of the Lane-Riesenfeld algorithms $\mathcal{M}_1$ (left) and $\mathcal{M}_2$ (right).

Remark 4: The refinement operator $\mathcal{M}_1$ is described by four masks, whereas the averaging operator $\mathcal{A}$ is described by a single mask only. Figures 15.3 and 15.4 show the masks for $\mathcal{A}$ and $\mathcal{A}^2$.

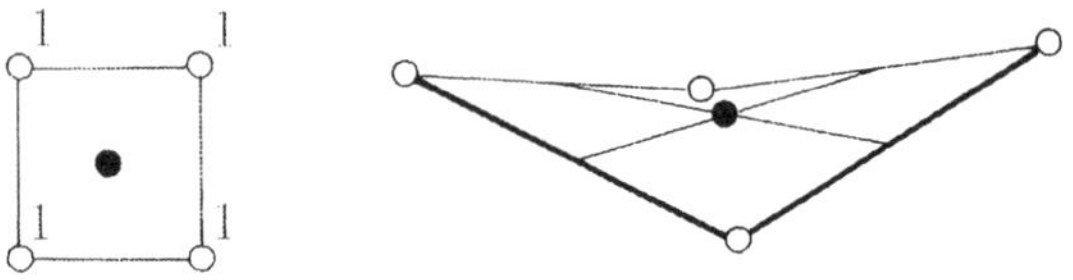

Figure 15.3: The mask of the averaging operator $\mathcal{A}$ (left) and its application (right).

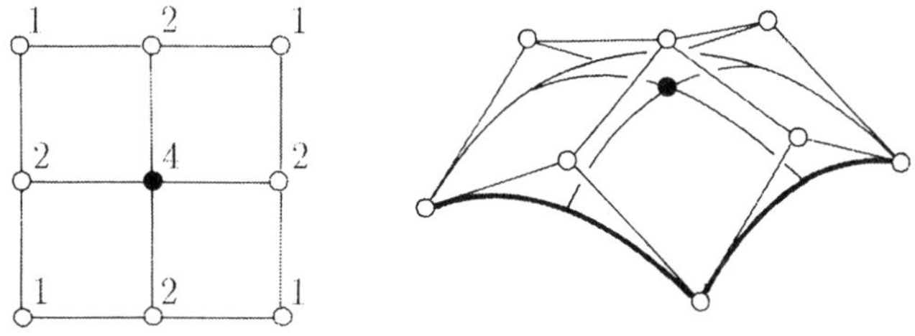

Figure 15.4: The mask of the averaging operator $\mathcal{A}^2$ (left) and its application (right).

15.3 Convergence theorems

A sequence of control nets $C_m = [\mathbf{c}_\mathbf{i}^m]$ is obtained under a stationary subdivision scheme if there is a finite sequence $\gamma_\mathbf{i}$ such that

$$\mathbf{c}_\mathbf{i}^{m+1} = \sum_\mathbf{j} \mathbf{c}_\mathbf{j}^m \gamma_{\mathbf{i}-2\mathbf{j}} \ .$$

The sequence C_m is said to converge uniformly towards a function $\mathbf{c}(x, y)$ if the maximum distance

$$\sup_\mathbf{i} \|\mathbf{c}_\mathbf{i}^m - \mathbf{c}(\mathbf{i}/2^m)\|$$

converges to zero, as m tends to infinity.

If the limiting function $\mathbf{c}(\mathbf{x})$ *is continuous and non-zero, then the weights* $\gamma_{\mathbf{i}-2\mathbf{j}}$, $\mathbf{j} \in \mathbb{Z}^2$, *of each mask sum to one.*

Namely, without loss of generality, let $\mathbf{c}(\mathbf{o}) \neq \mathbf{o}$. Then, the points

$$\mathbf{c}_\mathbf{i}^m = \mathbf{c}(\mathbf{i}/2^m) + (\mathbf{c}_\mathbf{i}^m - \mathbf{c}(\mathbf{i}/2^m))$$

converge to $\mathbf{c}(\mathbf{o})$, as m goes to infinity for any $\mathbf{i}$. Consequently, the finite sum

$$\mathbf{c}_\mathbf{i}^m = \sum_\mathbf{j} \mathbf{c}_\mathbf{j}^{m-1} \gamma_{\mathbf{i}-2\mathbf{j}}$$

converges to

$$\mathbf{c}(\mathbf{o}) = \sum_\mathbf{j} \mathbf{c}(\mathbf{o}) \gamma_{\mathbf{i}-2\mathbf{j}} \ ,$$

which concludes the proof. ◇

Whether a sequence of control nets C_m converges depends on the two sequences of difference polygons

$$\nabla_k C_m = [\mathbf{c}_\mathbf{i}^m - \mathbf{c}_{\mathbf{i}-\mathbf{e}_k}^m] \ , \quad k = 1, 2 \ .$$

Namely, the following holds.

A sequence C_m obtained under stationary subdivision converges uniformly towards a uniformly continuous function $\mathbf{c}(x, y)$ if and only if the difference polygons $\nabla_1 C_m$ and $\nabla_2 C_m$ converge uniformly to zero.

Obviously, if C_m converges to a continuous function, then the differences go to zero. To simplify the proof of the converse, let the maximum difference

$$\delta_m = \max_{k=1,2} \sup_{\mathbf{i}} \|\nabla_k \mathbf{c}_\mathbf{i}^m\|$$

converge geometrically to zero. Further, let

$$\mathbf{c}^m(x, y) := \sum_{i,j} \mathbf{c}_{ij}^m N_i(2^m x) N_j(2^m y)$$

be a piecewise linear interpolant of the control net C_m, where $N_i(x)$ is the piecewise linear B-spline with the knots $i-1, i$ and $i+1$. Thus, $\mathbf{c}^m(\mathbf{i}/2^m) = \mathbf{c}_\mathbf{i}^m$.

Since

$$\begin{aligned} \|\mathbf{c}_{2\mathbf{j}}^{m+1} - \mathbf{c}_\mathbf{j}^m\| &\leq \sum_\mathbf{k} \|\mathbf{c}_\mathbf{k}^m - \mathbf{c}_\mathbf{j}^m\| \cdot |\gamma_{2\mathbf{j}-2\mathbf{k}}| \\ &\leq \delta_m \gamma \ , \end{aligned}$$

where γ is some multiple of $\sum_\mathbf{i} |\gamma_\mathbf{i}|$, depending on the size of the mask $[\gamma_{2k}]$, we obtain for $\mathbf{i} \in 2\mathbf{j} + \{0, 1\}^2$

$$\begin{aligned} &\|\mathbf{c}_\mathbf{i}^{m+1} - \mathbf{c}_{2\mathbf{j}}^{m+1} + \mathbf{c}_{2\mathbf{j}}^{m+1} - \mathbf{c}^m(\mathbf{j}/2^m) + \mathbf{c}^m(\mathbf{j}/2^m) - \mathbf{c}^m(\mathbf{i}/2^{m+1})\| \\ &\qquad \leq 2\delta_{m+1} + \delta_m \gamma + \delta_m \ . \end{aligned}$$

This implies that

$$\sup \|\mathbf{c}^{m+1}(\mathbf{x}) - \mathbf{c}^m(\mathbf{x})\| \leq 2\delta_{m+1} + (\gamma + 1)\delta_m \ .$$

Consequently, the linear splines $\mathbf{c}^m(\mathbf{x})$ and their nets C_m converge uniformly to a uniformly continuous function $\mathbf{c}(\mathbf{x})$, as claimed. ◇

Furthermore, let the polygons C_m and the divided difference polygons $2^m \nabla_\mathbf{v} C_m = 2^m [\mathbf{c}_\mathbf{i}^m - \mathbf{c}_{\mathbf{i}-\mathbf{v}}^m]$ converge uniformly to uniformly continuous functions $\mathbf{c}(\mathbf{x})$ and $\mathbf{d}(\mathbf{x})$, respectively, where $\mathbf{v} \in \mathbb{Z}^2$.

Then $\mathbf{d}(\mathbf{x})$ *is the directional derivative of* $\mathbf{c}(\mathbf{x})$ *with respect to* $\mathbf{v}$.

For a proof, choose $\mathbf{u} \in \mathbb{Z}^2$ be such that $\mathbf{u}$ and $\mathbf{v}$ are linearly independent. Clearly, the control nets

$$[c_{ij}^m] = [\mathbf{c}_{i\mathbf{u}+j\mathbf{v}}^m]$$

converge to $\mathbf{c}(x\mathbf{u}+y\mathbf{v})$ and the difference polygons $2^m\nabla_{\mathbf{v}}[c_{ij}^m]$ to $\mathbf{d}(x\mathbf{u}+y\mathbf{v})$. So, without loss of generality, we can and do assume $\mathbf{v} = \mathbf{e}_2$. Obviously the piecewise constant splines

$$\mathbf{d}_m(x,y) := \sum_{i,j} 2^m \nabla_{\mathbf{v}} \mathbf{c}_{ij}^m N_i^0(2^m x) N_j^0(2^m y)$$

and the splines

$$\mathbf{c}_m(x,y) := \int \mathbf{d}_m(x,y)dy = \sum_{i,j} \mathbf{c}_{ij}^m N_i^0(2^m x) N_j^1(2^m y)$$

converge uniformly to $\mathbf{d}(\mathbf{x})$ and $\mathbf{c}(\mathbf{x})$, respectively. Hence, $\mathbf{c}(x,y) = \int \mathbf{d}(x,y)dy$, which concludes the proof. $\diamond$

15.4 Increasing averages

In 15.3, divided differences of control nets are discussed. In contrast, we now study averaged nets.

If the polygons $C_m = [\mathbf{c_i}]_{\mathbf{i}\in\mathbb{Z}^2}$ *converge uniformly to a Riemann integrable function* $\mathbf{c}(\mathbf{x})$ *with compact support, then the increasing averages*

$$\mathbf{a}_{ij}^m = \frac{1}{4^m} \sum_{k,l=0}^{2^m-1} \mathbf{c}_{i-k,j-l}^m$$

converge uniformly to the uniformly continuous function

$$\mathbf{a}(\mathbf{x}) = \int_{[0,1]^2} \mathbf{c}(\mathbf{x}-\mathbf{t})dt \ .$$

For a proof, let Ω be the interval $(\mathbf{i} - [0,1]^2)/2^m$, which depends on $\mathbf{i}$ and m. Since $\mathbf{c}(\mathbf{x})$ is Riemann integrable and has compact support, the sums

$$\sum_{\mathbf{i}\in\mathbb{Z}^2} 4^{-m}(\sup_{\mathbf{x}\in\Omega} \mathbf{c}(\mathbf{x}) - \inf_{\mathbf{x}\in\Omega} \mathbf{c}(\mathbf{x}))$$

converge to zero. Thus, the Riemann sums

$$\mathbf{r}_\mathbf{i}^m = 4^{-m} \sum_{k,l=0}^{2^m-1} \mathbf{c}((i-k, j-l)/2^m)$$

converge to $\mathbf{a}(\mathbf{i}/2^m)$ uniformly for all $\mathbf{i}$, as m goes to infinity.

Since the $\mathbf{c}_\mathbf{i}^m$ converge uniformly to $\mathbf{c}(\mathbf{x})$, the averages $\mathbf{a}_\mathbf{i}^m$ converge uniformly to the Riemann sums $\mathbf{r}_\mathbf{i}^m$, which concludes the proof. ◇

Similarly, one can prove the next result.

If the polygons $C_m = [\mathbf{c}_\mathbf{i}]_{\mathbf{i}\in\mathbb{Z}^2}$ converge uniformly over any compact set to a continuous function $\mathbf{c}(\mathbf{x})$, then the **increasing line averages**

$$\mathbf{b}_\mathbf{i}^m = \frac{1}{2^m} \sum_{k=0}^{2^m-1} \mathbf{c}_{\mathbf{i}-k\mathbf{v}}^m \ , \qquad \mathbf{v} \in \mathbb{Z}^2 \ ,$$

converge uniformly over any compact set to the uniformly continuous function

$$\mathbf{b}(\mathbf{x}) = \int_0^1 \mathbf{c}(\mathbf{x} - t\mathbf{v})dt \ .$$

Remark 5: The averages $\mathbf{a}_\mathbf{i}^m$ are line averages of line averages, since

$$\mathbf{a}_\mathbf{i}^m = \frac{1}{2^m} \sum_{k=0}^{2^m-1} \mathbf{b}_{\mathbf{i}-k\mathbf{e}_1}^m \ , \qquad \mathbf{e}_1 = [1 \ 0] \ ,$$

where

$$\mathbf{b}_\mathbf{i}^m = \frac{1}{2^m} \sum_{l=0}^{2^m-1} \mathbf{c}_{\mathbf{i}-l\mathbf{e}_2}^m \ , \qquad \mathbf{e}_2 = [0 \ 1] \ .$$

15.5 Computing the difference schemes

A stationary subdivision scheme can also be described by means of generating functions. As in 8.8, we multiply the refinement equation by the monomial $\mathbf{x}^\mathbf{i} = x^i y^j$ and sum over all $\mathbf{i}$. This results in

$$\begin{aligned}\sum_\mathbf{i} \mathbf{c}_\mathbf{i}^{m+1}\mathbf{x}^\mathbf{i} &= \sum_\mathbf{i}\sum_\mathbf{j} \mathbf{c}_\mathbf{j}^m \gamma_{\mathbf{i}-2\mathbf{j}} \mathbf{x}^{2\mathbf{j}}\mathbf{x}^{\mathbf{i}-2\mathbf{j}} \\ &= \sum_\mathbf{j} \mathbf{c}_\mathbf{j}^m \mathbf{x}^{2\mathbf{j}} \sum_\mathbf{k} \gamma_\mathbf{k}\mathbf{x}^\mathbf{k} \ ,\end{aligned}$$

which we abbreviate by

$$\mathbf{c}^{m+1}(\mathbf{x}) = \mathbf{c}^m(\mathbf{x}^2)\gamma(\mathbf{x}) \ .$$

The factor

$$\gamma(\mathbf{x}) = \sum_{\mathbf{k}} \gamma_{\mathbf{k}} \mathbf{x}^{\mathbf{k}}$$

represents the subdivision scheme, and it is called its **symbol** or **characteristic polynomial**. For a tensor product scheme, the characteristic polynomial is the product of two univariate polynomials $\alpha(x)$ and $\beta(y)$ representing two curve schemes.

Any stationary curve scheme has an underlying difference scheme, but for surface schemes this is not true in general. To study when a subdivision scheme has a difference scheme, we identify control nets and subdivision schemes with their generating polynomials.

Thus, given a control net

$$\mathbf{c}(\mathbf{x}) = \sum \mathbf{c}_{\mathbf{i}} \mathbf{x}^{\mathbf{i}} \ ,$$

its refinement under a stationary scheme $\gamma(\mathbf{x})$ is given by

$$\mathbf{b}(\mathbf{x}) = \mathbf{c}(\mathbf{x}^2)\gamma(\mathbf{x}) \ ,$$

and the differences $\nabla_{\mathbf{v}} \mathbf{c}_{\mathbf{i}} = \mathbf{c}_{\mathbf{i}} - \mathbf{c}_{\mathbf{i}-\mathbf{v}}, \mathbf{v} \in \mathbb{Z}^2$, form the polygon

$$\nabla_{\mathbf{v}} \mathbf{c}(\mathbf{x}) = \mathbf{c}(\mathbf{x})(1 - \mathbf{x}^{\mathbf{v}}) \ .$$

Hence, the differences of the refined polygon $\mathbf{b}(\mathbf{x}) = \mathbf{c}(\mathbf{x}^2)\gamma(\mathbf{x})$ are given by

$$\nabla_{\mathbf{v}} \mathbf{b}(\mathbf{x}) = \nabla_{\mathbf{v}} \mathbf{c}(\mathbf{x}^2)\gamma(\mathbf{x}) \frac{1 - \mathbf{x}^{\mathbf{v}}}{1 - \mathbf{x}^{2\mathbf{v}}} \ .$$

Thus, there exists a stationary scheme, the **$\nabla_{\mathbf{v}}$-difference scheme**, mapping $\nabla_{\mathbf{v}} \mathbf{c}$ onto $\nabla_{\mathbf{v}} \mathbf{b}$ if and only if

$$\delta(\mathbf{x}) = \gamma(\mathbf{x})/(1 + \mathbf{x}^{\mathbf{v}})$$

is a polynomial. In case $\delta(\mathbf{x})$ is a polynomial, it is the **characteristic polynomial of the difference scheme**.

Remark 6: Given a control net $C = [\mathbf{c}_{\mathbf{i}}]$, let ∇C be the control net whose "vertices" are the matrices

$$\nabla \mathbf{c}_{\mathbf{i}} = [\nabla_{\mathbf{e}_1} \mathbf{c}_{\mathbf{i}} \ \nabla_{\mathbf{e}_2} \mathbf{c}_{\mathbf{i}}] \ .$$

If the control net B is obtained application of a stationary subdivision scheme from C, then ∇B is obtained from ∇C using a stationary scheme whose weights are 2×2 matrices, see [Kobbelt '00, Cavaretta et al. '91, Thm. 2.3].

Remark 7: The Lane-Riesenfeld scheme $\mathcal{M}_n$, see 15.1, has the characteristic polynomial

$$\gamma(x, y) = 4^{-n}(1 + x)^{n+1}(1 + y)^{n+1} .$$

This follows directly from Remark 5 in 8.8.

15.6 Computing the averaging schemes

Using characteristic polynomials, it can be seen that for any stationary subdivision scheme there exists a stationary scheme for the averages considered in 15.4.

Let

$$\mathbf{c}^m(\mathbf{x}) = \gamma(\mathbf{x})\mathbf{c}^{m-1}(\mathbf{x}^2)$$

represent a sequence of control nets $[\mathbf{c}_\mathbf{i}^m]$ obtained under a stationary subdivision scheme γ. Using the variables

$$\mathbf{x}_k := \mathbf{x}^{2^k} ,$$

this sequence can be written as

$$\mathbf{c}^m(\mathbf{x}) = \gamma(\mathbf{x}_0) \ldots \gamma(\mathbf{x}_{m-1})\mathbf{c}^0(\mathbf{x}_m) .$$

Further, for any $\mathbf{v} \in \mathbb{Z}^2$, let the polynomial $\mathbf{b}^m(\mathbf{x}) = \sum \mathbf{b}_\mathbf{i}^m \mathbf{x}^\mathbf{i}$ represent the line averages

$$\mathbf{b}_\mathbf{i}^m = \frac{1}{2^m} \sum_{k=0}^{2^m-1} \mathbf{c}_{\mathbf{i}-k\mathbf{v}}^m .$$

With $\mathbf{y}^k := \mathbf{x}^{k\mathbf{v}}$, this can be written as

$$\begin{aligned} \mathbf{b}^m(\mathbf{x}) &= 2^{-m}(1 + \mathbf{y} + \mathbf{y}^2 + \mathbf{y}^3 + \cdots + \mathbf{y}^{2^m-1})\mathbf{c}^m(\mathbf{x}) \\ &= 2^{-m}(1 + \mathbf{y})(1 + \mathbf{y}^2)(1 + \mathbf{y}^4) \ldots (1 + \mathbf{y}^{2^m-1})\mathbf{c}^m(\mathbf{x}) \\ &= \beta(\mathbf{x}_0) \ldots \beta(\mathbf{x}_{m-1})\mathbf{c}^0(\mathbf{x}) , \end{aligned}$$

where

$$\beta(\mathbf{x}) = \gamma(\mathbf{x})(1 + \mathbf{x}^\mathbf{v})/2$$

is the characteristic polynomial of the **averaging scheme** obtained from the scheme γ.

This means that the scheme β is described by the following algorithm.

Given control points $\mathbf{b_i}$, $\mathbf{i} \in \mathbb{Z}^2$, and a vector $\mathbf{v} \in \mathbb{Z}^2$
repeat
1 For all **i**, subdivide by the scheme γ, i.e.,
$$\mathbf{d_i} := \sum_{\mathbf{j}} \mathbf{b_j} \gamma_{\mathbf{i}-2\mathbf{j}} \ .$$

2 For all **i**, compute the line averages
$$\mathbf{b_i} := \tfrac{1}{2}(\mathbf{d_i} + \mathbf{d_{i-v}}) \ .$$

Similarly, it follows that the averages

$$\mathbf{a}_{\mathbf{i}}^m = \frac{1}{4^m} \sum_{k,l=0}^{2^m-1} \mathbf{c}_{i-k,j-l}^m$$

are obtained from the points $\mathbf{a}_{\mathbf{i}}^0 = \mathbf{c}_{\mathbf{i}}^0$ under the stationary scheme represented by

$$\alpha(\mathbf{x}) = \gamma(\mathbf{x})(1+x)(1+y)/4.$$

This scheme is described by the following algorithm.

Given control points $\mathbf{a_i}$, $\mathbf{i} \in \mathbb{Z}^2$
repeat
1 For all **i**, subdivide by the scheme γ, i.e.,
$$\mathbf{d_i} := \sum_{\mathbf{j}} \mathbf{a_j} \gamma_{\mathbf{i}-2\mathbf{j}} \ .$$
2 For all **i**, compute the line averages
$$\mathbf{f_i} := \tfrac{1}{2}(\mathbf{d_i} + \mathbf{d_{i-e_1}}) \ .$$
3 For all **i**, compute the line averages
$$\mathbf{a_i} := \tfrac{1}{2}(\mathbf{f_i} + \mathbf{f_{i-e_2}}) \ .$$

15.7 Subdivision for triangular nets

Every regular quadrilateral net can be transformed into a regular triangular net and vice versa by adding or deleting "diagonal" edges, as illustrated in Figure 15.5. Thus, we can represent any regular triangular net by a biinfinite matrix

$$C = \begin{bmatrix} & \vdots & \iddots \\ \cdots & \mathbf{c}_{ij} & \cdots \\ \iddots & \vdots & \end{bmatrix}$$

whose entries are the vertices of the net.

In particular, the three vectors

$$\mathbf{e}_1 = \begin{bmatrix} 1 \\ 0 \end{bmatrix} , \quad \mathbf{e}_2 = \begin{bmatrix} 0 \\ 1 \end{bmatrix} , \quad \mathbf{e}_3 = \begin{bmatrix} -1 \\ -1 \end{bmatrix}$$

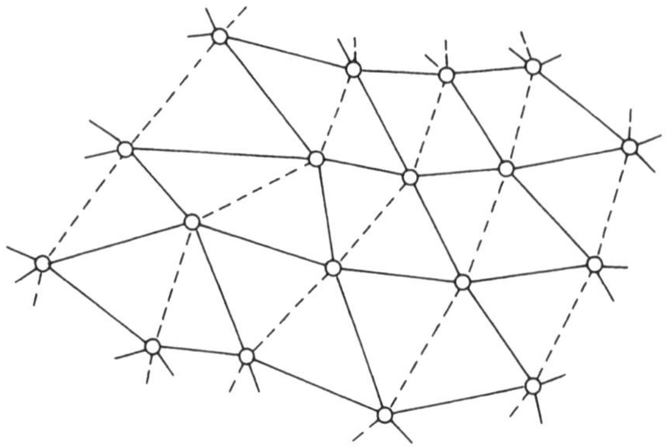

Figure 15.5: Transforming a regular quadrilateral net into a triangular net.

in $\mathbb{R}^2$ span a uniform regular triangular grid, as illustrated in Figure 15.6.

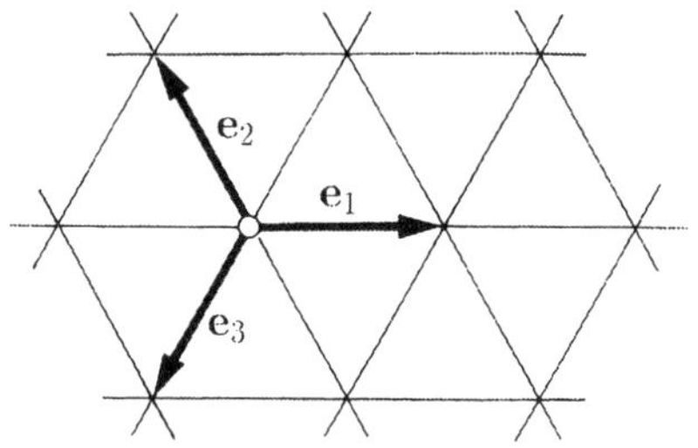

Figure 15.6: A uniform regular grid spanned by three vectors.

Using all three directions of a triangular net, we can generalize the Lane-Riesenfeld algorithm to a **three direction averaging** algorithm due to [Prautzsch '84b], see also [Boehm et al. '87].

Given a matrix $C = [\mathbf{c_i}]_{\mathbf{i}\in\mathbb{Z}^2}$ representing a regular triangular net, three directions averaging is described by four operators: a **doubling operator** $\mathcal{D}$ that quadruples all vertices,

$$\mathcal{D}(C) := [\mathbf{d_i}]_{\mathbf{i}\in\mathbb{Z}^2} \ , \text{ where } \mathbf{d_i} = \mathbf{c}_{\lfloor \mathbf{i}/2 \rfloor} \ ,$$

and three **averaging operators** $\mathcal{A}_k$, $k = 1, 2, 3$, that average the net with respect to the three directions $\mathbf{e}_k$,

$$\mathcal{A}_k(C) = [\mathbf{a_i}]_{\mathbf{i}\in\mathbb{Z}^2} \ , \text{ where } \mathbf{a_i} = \frac{1}{2}(\mathbf{c_i} + \mathbf{c}_{\mathbf{i}-\mathbf{e}_k}) \ .$$

For any $\mathbf{n} = (n_1, n_2, n_3) \in \mathbf{N}_0^3$, the composed operator

$$\mathcal{B}_{\mathbf{n}} = \mathcal{A}_1^{n_1}\mathcal{A}_2^{n_2}\mathcal{A}_3^{n_3}\mathcal{D}$$

represents a three direction averaging algorithm, and $C_m = \mathcal{B}_{\mathbf{n}}^m(C)$ represents the sequence of triangular nets obtained from C under the averaging algorithm $\mathcal{B}_{\mathbf{n}}$.

Remark 8: The doubling operator $\mathcal{D}$ is the Lane-Riesenfeld operator $\mathcal{M}_0$ given in Remark 8. Its characteristic polynomial is

$$\delta(x, y) = (1 + x)(1 + y) \ .$$

Remark 9: In particular, $\mathcal{B}_{001}$ represents the **refinement operator** $\mathcal{R}$ that subdivides all triangles of a regular net uniformly into four congruent triangles, as illustrated in Figure 15.7. The figure also depicts the four masks representing $\mathcal{B}_{001}$. The weights of these four masks form the coefficients of the characteristic polynomial of $\mathcal{B}_{001}$, see 15.2. This polynomial is

$$\begin{aligned}\gamma(x, y) &= (1 + x)(1 + y)(1 + \mathbf{x}^{\mathbf{e}_3})/2 \\ &= \frac{1}{2}\left[x^{-1}\ 1\ x\right]\begin{bmatrix} 1 & 1 & 0 \\ 1 & 2 & 1 \\ 0 & 1 & 1 \end{bmatrix}\begin{bmatrix} y^{-1} \\ 1 \\ y \end{bmatrix} .\end{aligned}$$

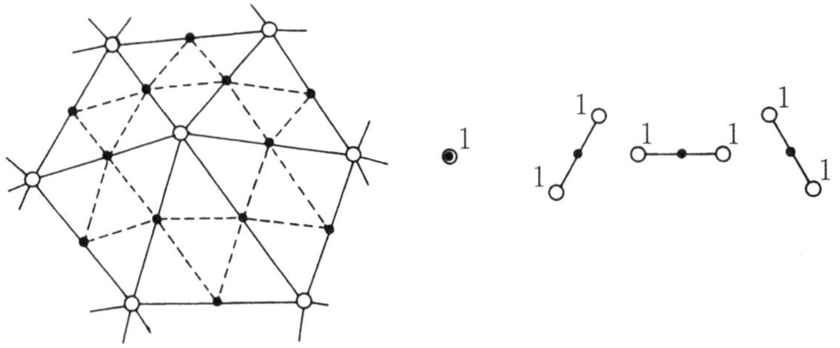

Figure 15.7: The refinement operator $\mathcal{R}$ applied to a regular triangular net and its four masks.

Remark 10: Any net obtained by successive application of the refinement operator $\mathcal{R}$ to a regular triangular net C represents the same continuous piecewise linear surface. Hence, a sequence of nets obtained under $\mathcal{R}$ converges.

Remark 11: The symmetric averaging operator $\mathcal{A}_{111} = \mathcal{A}_1\mathcal{A}_2\mathcal{A}_3$ is given by a single mask. The mask has been introduced in [Boehm '83] and is depicted in Figure 15.8. The polynomial representing $\mathcal{A}_{111}$ is

$$\gamma(x,y) = \frac{1}{8}\,[1\ x\ x^2] \begin{bmatrix} 0 & 1 & 1 \\ 1 & 2 & 1 \\ 1 & 1 & 0 \end{bmatrix} \begin{bmatrix} 1 \\ y \\ y^2 \end{bmatrix} .$$

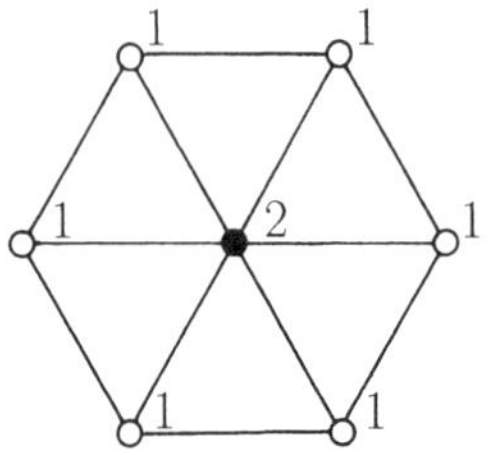

Figure 15.8: Boehm's mask of the symmetric averaging operator $\mathcal{A}_{111}$.

15.8 Box splines over triangular grids

Let C_m be a sequence of triangular nets obtained under repeated applications of the averaging operator $\mathcal{B}_{\mathbf{n}}$. If $\mathcal{B}_{\mathbf{n}}$ is the doubling operator, given by $\mathbf{n} = (0,0,0)$, or refinement operator, given by $\mathbf{n} = (0,0,1)$, then C_m converges to a piecewise constant or continuous piecewise linear spline, respectively.

In general, if

$$k := \min\{n_1{+}n_2,\ n_1{+}n_3{-}1,\ n_2{+}n_3{-}1\} \geq 0 \ ,$$

then, over every compact domain, C_m converges uniformly to a C^k spline, which is polynomial of total degree $|\mathbf{n}| = n_1 + n_2 + n_3$ over each triangle of the grid spanned by $\mathbf{e}_1, \mathbf{e}_2$ and $\mathbf{e}_3$.

These splines are three-direction box splines, see Chapter 17.

For a proof, we apply repeatedly the results in 15.4 and 15.6 and take into account the following fact. If $f(\mathbf{x})$ is continuous, then the integral

$$\int_0^1 \int_0^1 \int_0^1 f(\mathbf{x} - u\mathbf{e}_1 - v\mathbf{e}_2 - w\mathbf{e}_3)\,du\,dv\,dw$$

has continuous mixed partial derivatives with respect to any two distinct directions $\mathbf{e}_i$ and $\mathbf{e}_j$. Since $\mathbf{e}_i = \mathbf{e}_j - \mathbf{e}_k$ for all permutations (i,j,k) of $(1,2,3)$, all second partial derivatives exist. Consequently, iterated integration with respect to two different directions raises the smoothness order by one, but iterated integration with respect to three directions raises it by two. ◇

Remark 12: The second result in 15.4 also holds if we assume convergence over a compact domain to a piecewise polynomial spline over the triangular grid. Consequently, the theorem above also holds for $k = -1$.

Remark 13: If $n_3 = 0$, then $\mathcal{B}_{\mathbf{n}}$ represents the tensor product Lane-Riesenfeld subdivision algorithm for uniform tensor product splines of bidegree (n_1, n_2).

15.9 Subdivision for hexagonal nets

Any regular triangular net determines a regular hexagonal net that connects the midpoints of all adjacent triangles with a common edge and vice versa. Figure 15.9 illustrates this for the special case of uniform regular grids.

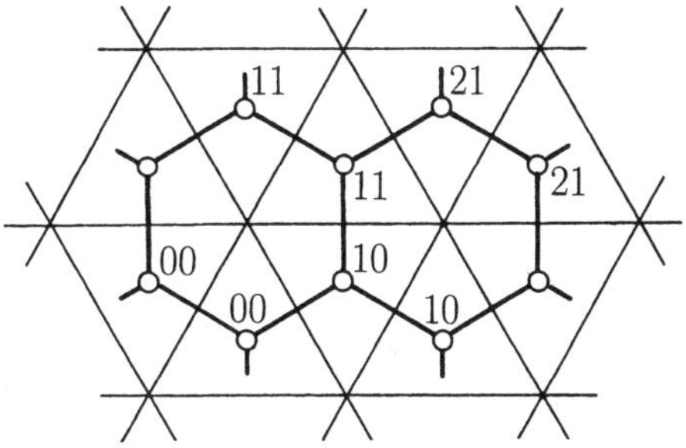

Figure 15.9: Regular triangular and hexagonal grids.

Further, any regular hexagonal net can be split into two regular triangular nets with distinct vertices, as illustrated in Figure 15.10 for a uniform regular grid. Thus, we can represent any regular hexagonal net by two biinfinite matrices,

$$C^{\Delta} = \begin{bmatrix} & \vdots & \cdot^{\cdot^{\cdot}} \\ \cdots & \mathbf{c}_{ij}^{\Delta} & \cdots \\ \cdot^{\cdot^{\cdot}} & \vdots & \end{bmatrix}, \quad \text{and} \quad C^{\nabla} = \begin{bmatrix} & \vdots & \cdot^{\cdot^{\cdot}} \\ \cdots & \mathbf{c}_{ij}^{\nabla} & \cdots \\ \cdot^{\cdot^{\cdot}} & \vdots & \end{bmatrix},$$

or by two polynomials

$$\mathbf{c}^{\Delta}(\mathbf{x}) = \sum \mathbf{c}_{\mathbf{i}}^{\Delta}\mathbf{x}^{\mathbf{i}} \quad \text{and} \quad \mathbf{c}^{\nabla}(\mathbf{x}) = \sum \mathbf{c}_{\mathbf{i}}^{\nabla}\mathbf{x}^{\mathbf{i}} \ .$$

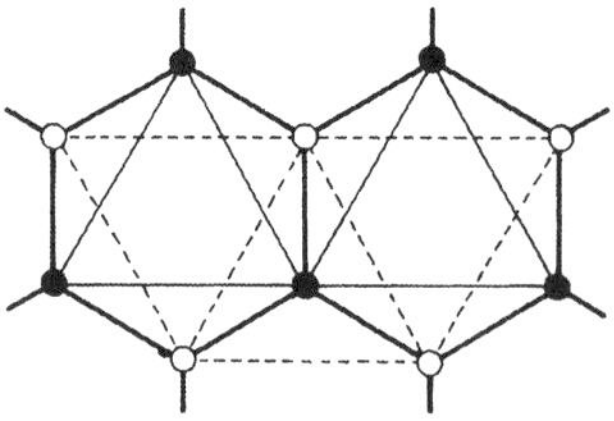

Figure 15.10: Decomposition of a regular hexagonal grid into two triangular grids.

The three direction averaging algorithm in 15.7 can be modified for hexagonal nets. The doubling operator $\mathcal{D}_{\bigcirc}$ for hexagonal nets is given by

$$\mathcal{D}_{\bigcirc}[C^{\triangle}, C^{\nabla}] = [[\mathbf{d}_{\mathbf{i}}^{\triangle}], [\mathbf{d}_{\mathbf{i}}^{\nabla}]] \ ,$$

where

$$\mathbf{d}_{2i,2j}^{\triangle} = \mathbf{d}_{2i+1,2j}^{\triangle} = \mathbf{d}_{2i+1,2j}^{\nabla} = \mathbf{d}_{2i+1,2j+1}^{\triangle} = \mathbf{c}_{ij}^{\triangle}$$

and

$$\mathbf{d}_{2i,2j}^{\nabla} = \mathbf{d}_{2i,2j+1}^{\nabla} = \mathbf{d}_{2i,2j+1}^{\triangle} = \mathbf{d}_{2i+1,2j+1}^{\nabla} = \mathbf{c}_{ij}^{\nabla} \ .$$

The labelling of the indices is illustrated in Figure 15.9. The three averaging operators $\mathcal{A}_k, k = 1, 2, 3$, defined in 15.7 are extended to

$$\mathcal{A}_k[C^{\triangle}, C^{\nabla}] = [\mathcal{A}_k C^{\triangle}, \mathcal{A}_k C^{\nabla}] \ .$$

For any $\mathbf{n} = (n_1, n_2, n_3) \in \mathbf{N}_0^3$, the composed operator

$$\mathcal{H}_{\mathbf{n}} = \mathcal{A}_1^{n_1} \mathcal{A}_2^{n_2} \mathcal{A}_3^{n_3} \mathcal{D}_{\bigcirc}$$

represents the **three direction averaging algorithm** for hexagonal nets due to [Prautzsch '84b].

The subdivision operators $\mathcal{H}_{\mathbf{n}}$ can also be represented by 2×2 **characteristic matrices** with polynomial entries. In particular, the doubling operation

$$[C^{\triangle}, C^{\nabla}] := \mathcal{D}_{\bigcirc}[C^{\triangle}, C^{\nabla}]$$

is described by the polynomial assignment

$$\begin{bmatrix} \mathbf{c}^{\triangle}(\mathbf{x}) \\ \mathbf{c}^{\nabla}(\mathbf{x}) \end{bmatrix} := \begin{bmatrix} 1+x+xy & y \\ x & 1+y+xy \end{bmatrix} \begin{bmatrix} \mathbf{c}^{\triangle}(\mathbf{x}^2) \\ \mathbf{c}^{\nabla}(\mathbf{x}^2) \end{bmatrix} ,$$

which we abbreviate by

$$\mathbf{c}(\mathbf{x}) := D(\mathbf{x})\mathbf{c}(\mathbf{x}^2) \ .$$

Figure 15.11 shows the associated four masks, for the net C^Δ on the top and for the net C^∇ on the bottom.

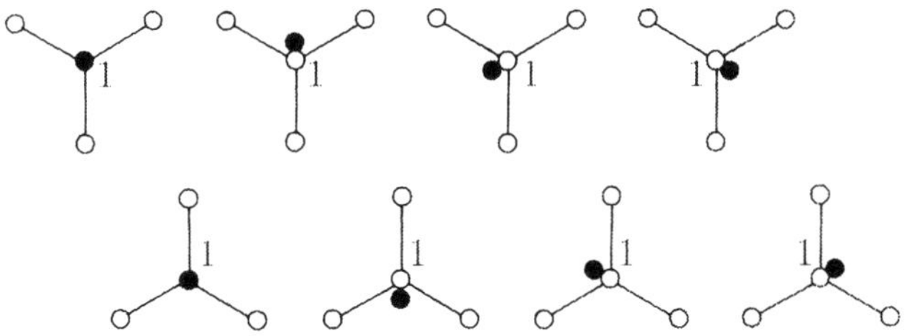

Figure 15.11: The eight masks of the doubling operator.

Furthermore, an averaging operation $[C^\Delta, C^\nabla] := \mathcal{A}^k[C^\Delta, C^\nabla]$ is described by the polynomial assignment

$$\begin{bmatrix} \mathbf{c}^\Delta(\mathbf{x}^2) \\ \mathbf{c}^\nabla(\mathbf{x}^2) \end{bmatrix} := \frac{1}{2}(1+\mathbf{x}^{\mathbf{e}_k}) \begin{bmatrix} \mathbf{c}^\Delta(\mathbf{x}^2) \\ \mathbf{c}^\nabla(\mathbf{x}^2) \end{bmatrix} ,$$

which we abbreviate by

$$\mathbf{c}(\mathbf{x}) := \alpha_k(\mathbf{x})\mathbf{c}(\mathbf{x}) \ .$$

Hence, $\mathcal{H}_\mathbf{n}$ is represented by the polynomial 2×2 matrix

$$\alpha_1^{n_1}\alpha_2^{n_2}\alpha_3^{n_3}D \ .$$

Figure 15.12 shows the single mask associated with the symmetric averaging operator $\mathcal{A}_{111} := \mathcal{A}_1\mathcal{A}_2\mathcal{A}_3$ for C^Δ on the left and for C^∇ on the right.

15.10 Half–box splines over triangular grids

In analogy to triangular and quadrilateral nets, we say that a sequence of hexagonal nets $C_m = [C_m^\Delta, C_m^\nabla]$ converges uniformly to a function $\mathbf{c}(\mathbf{x})$ if the suprema

$$\sup_{\mathbf{i}\in\mathbb{Z}^2} \max\{\|\mathbf{c}_{\mathbf{i}m}^\Delta - \mathbf{c}((\mathbf{i}+(\mathbf{e}_1-\mathbf{e}_3)/3)/2^m)\|, \|\mathbf{c}_{\mathbf{i}m}^\nabla - \mathbf{c}((\mathbf{i}+(\mathbf{e}_2-\mathbf{e}_3)/3)/2^m)\|\}$$

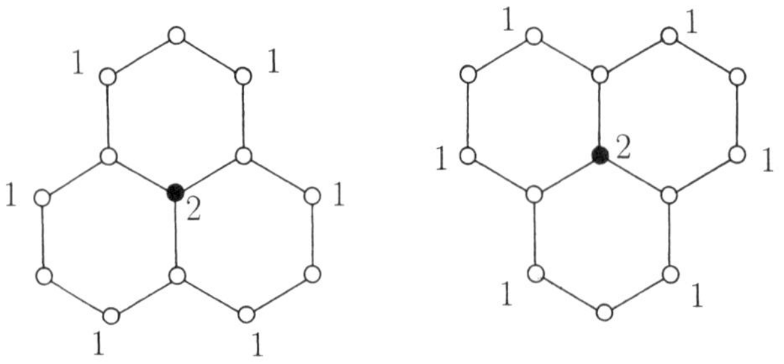

Figure 15.12: The mask of the symmetric averaging operator $\mathcal{A}_{111}$.

converge to zero, as m tends to infinity. Thus, C_m converges to a continuous function $\mathbf{c}(\mathbf{x})$ if both sequences of triangular nets C_m^Δ and C_m^∇ converge to $\mathbf{c}(\mathbf{x})$.

Let $C_m = [C_m^\Delta, C_m^\nabla]$ be a sequence of hexagonal nets obtained under repeated application of the averaging operator $\mathcal{H}_\mathbf{n}$ defined in 15.9. If $\mathcal{H}_\mathbf{n}$ is the doubling operator, i.e., if $\mathbf{n} = (0, 0, 0)$, then C_m converges to a piecewise constant spline over the triangular grid.

Moreover, using the arguments in 15.6, we see that any sequence $(\mathcal{A}_k \mathcal{H}_\mathbf{n})^m C$ can be obtained from $\mathcal{H}_\mathbf{n}^m C$ by increased averaging, as described in 15.4. Since we average a hexagonal net C by averaging the triangular subnets C^Δ and C^∇ separately, we obtain as in 15.8 that

> *for* $k = \min\{n_i + n_j | i \neq j\} - 1$, *the sequence* C_m *converges over every compact domain to a* C^k *spline, which is polynomial of total degree* $|\mathbf{n}| = n_1 + n_2 + n_3$ *over each triangle of the grid spanned by* $\mathbf{e}_1, \mathbf{e}_2$ *and* $\mathbf{e}_3$.

These splines are half-box spline surfaces, see Chapter 17. In particular, the symmetric operators $\mathcal{H}_{nnn}$ generate piecewise polynomial C^{2n-1} splines of degree $3n$.

15.11 Problems

1 The **butterfly algorithm** devised in [Dyn et al. '90] is an interpolatory subdivision scheme for triangular nets. It is given by the two masks shown in Figure 15.13, where the second mask stands for three symmetric masks. Show that any sequence of regular triangular nets obtained under repeated application of the butterfly algorithm converges to a C^1 surface for $0 < \omega < 1/2$, see [Gregory '91, Shenkman et al. '99].

2 Let $C_m = [\mathbf{c}_\mathbf{i}^m]$ be a sequence of control nets obtained by the refinement

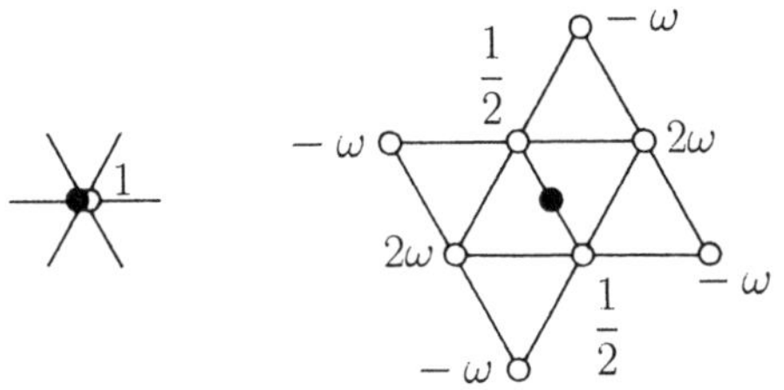

Figure 15.13: The masks of the butterfly algorithm.

rule

$$\mathbf{c}_{\mathbf{i}}^{m+1} = \sum \mathbf{c}_{\mathbf{j}}^{m} \gamma_{2\mathbf{i}-\mathbf{j}}$$

from some initial net C_0. Assume that for all nets C_0 the sequence C_m converges to some function $\mathbf{c}(\mathbf{x})$. Show that a refinable function $N(\mathbf{x})$ exists, i.e.,

$$N(\mathbf{x}) = \sum \gamma_{\mathbf{i}} N(2\mathbf{x} - \mathbf{i}) \ ,$$

and that it defines a basis, i.e., show that for all nets C_0 we can express the limiting function as

$$\mathbf{c}(\mathbf{x}) = \sum \mathbf{c}_{\mathbf{i}}^{m} N(2^{-m}\mathbf{x} - \mathbf{i}) \ .$$

3 Let $N(\mathbf{x})$ be a continuous basis function for the subdivision scheme considered in Problem 2. Let $\mathbf{v} \in \mathbb{Z}^2$ and show that

$$\overline{N}(\mathbf{x}) = \int_0^1 N(\mathbf{x} - t\mathbf{v})dt$$

is the basis function for the subdivision scheme with mask coefficients

$$\overline{\gamma}_{\mathbf{i}} = \frac{1}{2}(\gamma_{\mathbf{i}} + \gamma_{\mathbf{i}-\mathbf{v}}) \ .$$

4 In 15.5, we defined the ∇_1-, ∇_2- and ∇_3-difference schemes. Show that the $\nabla_{\mathbf{v}}$-difference scheme exists for any $\mathbf{v} \in \mathbb{Z}^2$ provided that the ∇_1- and ∇_2-scheme exist.

5 Describe a stationary subdivision scheme without stationary ∇_1- and ∇_2-difference schemes.

6 Prove the second theorem in 15.4 where convergence to uniformly continuous functions is replaced by convergence over compact domains to a piecewise polynomial spline over the regular triangular grid spanned by $\mathbf{e}_1, \mathbf{e}_2$ and $\mathbf{e}_3$.

16 Stationary subdivision for arbitrary nets

16.1 The midpoint scheme — 16.2 The limiting surface — 16.3 The standard parametrization — 16.4 The subdivision matrix — 16.5 Continuity of subdivision surfaces — 16.6 The characteristic map — 16.7 Higher order smoothness — 16.8 Triangular and hexagonal nets — 16.9 Problems

In 1978 Doo and Sabin presented a generalization of the subdivision algorithm for biquadratic tensor product splines. Simultaneously, Catmull and Clark presented a similar generalization for bicubic splines. Their algorithms can be applied to arbitrary quadrilateral control nets and yield sequences of control nets that converge to piecewise biquadratic or bicubic surfaces with finitely many so-called extraordinary points.

In contrast to the appealing simplicity of these algorithms, it had been difficult to analyze the smoothness properties of the limiting surfaces at their extraordinary points. It took 15 years till Ulrich Reif succeeded, after several attempts by others, in establishing a complete set of sufficient conditions under which the Catmull-Clark, Doo-Sabin, and similar algorithms generate tangent plane continuous regular surfaces.

16.1 The midpoint scheme

In 15.1 we described the Lane-Riesenfeld algorithm for tensor product splines by two operators.

The **refinement operator** $\mathcal{R}$ maps any control net C to the net $\mathcal{R}C$ that connects the midpoints of all edges in C with both their endpoints and the centroids of both abutting meshes.

The **averaging operator** $\mathcal{A}$ maps any net C to the net $\mathcal{A}C$ that connects the centroids of any two meshes with a common edge.

To apply these two operators we do not need to assume that the net C consists of quadrilateral only or that it is regular. The net C can be an arbitrary net, as the one illustrated in Figure 16.1. In this figure, the light edges form the net C, the light and broken edges form the net $\mathcal{R}C$ and the bold edges form the net $\mathcal{ARC}$.

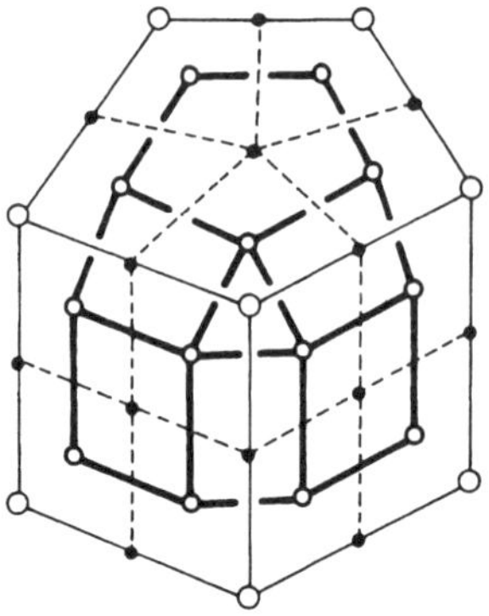

Figure 16.1: Refining and averaging a net.

We call the operator $\mathcal{M}_n = \mathcal{A}^{n-1}\mathcal{R}$, which refines a net and averages it $(n-1)$ times successively, the **midpoint operator** and say that any sequence of nets $\mathcal{M}_n^i C$ is obtained from C under the **midpoint** scheme $\mathcal{M}_n$.

In particular, if C is a regular quadrilateral net, then $\mathcal{M}_n$ represents the Lane-Riesenfeld algorithm for tensor product splines of bidegree n. Furthermore, for arbitrary nets, $\mathcal{M}_2$ and $\mathcal{M}_3$ represent specific instances of the **Doo-Sabin** [Doo & Sabin '78] and **Catmull-Clark algorithm** [Catmull & Clark '78], respectively. A sequence of nets obtained by application of $\mathcal{M}_3$ is shown in Figure 16.2.

For odd and even n, the midpoint schemes $\mathcal{M}_n$ have dual properties. If n is odd, all nets $\mathcal{M}_n^i C, i \geq 1$, are quadrilateral and if n is even, all nets $\mathcal{M}_n^i C, i \geq 1$, have interior vertices of valence four only. Non-quadrilateral meshes and interior vertices of valence is different from four are called **extraordinary meshes** and **extraordinary vertices**, respectively. In short, we refer to both types using the term **extraordinary elements**.

Furthermore, every extraordinary element of a net $\mathcal{M}_n^i C$ is obtained by an affine combination of a fixed number of vertices around a corresponding extraordinary element of the preceding net $\mathcal{M}_n^{i-1} C$. Since any extraordinary element in a net $\mathcal{M}_n^{i-1} C$ corresponds to at most one extraordinary element in $\mathcal{M}_n^i C$, the number of extraordinary elements in any net $\mathcal{M}_n^i C$ is bounded by the number of extraordinary elements in C. If C is a closed net, i.e., if C has no boundary, then the number of extraordinary elements is the same for all nets $\mathcal{M}_n^i C, i \geq 0$.

Remark 1: The distance between two extraordinary elements in some net $\mathcal{M}_n^i C$ is the number of edges of a shortest path connecting the two elements.

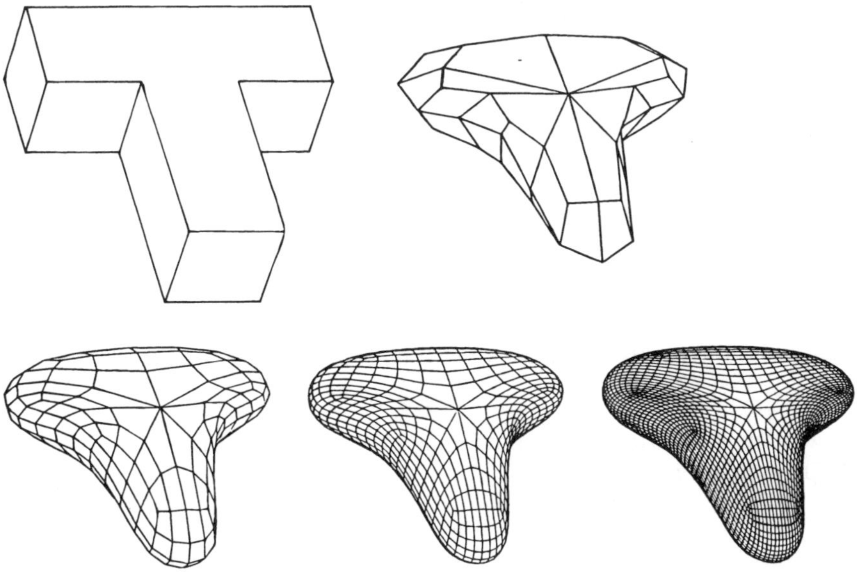

Figure 16.2: A sequence of nets obtained by the Catmull-Clark algorithm.

The distance between the corresponding extraordinary elements in $\mathcal{M}_n^{i+1}C$ is approximately twice as large.

16.2 The limiting surface

Any midpoint scheme $\mathcal{M}_n$ restricted to a regular quadrilateral subnet of an arbitrary net C is the Lane-Riesenfeld algorithm for splines of bidegree n. Thus, the sequence of nets $\mathcal{M}_n^i C, i \in \mathbb{N}$, converges to a piecewise polynomial surface $\mathbf{s}$. The sequences of corresponding extraordinary elements converge, see 16.5. Their limit points are called **extraordinary points** of $\mathbf{s}$. In every neighborhood of an extraordinary point, the surface $\mathbf{s}$ consists of infinitely many polynomial patches.

We now analyze the limiting surface $\mathbf{s}$ at an extraordinary point. Because of Remark 1 in 16.1, it suffices to consider a net C with just one extraordinary element surrounded by several rings of quadrilateral meshes, as illustrated in Figure 16.3.

Each regular subnet of a net $\mathcal{M}_n^i C$ consisting of $n \times n$ quadrilateral meshes forms the B-spline control net of some polynomial patch of $\mathbf{s}$. The patches defined by all these regular subnets of $\mathcal{M}_n^i C$ form an $(n-1)$ times differentiable surface $\mathbf{s}_i$, which is part of the limiting surface $\mathbf{s}$. Moreover, the surface

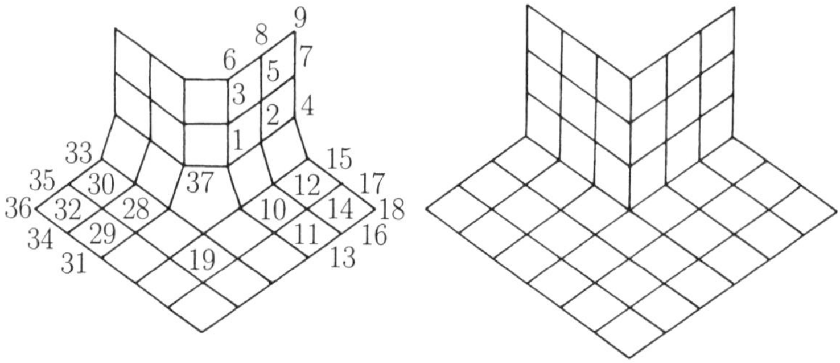

Figure 16.3: Control nets with one extraordinary element.

$\mathbf{s}_i$ without the surface $\mathbf{s}_{i-1}$ forms some m-sided surface ring $\mathbf{r}_i$ consisting of $3m\rho_n^2$ patches, where

$$\rho_n = \lfloor n/2 \rfloor := \max\{i \in \mathbb{N} | i \leq n/2\} \ .$$

The rings $\mathbf{r}_i$ together form the limiting surface $\mathbf{s}$. We can partition each surface ring $\mathbf{r}_i$ into $3m$ macro patches $\mathbf{r}_i^1, \ldots, \mathbf{r}_i^{3m}$, parametrized over $[0,1]^2$, where each macro patch consists of $\rho_n \times \rho_n$ patches. This is illustrated schematically in Figure 16.4, where $m = 5$. The dashed lines indicate the patches of $\mathbf{r}_1^5$ for $\rho_n = 3$.

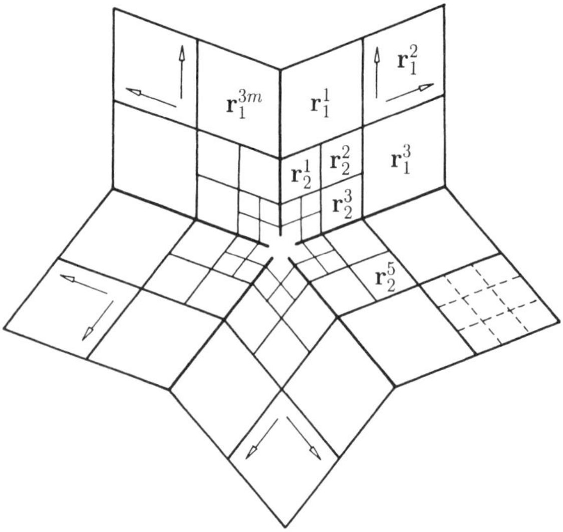

Figure 16.4: The adjacency of the patches $\mathbf{r}_i^j$ for $m = 5$.

16.3 The standard parametrization

An entire surface ring $\mathbf{r}_i$ is parametrized over $3m$ copies of $[0,1]^2$ or, more precisely, over

$$\Omega := \{1, \dots, 3m\} \times [0,1]^2 \ .$$

The rings $\mathbf{r}_i$ together form a surface with simple C^{n-1} joints. According to 9.7, this means that

$$\begin{aligned}
\frac{\partial^\kappa}{\partial u^\kappa}\mathbf{r}_i^j(1,v) &= \frac{\partial^\kappa}{\partial u^\kappa}\mathbf{r}_i^{j+1}(0,v) \ , \\
\frac{\partial^\kappa}{\partial v^\kappa}\mathbf{r}_i^{j+1}(u,0) &= \frac{\partial^\kappa}{\partial v^\kappa}\mathbf{r}_i^{j+2}(u,1) \ , \\
\frac{\partial^\kappa}{\partial v^\kappa}\mathbf{r}_i^{j+2}(u,0) &= -\frac{\partial^\kappa}{\partial u^\kappa}\mathbf{r}_i^{j+3}(0,u)
\end{aligned} \tag{1}$$

and

$$\begin{aligned}
\frac{\partial^\kappa}{\partial v^\kappa}\mathbf{r}_{i+1}^j(u,1) &= \frac{\partial^\kappa}{\partial v^\kappa}\mathbf{r}_i^j(\tfrac{u}{2},0) \ , \\
\frac{\partial^\kappa}{\partial v^\kappa}\mathbf{r}_{i+1}^{j+1}(u,1) &= \frac{\partial^\kappa}{\partial v^\kappa}\mathbf{r}_i^j(\tfrac{1}{2}+\tfrac{u}{2},0) \ , \\
\frac{\partial^\kappa}{\partial u^\kappa}\mathbf{r}_{i+1}^{j+1}(1,v) &= \frac{\partial^\kappa}{\partial u^\kappa}\mathbf{r}_i^{j+2}(0,\tfrac{1}{2}+\tfrac{v}{2}) \ , \\
\frac{\partial^\kappa}{\partial u^\kappa}\mathbf{r}_{i+1}^{j+2}(1,v) &= \frac{\partial^\kappa}{\partial u^\kappa}\mathbf{r}_i^{j+2}(1,\tfrac{v}{2}) \ ,
\end{aligned} \tag{2}$$

for all $u, v \in [0,1], i \in \mathbf{N}, \kappa = 0, 1, \dots, n-1$, and $j = 1, 4, 7, \dots, 3n-2$, where $\mathbf{r}_i^{3n+1} = \mathbf{r}_i^1$.

Moreover, if $\mathbf{c}_1, \dots, \mathbf{c}_p$ denote the control points of $\mathbf{r}_i$, then $\mathbf{r}_i$ can also be written as

$$\mathbf{r}_i(j,u,v) = \mathbf{r}_i^j(u,v) = \sum_{l=1}^{p} \mathbf{c}_l B_l^j(u,v) \ ,$$

where the $B_l^j(u,v)$ are certain tensor product B-splines of bidegree n.

Remark 2: Let N_{km} denote the tensor product $N_k^2(u)N_m^2(v)$ of the quadratic B-splines with the knots $k, \dots, k+3$ and $m, \dots, m+3$, respectively. Then, using the numbering scheme shown in Figures 16.3 and 16.4, we obtain, for example, for $n = 2$ and $j = 2, 3$

$$\begin{bmatrix} B_6^2 & B_8^2 & B_9^2 \\ B_3^2 & B_5^2 & B_7^2 \\ B_1^2 & B_2^2 & B_4^2 \end{bmatrix} = \begin{bmatrix} B_3^3 & B_5^3 & B_7^3 \\ B_1^3 & B_2^3 & B_4^3 \\ B_{10}^3 & B_{12}^3 & B_{15}^3 \end{bmatrix} = \begin{bmatrix} N_{0,-2} & \cdots & N_{0,0} \\ \vdots & & \vdots \\ N_{-2,-2} & \cdots & N_{-2,0} \end{bmatrix} .$$

16.4 The subdivision matrix

Any surface ring $\mathbf{r}_i$ has three coordinate functions, and the possible coordinate functions form the linear space

$$\mathcal{R} = \{\sum c_l B_l^j(u,v) \mid c_l \in \mathbb{R}\} \ .$$

Let

$$\mathbf{r}_i(j,u,v) = \sum \mathbf{c}_l^i B_l^j(u,v) \in \mathbb{R}^3$$

be surface rings obtained under repeated applications of some midpoint scheme $\mathcal{M}_n$ from initial control points $\mathbf{c}_1^0, \dots, \mathbf{c}_p^0$. Since the points $\mathbf{c}_l^i$ are affine combinations of the points $\mathbf{c}_l^{i-1}$, there is a $p \times p$ matrix S, the **subdivision matrix**, such that, for all $i \geq 1$,

$$[\mathbf{c}_1^i \dots \mathbf{c}_p^i]^t = S[\mathbf{c}_1^{i-1} \dots \mathbf{c}_p^{i-1}]^t \ .$$

The subdivision matrix S is **stochastic**, which means that it has non-negative entries summing to one in each row. Therefore, the leading eigenvalue of S is one with $\mathbf{e} = [1 \dots 1]^t$ as associated eigenvector. All points $\mathbf{c}_l^2$ depend on the extraordinary element in the initial net $\mathbf{c}_1^0 \dots \mathbf{c}_p^0$. Thus, S^2 has a strictly positive row, which implies that one is a simple eigenvalue of S^2, see [Micchelli & Prautzsch '89].

Consequently, the control nets $[\mathbf{c}_1^i \dots \mathbf{c}_p^i]$ converge to a p-fold point $[\mathbf{c} \dots \mathbf{c}]$, which represents the extraordinary point $\mathbf{c}$ of the subdivision surface $\mathbf{s}$.

Remark 3: Using the numbering scheme shown in Figure 16.3 left, the subdivision matrix S for the midpoint scheme $\mathcal{M}_2$ around an m-sided mesh is the cyclic $9m \times 9m$ matrix

$$S = \frac{1}{16}\begin{bmatrix} S_1 & S_2 & \cdots & S_m \\ S_m & S_1 & \cdots & S_{m-1} \\ \vdots & & \ddots & \vdots \\ S_2 & \cdots & S_m & S_1 \end{bmatrix} .$$

The S_i are 9×9 blocks of the form

$$S_i = \begin{bmatrix} A_i & O \\ O & O \end{bmatrix} ,$$

where O denotes zero matrices of different sizes,

$$A_3 = \dots = A_{m-1} = [4/m] \ ,$$

$$A_1 = \begin{bmatrix} a & . & . & . & . \\ 9 & 3 & . & . & . \\ 9 & . & 3 & . & . \\ 3 & 9 & . & . & . \\ 3 & 9 & 1 & 3 & . \\ 3 & . & 9 & . & . \\ 3 & 9 & 1 & . & 3 \\ 3 & 1 & 9 & . & 3 \\ 1 & 3 & 3 & . & 9 \end{bmatrix}, \quad A_2 = \begin{bmatrix} b & . & . \\ 3 & . & 1 \\ . & . & . \\ 1 & . & 3 \end{bmatrix} \quad \text{and} \quad A_m = \begin{bmatrix} b & . \\ . & . \\ 3 & 1 \\ . & . \\ . & . \\ 1 & 3 \end{bmatrix},$$

with $a = 8 + 4/m$, $b = 4 + 4/m$ and zero entries are marked by a dot.

16.5 Continuity of subdivision surfaces

From now on let $\mathcal{R}$ denote an arbitrary linear space of functions $r(j, u, v)$, defined for $j \in \{1, \ldots, 3n\}$ and $u, v \in [0, 1]$ that are k times differentiable in u and v. Further, let S be an arbitrary linear map on $\mathcal{R}$ and assume that any function $r_i \in \mathcal{R}$ and its image $r_{i+1} = Sr_i$ satisfy the smoothness conditions (1) and (2) in 16.3 for all $\kappa = 1, \ldots, k$.

Let $x_1, \ldots, x_p$ be eigenvectors and generalized eigenvectors of S forming a basis of $\mathcal{R}$, and let $\lambda_1, \ldots, \lambda_p$ be the associated eigenvalues. We assume that their moduli are in decreasing order, i.e., $|\lambda_1| \geq |\lambda_2| \geq \ldots \geq |\lambda_p|$.

Any surface ring whose coordinates are functions of $\mathcal{R}$ is an image of the normal ring

$$\mathbf{r}_0 = \begin{bmatrix} x_1 \\ \vdots \\ x_p \end{bmatrix},$$

and any subdivision surface obtained by repeated application of S is an image of the surface $\mathbf{r}$ composed of the rings

$$\mathbf{r}_m = S^m \mathbf{r}_0 = \begin{bmatrix} S^m x_1 \\ \vdots \\ S^m x_p \end{bmatrix}, \quad m \in \mathbf{N}_0,$$

under some linear map.

Hence, in order to study the smoothness of these subdivision surfaces, it suffices to analyze just $\mathbf{r}$. First, we discuss convergence. The iterates $\mathbf{r}_m$ converge to the origin if and only if $|\lambda_1| < 1$, and they converge to some other point, i.e., a non-zero constant function, if $\lambda_1 = 1$ and $|\lambda_2| < 1$ and x_1 is a constant function. Furthermore, the functions x_i are linearly independent. Thus, only one coordinate function x_i can be constant.

If x_1 is constant and $\lambda_1 = 1$, then the subdivision surface $\mathbf{r}$ is a translate of the subdivision surface formed by the "shorter" iterates $S^m[x_2 \ldots x_p]^t$. Hence, for the smoothness analysis it suffices to assume $|\lambda_1| < 1$.

Any affinely independent subdivision scheme has a subdivision matrix S with eigenvalue one associated with the constant functions as eigenvectors, see, for example, the midpoint scheme in 16.4.

16.6 The characteristic map

We continue to analyze the smoothness properties of $\mathbf{r}$ at the origin assuming that $|\lambda_1| < 1$. Moreover, we assume that $\lambda_1 = \lambda_2$ (which implies that λ_1 is real) and $|\lambda_2| > |\lambda_3|$ and suppose, for simplicity, that all x_i are (non-generalized) eigenvectors.

Under these conditions, $\mathbf{r}_m/\lambda_1^m$ converges to $[x_1\, x_2\, 0 \ldots 0]^t$. Consequently, if $\mathbf{r}$ has a tangent plane at the origin, this plane is the x_1x_2-plane. By the way, this argument also shows that $\mathbf{r}$ has no tangent plane if there are more than two dominant eigenvectors x_i.

If $\mathbf{r}$ is a regular surface at the origin, then it can be viewed as the graph of a function over the x_1x_2-plane. In particular, this means that for each $j = 1, \ldots, 3n$ the patch

$$\mathbf{c}_j(u,v) = [x_1(j,u,v)\; x_2(j,u,v)]$$

is invertible and that the only overlapping parts of these patches are their common boundary curves, according to conditions (1) and (2) in 16.3. More concisely we say that the map $\mathbf{c} : \{1, \ldots, 3n\} \times \Omega \to \mathbb{R}^2$ is invertible. The map $\mathbf{c}$ is introduced in [Reif '93] and called the **characteristic map** of the subdivision scheme S. Its image is a planar surface ring R without foldings, and the iterates $\lambda^m R$ fill the hole enclosed by R without overlapping each other, see Figure 16.5.

So, any coordinate r_i of $\mathbf{r}$ can be viewed as a function over $U = \bigcup_{m=0}^{\infty} \lambda^m R$, i.e., the i-th coordinate is the function

$$r_i(\mathbf{x}) = \lambda_i^m x_i(\mathbf{c}^{-1}(\mathbf{x}/\lambda^m))\ , \text{ where } \mathbf{x} \in \lambda^m R\ . \tag{3}$$

16.7 Higher order smoothness

If the subdivision surface $\mathbf{r}$ is k times continuously differentiable and regular, it is a function in x_1 and x_2, with a Taylor expansion of order k at the origin. Comparing this expansion with (3), we see that λ_i must either be a power of λ or satisfy $|\lambda_i| < |\lambda|^k$. Further, if $\lambda_i = \lambda^\kappa$, we see that the eigenvector x_i is a homogeneous polynomial of order κ in x_1 and x_2.

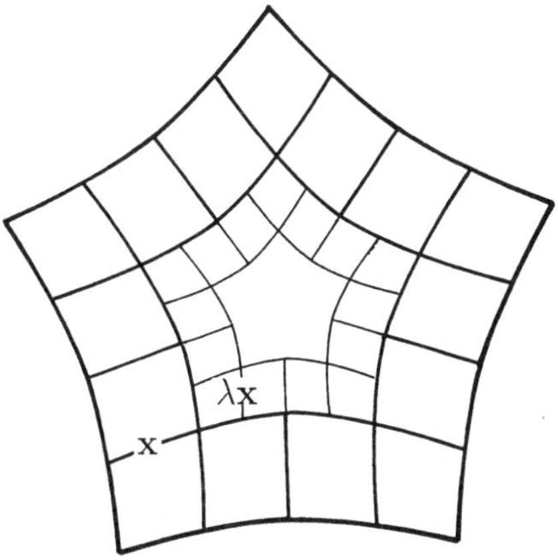

Figure 16.5: A characteristic map $\mathbf{x}$ and its scaled version $\lambda_{\mathbf{x}}$.

Conversely, let the eigenvalues and eigenvectors of S have these properties. If $\lambda_i = \lambda^\kappa$, $\kappa < k$, then, r_i is a polynomial in x_1 and x_2. If $|\lambda_i| < |\lambda^k|$, then for any mixed partial derivative ∂ of order $\kappa \leq k$, we get that

$$\partial r_i(\mathbf{x}) = \left(\frac{\lambda_i}{\lambda^\kappa}\right)^m \partial(x_i \circ \mathbf{c}^{-1})(\mathbf{x}/\lambda^m)$$

converges to zero, as $\mathbf{x}$ goes to zero, or, equivalently, as m goes to infinity.

Hence, we arrive at the following theorem from [Prautzsch '98].

Assume that $1 > |\lambda| > |\lambda_3|$ *and* λ *is associated with two eigenvectors* x_1 *and* x_2 *defining a regular characteristic map. The subdivision surface* $\mathbf{r}$ *is a regular* C^k *surface if and only if*

- *the characteristic map* $\mathbf{x}$ *is invertible*
- *and for all* $i = 1, \ldots, p$ *it holds that either* $\lambda_i = \lambda^\kappa$ *and* $x_i \in \operatorname{span}\{x_1^\alpha x_2^\beta \mid \alpha + \beta = \kappa\}$ *or* $|\lambda_i| < |\lambda|^k$.

Remark 4: This theorem also holds if one is an eigenvalue of S with any constant function as associated eigenvector, see 16.5.

Remark 5: The theorem takes on a similar form if λ_1 and λ_2 are complex conjugate or different real eigenvalues, see [Prautzsch '98, Reif '95b].

16.8 Triangular and hexagonal nets

The analysis given in this chapter applies also to stationary subdivision schemes for triangular and hexagonal nets.

An arbitrary triangular net consists only of triangular meshes, but it may have extraordinary vertices, which are vertices with valence different from six, see Figure 16.6.

Dually to this, an arbitrary hexagonal net has only regular vertices of valence three, but it may possess extraordinary, i.e., non-hexagonal, meshes, see Figure 16.6.

As already observed in 15.9, we obtain a hexagonal net from any triangular net by connecting the centroids of adjacent triangles. Similarly, we obtain triangular nets from arbitrary hexagonal nets see Figure 16.6.

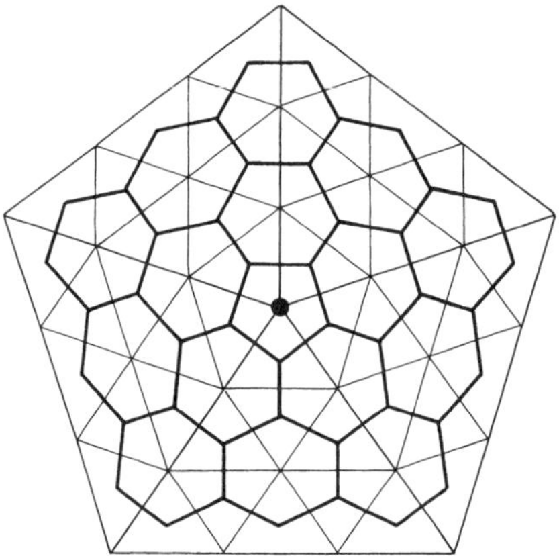

Figure 16.6: Converting a triangular net into a hexagonal net and vice versa.

As discussed in 15.8 and 15.10, any triangular and any hexagonal net controls a surface composed of triangular patches. In general, these patches need not to be polynomial. Around an extraordinary point, the patch arrangement looks like the one shown schematically in Figure 16.7. Moreover, around an extraordinary point, it is possible to merge adjacent triangular into quadrilateral patches, which then form the configuration shown in Figure 16.4.

Therefore, the smoothness analysis and the Theorem in 16.7 also apply to triangular and hexagonal nets.

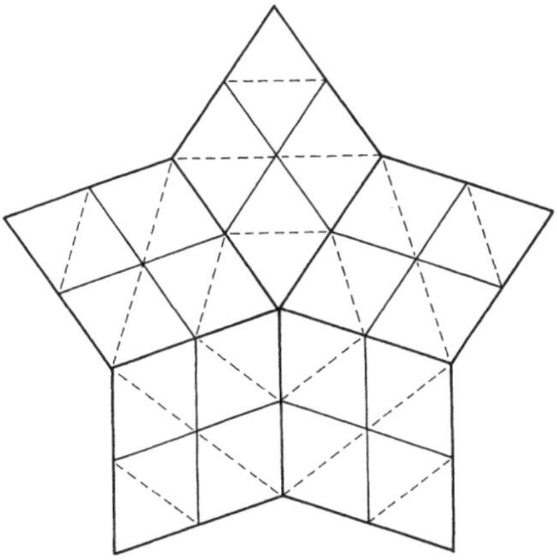

Figure 16.7: Triangular (quadrilateral) patches around an extraordinary point.

16.9 Problems

1 Let S be a stochastic $n \times n$ matrix with a strictly positive column. Show that there is some constant $\gamma \in (0,1)$ such that for all vectors $\mathbf{v} = [v_1 \dots v_n]^t$

$$\text{diam}(S\mathbf{v}) \leq \gamma \text{diam}(\mathbf{v}) \ ,$$

where $\text{diam}(\mathbf{v})$ denotes the diameter $\max v_i - \min v_i$, see also [Micchelli & Prautzsch '89].

2 Implement the butterfly algorithm given by the mask in Figure 15.13 for arbitrary triangular nets.

3 In 1987, Loop [Loop '87] generalized Boehm's masks for Prautzsch' quartic box spline subdivision algorithm, which is given in 15.7. The generalized masks by Loop can be applied to arbitrary triangular control nets and are shown in Figure 16.8, where n denotes the valence of the vertex and α is a free parameter depending on n. Verify Loop's result that the limiting surface is tangent plane continuous at extraordinary points if

$$\alpha(6) = 5/8 \quad \text{and} \quad -\frac{1}{4}\cos\frac{2\pi}{n} < \alpha(n) < \frac{3}{4} + \frac{1}{4}\cos\frac{2\pi}{n} \ .$$

4 Refine an arbitrary triangular net by subdividing each triangle uniformly into four subtriangles. Then compute the centroids of all subtriangles. These centroids are the vertices of a hexagonal net. Finally compute the centroids of all ordinary and extraordinary meshes of this hexagonal net. These operations comprise one step of a **midpoint scheme** for

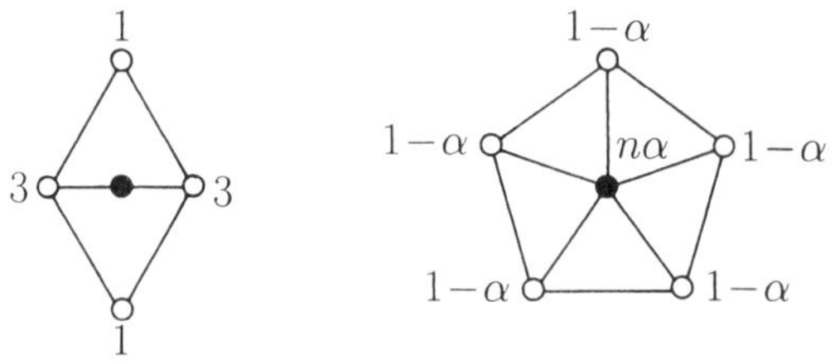

Figure 16.8: The masks of Loop's algorithm.

triangular nets. Compute the masks of this scheme and compare them with Loop's masks.

5 Refine an arbitrary hexagonal net by quadrupling all vertices, see Figure 15.11. The new vertices define a hexagonal net with degenerate meshes. Compute the centroids of all meshes. They form a triangular net. Finally, compute the centroids of all triangles. These operations comprise one step of a **midpoint scheme** for hexagonal nets. Compute the masks of this scheme and compare them with the masks of the subdivision algorithm for cubic half-box splines in 15.9.

6 A piecewise polynomial regular and invertible characteristic map $\mathbf{x}$ that belongs to a subdivision matrix S around an extraordinary point and that is k times differentiable has polynomial degree $\geq k+1$, see [Reif '96] and [Prautzsch & Reif '99]. Use this fact to show that a piecewise polynomial C^2-subdivision surface with positive Gaussian curvature at its extraordinary point must be of degree ≥ 6.

Part III

Multivariate Splines

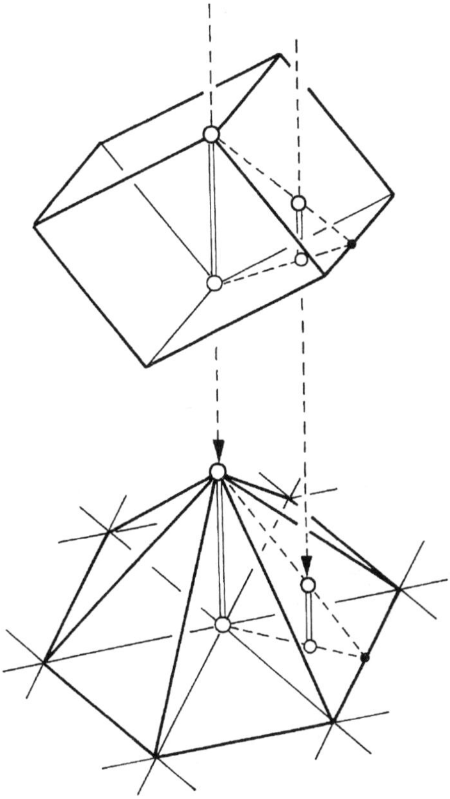

17 Box splines

17.1 Definition of box splines — 17.2 Box splines as shadows — 17.3 Properties of box splines — 17.4 Derivatives of box splines — 17.5 Box spline surfaces — 17.6 Subdivision for box spline surfaces — 17.7 Convergence under subdivision — 17.8 Half-box splines — 17.9 Half-box spline surfaces — 17.10 Problems

Box splines are density functions of the shadows of higher dimensional polyhedra, namely boxes. For example, B-splines with equidistant knots are special univariate box splines, and the surfaces obtained by the averaging algorithm described in Section 15.7 are box spline surfaces over a regular triangular grid. This chapter (an abbreviated version of [Prautzsch & Boehm '02]) provides a brief introduction to general box splines. It also covers half-box splines. Symmetric half-box splines of degree $3n$ are more suitable than box splines for the construction of arbitrary G^{2n-1} free-form surfaces with triangular patches than box splines.

17.1 Definition of box splines

An s-variate **box spline** $B(\mathbf{x}|\mathbf{v}_1 \ldots \mathbf{v}_k)$ is determined by k directions $\mathbf{v}_i$ in $\mathbb{R}^s$. For simplicity, we assume that $k \geq s$ and that $\mathbf{v}_1, \ldots, \mathbf{v}_s$ are linearly independent. Under these assumptions, the box splines $B_k(\mathbf{x}) = B(\mathbf{x}|\mathbf{v}_1 \ldots \mathbf{v}_k)$, $k = s, s+1, \ldots$, are defined by successive convolutions, similarly to the definition given in 8.1,

$$B_s(\mathbf{x}) = \begin{cases} 1/\det[\mathbf{v}_1 \ldots \mathbf{v}_s] & \text{if } \mathbf{x} \in [\mathbf{v}_1 \ldots \mathbf{v}_s][0,1)^s \\ 0 & \text{otherwise} \end{cases}$$

$$B_k(\mathbf{x}) = \int_0^1 B_{k-1}(\mathbf{x} - t\mathbf{v}_k)dt, \quad k > s \ .$$

This recursive definition is illustrated in Figure 17.1 for $s = 2$ and

$$[\mathbf{v}_1 \ldots \mathbf{v}_4] = \begin{bmatrix} 1 & 0 & 1 & 1 \\ 0 & 1 & 1 & 0 \end{bmatrix} ,$$

see also Figure 8.1.

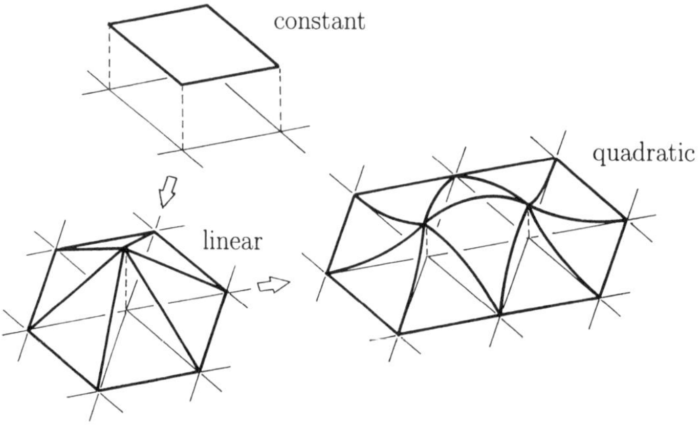

Figure 17.1: Bivariate box splines over the triangular grid.

These box splines are normalized such that

$$\int_{\mathbb{R}^s} B_k(\mathbf{x})dx = 1 ,$$

which can easily be verified for $k = s$ and by induction over k:

$$\int_{\mathbb{R}^s} \int_0^1 B_{k-1}(\mathbf{x} - t\mathbf{v}_k)dtd\mathbf{x} = \int_0^1 \int_{\mathbb{R}^s} B_{k-1}(\mathbf{x} - t\mathbf{v}_k)d\mathbf{x}dt = \int_0^1 dt = 1 .$$

17.2 Box splines as shadows

A box spline $B_k(\mathbf{x}) = B(\mathbf{x}|\mathbf{v}_1 \ldots \mathbf{v}_k)$ can also be constructed geometrically. Let π be the orthogonal projection

$$\pi : [t_1 \ldots t_k]^t \mapsto [t_1 \ldots t_s]^t ,$$

and let

$$\beta_k = [\mathbf{u}_1 \ldots \mathbf{u}_k][0,1)^k$$

be a parallelepiped such that $\mathbf{v}_i = \pi\mathbf{u}_i$.

Then, $B_k(\mathbf{x})$ represents the density of the "shadow" of β_k, i.e.,

(1) $$B_k(\mathbf{x}) = \frac{1}{\mathrm{vol}_k \beta_k} \mathrm{vol}_{k-s} \beta_k(\mathbf{x}) \ ,$$

where

$$\beta_k(\mathbf{x}) = \pi^{-1}\mathbf{x} \cap \beta_k \ .$$

For $k = 3$ and $s = 2$, the corresponding geometric construction is illustrated in Figure 17.2.

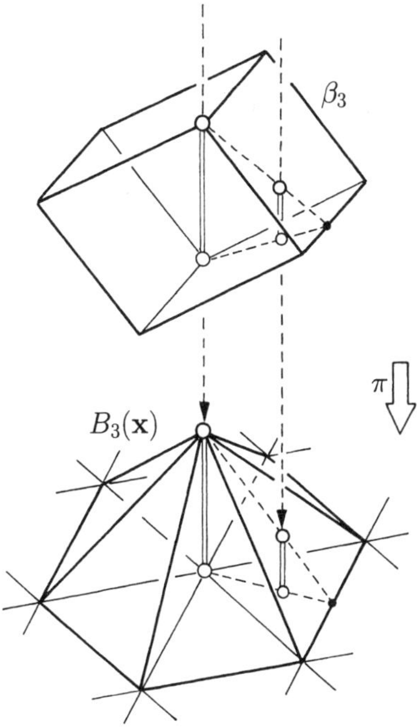

Figure 17.2: The geometric construction of a piecewise linear box spline over a triangular grid.

We prove this characterization of box splines by induction: For $k = s$, equation (1) is obvious, and, for greater values of k, we observe that

$$\beta_k(\mathbf{x}) = \bigcup_{s \in [0,1)} \Big((\beta_{k-1} + s\mathbf{u}_k) \cap \pi^{-1}\mathbf{x} \Big) \ .$$

Hence, if h measures the distance between β_{k-1} and $\mathbf{u}_k + \beta_{k-1}$ along the kth unit vector in $\mathbb{R}^k$, see Figure 17.3, then it follows that

$$\mathrm{vol}_{k-s} \beta_k(\mathbf{x}) = \int_0^1 h \ \mathrm{vol}_{k-s-1}\Big(\beta_{k-1}(\mathbf{x} - s\mathbf{v}_k)\Big) ds \ ,$$

which corresponds, up to a constant factor, to the inductive definition of box splines. Consequently, $\mathrm{vol}_{k-s}\beta_k(\mathbf{x})$ is a multiple of the box spline $B_k(\mathbf{x})$, and, since

$$\int_{\mathbb{R}^s} \mathrm{vol}_{k-s}\beta_k(\mathbf{x})\, d\mathbf{x} = \mathrm{vol}_k\beta_k \quad \text{and} \quad \int_{\mathbb{R}^s} B_k(\mathbf{x}) d\mathbf{x} = 1 \ ,$$

equation (1) follows. ◇

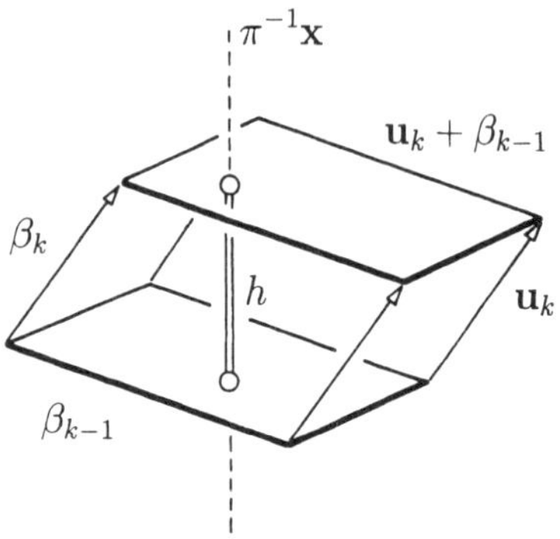

Figure 17.3: Measurements of the box β_k.

Remark 1: From the geometric definition (1), it follows that a box spline solves the functional equation

$$\int_{\mathbb{R}^s} B(\mathbf{x}|\mathbf{v}_1 \ \ldots \ \mathbf{v}_k) f(\mathbf{x}) d\mathbf{x} = \int_{[0,1)^k} f([\mathbf{v}_1 \ \ldots \ \mathbf{v}_k]\mathbf{t}) d\mathbf{t}$$

for all continuous test functions $f(\mathbf{x})$.

17.3 Properties of box splines

From the geometric construction (1) of box splines, it follows that $B(\mathbf{x}) = B(\mathbf{x}|\mathbf{v}_1 \ \ldots \ \mathbf{v}_k)$

- *does not depend on the* **ordering** *of the directions* $\mathbf{v}_i$,
- *is* **positive** *over the convex set* $[\mathbf{v}_1 \ldots \mathbf{v}_k][0,1)^k$,
- *has the* **support** $\mathrm{supp}B(\mathbf{x}) = [\mathbf{v}_1 \ \ldots \ \mathbf{v}_k][0,1]^s$,
- *is* **symmetric** *with respect to the center of its support.*

Further, let $B(\mathbf{x})$ be the shadow of a box β as in 17.2. The $(s-1)$-dimensional faces of β projected into $\mathbb{R}^s$ form a tessellation of the support. It is illustrated in Figure 17.4 for

$$[\mathbf{v}_1 \ldots \mathbf{v}_k] = \begin{bmatrix} 1 & 1 & 1 & 0 \\ -1 & 0 & 1 & 1 \end{bmatrix} \quad \text{and} \quad [\mathbf{v}_1 \ldots \mathbf{v}_k] = \begin{bmatrix} 1 & 1 & 1 & 1 & 0 \\ 0 & 0 & 1 & 1 & 1 \end{bmatrix} .$$

- *The box spline $B(\mathbf{x})$ is* **polynomial** *of degree $\leq k-s$ over each tile of this tessellation.*

For a proof, we observe that the extreme points of the convex sets $\pi^{-1}\mathbf{x} \cap \beta$ lie in s-dimensional faces of β. Hence, an extreme point is of the form $[\mathbf{x}^t \mathbf{e}^t]^t$, where $\mathbf{e} \in \mathbb{R}^{k-s}$ depends linearly on $\mathbf{x}$ over the projection of the corresponding s-dimensional face. The volume of $\pi^{-1}\mathbf{x} \cap \beta$ can be expressed as a linear combination of determinants of $k-s \times k-s$ matrices whose columns represent differences of extreme points $\mathbf{e}$. Hence, the volume is a polynomial of degree $\leq k-s$ in $\mathbf{x}$ in each tile of the tessellation. $\diamond$

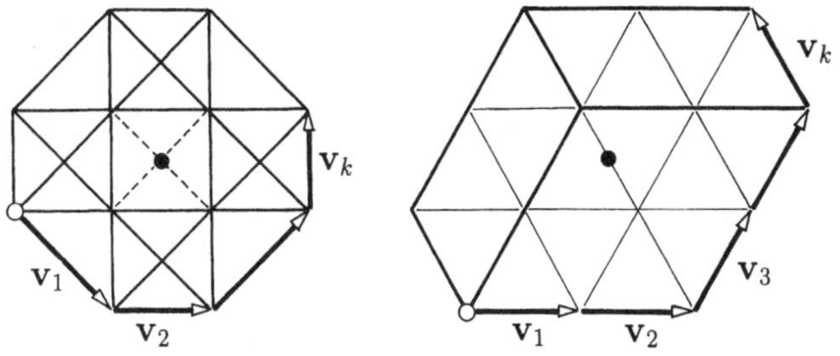

Figure 17.4: The supports of a quadratic and a cubic box spline over the criss-cross and the regular triangular grid.

17.4 Derivatives of box splines

From the inductive definition of box splines, it follows that the restricted box spline $B(y) = B(\mathbf{x} + y\mathbf{v}_r)$ is piecewise constant in y if $\mathbf{v}_r \notin \operatorname{span}\{\mathbf{v}_1, \ldots, \mathbf{v}_r^*, \ldots, \mathbf{v}_k\}$. If $\mathbf{v}_r \in \operatorname{span}\{\mathbf{v}_1, \ldots, \mathbf{v}_r^*, \ldots, \mathbf{v}_k\}$, then $B(y)$ is continuous since it can be obtained by a convolution from

$$B^*(y) = B(\mathbf{x} + y\mathbf{v}_r | \mathbf{v}_1 \ldots \mathbf{v}_r^* \ldots \mathbf{v}_k) \ ,$$

i.e.,

$$\begin{aligned} B(y) &= \int_0^1 B^*(y-t)dt = \int_{y-1}^{y} B^*(t)dt \\ &= \int_{-\infty}^{y} B^*(t) - B^*(t-1)dt \ . \qquad (2) \end{aligned}$$

Further, the **directional derivative** with respect to $\mathbf{v}_r$ is given by

$$(3) \qquad B'(y)|_{y=0} = B^*(0) - B^*(-1) \ .$$

If $\mathbf{v}_1, \ldots, \mathbf{v}_r^*, \ldots, \mathbf{v}_k$ span $\mathbb{R}^s$ for $r = 1, \ldots, s$, then $B(\mathbf{x}|\mathbf{v}_1 \ldots \mathbf{v}_k)$ is continuous and its directional derivatives can be written as linear combinations of translates of the box splines $B(\mathbf{x}|\mathbf{v}_1 \ldots \mathbf{v}_r^* \ldots \mathbf{v}_k)$. Applying this argument repeatedly, we see that

> *$B(\mathbf{x})$ is r times continuously differentiable if all subsets of $\{\mathbf{v}_1, \ldots, \mathbf{v}_k\}$ obtained by deleting $r+1$ vectors $\mathbf{v}_i$ span $\mathbb{R}^s$.*

Remark 2: Another and inductive proof of the polynomial properties of $B(\mathbf{x})$ is based on equation (3): If $B^*(\mathbf{x})$ is a polynomial of degree $\leq k-s-1$ over each tile of the tessellation above, then $B^*(\mathbf{x}-\mathbf{v}_r)$ has the same property. Due to (3), the property, but with degree $k-s$, is passed on to $B(y)$. Consequently, $B(\mathbf{x})$ is piecewise polynomial of degree $k-s$ in each direction $\mathbf{v}_r$ over this tessellation above and hence in $\mathbf{x}$.

Remark 3: The piecewise quadratic C^1 box spline whose support is shown on the left side of Figure 17.4 is called a **Zwart-Powell element**. Because of its symmetry, it is quadratic, but not piecewise quadratic, over the inner square with the dashed diagonals.

17.5 Box spline surfaces

In the following we assume that the directions $\mathbf{v}_1, \ldots, \mathbf{v}_k$ are in $\mathbb{Z}^s$ and that $\mathbf{v}_1, \ldots, \mathbf{v}_s$ span $\mathbb{R}^s$.

Obviously, the translates $B(\mathbf{x}-\mathbf{j}|\mathbf{v}_1 \ldots \mathbf{v}_s)$, $\mathbf{j} \in [\mathbf{v}_1 \ldots \mathbf{v}_s]\mathbb{Z}^s$, of the piecewise constant box spline sum to

$$(4) \qquad \gamma = 1/|\det[\mathbf{v}_1 \ldots \mathbf{v}_s]| \ .$$

Since $\mathbb{Z}^s$ can be decomposed into some m sets $\mathbf{i}+[\mathbf{v}_1 \ldots \mathbf{v}_s]\mathbb{Z}^s$, see Figure 17.5, it follows that

$$\sum_{\mathbf{i} \in \mathbb{Z}^s} B(\mathbf{x}-\mathbf{i}|\mathbf{v}_1 \ldots \mathbf{v}_s) = m\gamma \ .$$

Further, since $\int_0^1 m\gamma dt = m\gamma$, we obtain by $k-s$ and again s successive

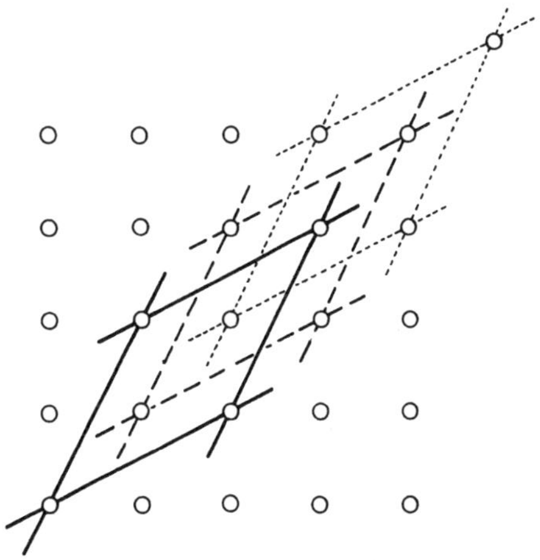

Figure 17.5: Decomposition of $\mathbb{Z}^s$ into translates of coarser grids.

convolutions the equation

$$\begin{aligned} m\gamma &= \sum_{\mathbf{i}\in\mathbb{Z}^s} B(\mathbf{x}-\mathbf{i}|\mathbf{v}_1 \ldots \mathbf{v}_k) \\ &= \sum_{\mathbf{i}\in\mathbb{Z}^s} B(\mathbf{x}-\mathbf{i}|\mathbf{e}_1 \ldots \mathbf{e}_s\, \mathbf{v}_1 \ldots \mathbf{v}_k) \ , \end{aligned}$$

where $\mathbf{e}_i$ is the ith unit vector. Since the last sum is identically one, due to (4), we have shown that

the (integer) shifts of any box spline $B(\mathbf{x}) = B(\mathbf{x}|\mathbf{v}_1 \ldots \mathbf{v}_k)$ form a **partition of unity**.

Consequently, any **box spline surface**

$$\mathbf{s}(\mathbf{x}) = \sum_{\mathbf{i}\in\mathbb{Z}^s} \mathbf{c}_\mathbf{i} B(\mathbf{x}-\mathbf{i})$$

is an affine combination of its **control points** $\mathbf{c}_\mathbf{i}$. This surface representation is **affinely invariant**, i.e., that under any affine map the control point images control the surface image.

Since the box splines are non-negative, $\mathbf{s}(\mathbf{x})$ even is a convex combination of its control points and, hence, it lies in their **convex hull**.

Apparently, the sequence

$$B(\mathbf{x}-\mathbf{i}|\mathbf{v}_1 \ldots \mathbf{v}_k), \quad \mathbf{i} \in \mathbb{Z}^s \ ,$$

is linearly dependent if
$|\det[\mathbf{v}_1 \dots \mathbf{v}_s]| \neq 1$. Since the ordering of the $\mathbf{v}_i$ does not matter, this sequence is also linearly dependent if there is any independent subsequence $\mathbf{v}_{i_1} \dots \mathbf{v}_{i_s}$ with

$$|\det[\mathbf{v}_{i_1} \dots \mathbf{v}_{i_s}]| \neq 1 \ .$$

The converse is also true [Dahmen & Micchelli '83, '85]. One can prove it, for example, by induction, see [Jia '83, '85]. We leave it as an exercise. In summary, we have the following theorem.

> $B(\mathbf{x}-\mathbf{i}|\mathbf{v}_1 \dots \mathbf{v}_k), \mathbf{i} \in \mathbb{Z}^s$, *is linearly independent over each open subset of* $\mathbb{R}^s$ *if and only if* $[\mathbf{v}_1 \dots \mathbf{v}_k]$ *is* **unimodular**,

which means that the determinant of any submatrix $[\mathbf{v}_{i_1} \dots \mathbf{v}_{i_s}]$ is $1, 0$ or -1.

If the directions $\mathbf{v}_1, \dots, \mathbf{v}_{k-1}$ span $\mathbb{R}^s$, then we can compute the **directional derivative** $D_{\mathbf{v}_k}\mathbf{s}$ of $\mathbf{s}$ with respect to $\mathbf{v}_k$. Using derivative formula (3) from 17.3, we obtain

$$D_{\mathbf{v}_k}\mathbf{s}(\mathbf{x}) = \sum_{\mathbf{i}\in\mathbb{Z}^s} \nabla_{\mathbf{v}_k}\mathbf{c}_\mathbf{i} B(\mathbf{x}-\mathbf{i}|\mathbf{v}_1 \dots \mathbf{v}_{k-1}) \ , \tag{5}$$

where $\nabla_\mathbf{v}\mathbf{c}_\mathbf{i} = \mathbf{c}_\mathbf{i} - \mathbf{c}_{\mathbf{i}-\mathbf{v}}$. Further, if, for all $j = 1, \dots, k$ the $k-1$, directions $\mathbf{v}_1, \dots, \mathbf{v}_j^*, \dots, \mathbf{v}_k$ span $\mathbb{R}^s$, then $B(\mathbf{x})$ is continuous, as shown in 17.3, and the span of its shifts contains the linear polynomials. In particular, if

$$\mathbf{m}_\mathbf{i} = \mathbf{i} + \frac{1}{2}(\mathbf{v}_1 + \cdots + \mathbf{v}_k)$$

is the **center** of supp$B(\mathbf{x}-\mathbf{i})$, then

$$\sum_{\mathbf{i}\in\mathbb{Z}^s} \mathbf{m}_\mathbf{i} B(\mathbf{x}-\mathbf{i}) = \mathbf{x} \ . \tag{6}$$

Namely, because of symmetry, this equation holds for $\mathbf{x} = \mathbf{o}$, and for all $j = 1, \dots, s$, we obtain

$$D_{\mathbf{v}_j} \sum \mathbf{m}_\mathbf{i} B(\mathbf{x}-\mathbf{i}) = \mathbf{v}_j \ .$$

Since the box spline representation is affinely invariant, we obtain for any linear polynomial $l(\mathbf{x})$ that

$$l(\mathbf{x}) = \sum l(\mathbf{m}_\mathbf{i}) B(\mathbf{x}-\mathbf{i}) \ .$$

This property is referred to as the **linear precision** of the box spline representation, see also Problems 1 and 2.

17.6 Subdivision for box spline surfaces

Any box $\beta = [\mathbf{u}_1 \ \dots \ \mathbf{u}_k][0,1)^k$ in $\mathbb{R}^k$ can be partitioned into 2^k translates of the scaled box $\hat{\beta} = \beta/2$ spanned by the half directions $\hat{\mathbf{u}}_i = \mathbf{u}_i/2$. Based on this observation, it is concluded in [Prautzsch '84a] that the non-normalized "shadow"

$$M_\beta(\mathbf{x}) = \mathrm{vol}_{k-s}(\pi^{-1}\mathbf{x} \cap \beta)$$

of β under the projection

$$\pi : [t_1 \dots t_k]^t \mapsto [t_1 \dots t_s]^t$$

can be written as a linear combination of translates of the scaled box spline

$$M_{\hat{\beta}}(\mathbf{x}) = 2^{s-k} M_\beta(2\mathbf{x}) \ .$$

Consequently, if the projections $\mathbf{v}_i = \pi \mathbf{u}_i$ lie in $\mathbb{Z}^s$, then any box spline surface

$$\mathbf{s}(\mathbf{x}) = \sum_{\mathbf{i} \in \mathbb{Z}^s} \mathbf{c}_\mathbf{i}^1 B(\mathbf{x} - \mathbf{i}) \ ,$$

with $B(\mathbf{x}) = B(\mathbf{x}|\mathbf{v}_1 \ \dots \ \mathbf{v}_k)$, has also a "finer" representation

$$\mathbf{s}(\mathbf{x}) \ = \ \sum_{\mathbf{i} \in \mathbb{Z}^s} \mathbf{c}_\mathbf{i}^2 B(2\mathbf{x} - \mathbf{i}) \ .$$

The new control points $\mathbf{c}_\mathbf{i}^2$ can be computed iteratively from the initial control points $\mathbf{c}_\mathbf{i}^1$ by the recursion formula

$$\begin{aligned}
\mathbf{d}^0(\mathbf{i}) &= \begin{cases} \mathbf{o} & \text{if } \ \mathbf{i}/2 \notin \mathbb{Z}^s \\ \mathbf{c}_{\mathbf{i}/2}^1 & \phantom{\text{if } \ \mathbf{i}/2} \in \mathbb{Z}^s \end{cases} \ , \\
\mathbf{d}^r(\mathbf{i}) &= \frac{1}{2}\Big(\mathbf{d}^{r-1}(\mathbf{i}) + \mathbf{d}^{r-1}(\mathbf{i} - \mathbf{v}_r)\Big) \ , \quad r = 1, \dots, k \ , \\
\mathbf{c}_\mathbf{i}^2 &= 2^s \mathbf{d}^k(\mathbf{i}) \ .
\end{aligned}$$

For a proof, we follow the ideas described in [Prautzsch '84a] and subdivide each box

$$\beta_{r-1} = [\hat{\mathbf{u}}_1 \ \dots \ \hat{\mathbf{u}}_{r-1} \, \mathbf{u}_r \ \dots \ \mathbf{u}_k][0,1)^k$$

into β_r and $\hat{\mathbf{u}}_r + \beta_r$. The associated shadows satisfy the equation

$$M_{\beta_{r-1}}(\mathbf{x}) = M_{\beta_r}(\mathbf{x}) + M_{\beta_r}(\mathbf{x} - \hat{\mathbf{v}}_r) \ ,$$

where $\hat{\mathbf{v}}_r = \mathbf{v}_r/2$. Dividing this equation by vol $\beta_{r-1} = 2\,\text{vol}\beta_r$, we obtain

$$B_{r-1}(\mathbf{x}) = \frac{1}{2}\Big(B_r(\mathbf{x}) + B_r(\mathbf{x} - \hat{\mathbf{v}}_r)\Big) \ ,$$

where $B_r(\mathbf{x}) = B(\mathbf{x}|\hat{\mathbf{v}}_1 \ldots \hat{\mathbf{v}}_r \mathbf{v}_{r+1} \ldots \mathbf{v}_k)$.
Using this identity repeatedly and making use of the relation

$$B(\mathbf{x}|\hat{\mathbf{v}}_1 \ldots \hat{\mathbf{v}}_k) = 2^s B(2\mathbf{x}|\mathbf{v}_1 \ldots \mathbf{v}_k) \ ,$$

we see that

$$\begin{aligned} \mathbf{s}(\mathbf{x}) &= \sum_{\mathbf{i}\in\mathbb{Z}^s} \mathbf{c}_\mathbf{i}^1 B_0(\mathbf{x} - \mathbf{i}) \\ &= \sum_{\mathbf{i}\in\mathbb{Z}^s} \mathbf{d}^r(\mathbf{i}) B_r(\mathbf{x} - \mathbf{i}/2) \ , \quad r = 0, 1, \ldots, k \ , \end{aligned}$$

which equals

$$\mathbf{s}(\mathbf{x}) = \sum_{\mathbf{i}\in\mathbb{Z}^s} \mathbf{c}_\mathbf{i}^2 B_0(2\mathbf{x} - \mathbf{i}) \ ,$$

and, thus, concludes the proof. ◇

Remark 4: The shifts $B_r(\mathbf{x} - \mathbf{j})$ sum to 2^s for $\mathbf{j} \in \mathbb{Z}^s/2$ and $r \geq s$ since

$$B_r(\mathbf{x} - \mathbf{j}) = 2^s B(2\mathbf{x} - 2\mathbf{j}|\mathbf{v}_1 \ldots \mathbf{v}_r\, 2\mathbf{v}_{r+1} \ldots 2\mathbf{v}_k) \ .$$

Remark 5: If $[\mathbf{v}_1 \ldots \mathbf{v}_s]\mathbb{Z}^s = \mathbb{Z}^s$, then $2^s \mathbf{d}_\mathbf{i}^s = \mathbf{c}_{\lfloor \mathbf{i}/2 \rfloor}^1$ and every point $\mathbf{c}_\mathbf{i}^2$ is a convex combination of the initial points $\mathbf{c}_\mathbf{i}^1$, see also Problem 7.

Remark 6: It is straightforward to generalize the subdivision algorithm to obtain, for any $m \in \mathbb{N}$, the finer representation

$$\mathbf{s}(\mathbf{x}) = \sum_{\mathbf{i}\in\mathbb{Z}^s} \mathbf{c}_\mathbf{i}^m B(m\mathbf{x} - \mathbf{i}) \ ,$$

where the "points" $\mathbf{c}_\mathbf{i}^m = m^s \mathbf{d}^k(\mathbf{i})$ can be computed iteratively by the formulae

$$\begin{aligned} \mathbf{d}^0(\mathbf{i}) &= \begin{cases} \mathbf{o} & \text{if } \ \mathbf{i}/m \notin \mathbb{Z}^s \\ \mathbf{c}_{\mathbf{i}/m}^1 & \text{if } \ \mathbf{i}/m \in \mathbb{Z}^s \end{cases} \quad \text{and} \\ \mathbf{d}^r(\mathbf{i}) &= \frac{1}{m}\sum_{l=0}^{m-1} \mathbf{d}^{r-1}(\mathbf{i} - l\mathbf{v}_r) \ , \quad r = 1, \ldots, k \ . \end{aligned}$$

In this form and in an even more general form, this algorithm can be found in [Cohen et al. '84] and [Dahmen & Micchelli '84], where it is derived algebraically.

Remark 7: The geometric derivation of the subdivision algorithm above shows that any new control point $\mathbf{c}_\mathbf{j}^m$ only depends on the control points $\mathbf{c}_\mathbf{i}^1$, where $\text{supp}B(m\mathbf{x}-\mathbf{j}) \subset \text{supp}B(\mathbf{x}-\mathbf{i})$. Their number is bounded by some h not depending on m and $\mathbf{j}$. Hence, we obtain

(7) $$\|\mathbf{c}_\mathbf{j}^m\| \le h \sup_{\mathbf{i}\in\mathbb{Z}^s} \|\mathbf{c}_\mathbf{i}^1\| \ .$$

Remark 8: The subdivision algorithm can be used a second time to compute control points of $\mathbf{s}(\mathbf{x})$ over any finer grid $\mathbb{Z}^s/(mn)$. Since the partition of a box β into translates of the scaled box $\beta/(mn)$ is unique, the same points $\mathbf{c}_\mathbf{i}^{mn}$ result that can be computed directly from the initial control points $\mathbf{c}_\mathbf{i}^1$ by one application of the subdivision algorithm.

Remark 9: Similarly, the points $\mathbf{c}_\mathbf{i}^m$ do not depend on the ordering of the directions $\mathbf{v}_1, \ldots, \mathbf{v}_k$, i.e., the ordering of the averaging steps.

17.7 Convergence under subdivision

Repeated subdivision by algorithm 17.6 produces the control points of the finer representation

$$\mathbf{s}(\mathbf{x}) = \sum_{\mathbf{i}\in\mathbb{Z}^s} \mathbf{c}_\mathbf{i}^m B(m\mathbf{x}-\mathbf{i}) \ , \quad \text{where} \quad B(\mathbf{x}) = B(\mathbf{x}|\mathbf{v}_1 \ldots \mathbf{v}_k) \ ,$$

of any surface $\mathbf{s}(\mathbf{x})$ over all scaled grids $\mathbb{Z}^s/m$, $m \in \mathbb{N}$. A major value of this procedure is that, under some reasonable conditions imposed on $\mathbf{v}_1, \ldots, \mathbf{v}_k$, the points $\mathbf{c}_\mathbf{i}^m$ converge to $\mathbf{s}(\mathbf{x})$. Let h be as in Remark (7) and

$$M = \sup\{\|\nabla_{\mathbf{v}_r}\mathbf{c}_\mathbf{i}^1\|, \text{ where } \mathbf{i} \in \mathbb{Z}^s \text{ and } \mathbf{v}_1, \ldots, \mathbf{v}_r^*, \ldots, \mathbf{v}_k \text{ span } \mathbb{R}^s\} \ .$$

Then, the following theorem holds.

If $[\mathbf{v}_1 \ldots \mathbf{v}_k]\mathbb{Z}^k = \mathbb{Z}^s$, *then* $\|\mathbf{c}_\mathbf{i}^m - \mathbf{s}(\mathbf{x})\| \le hM/m$ *for all* $\mathbf{i}$, *where* $B(m\mathbf{x}-\mathbf{i}) > 0$.

This result, in this generality, is due to de Boor et al. [de Boor et al. '93], who have also shown quadratic convergence when $[\mathbf{v}_1 \ldots \mathbf{v}_i^* \ldots \mathbf{v}_k]\mathbb{Z}^{k-1} = \mathbb{Z}^s$ and $B(\mathbf{x}|\mathbf{v}_1 \ldots \mathbf{v}_k)$ is differentiable.

For a proof, we enumerate the directions $\mathbf{v}_1, \ldots, \mathbf{v}_k$ such that $\mathbf{v}_1, \ldots, \mathbf{v}_r^*, \ldots, \mathbf{v}_k$ span $\mathbb{R}^s$ if and only if $r \ge q$ for some number $q \in \{1, \ldots, k+1\}$. Then,

$\mathbb{R}^s$ is the direct sum

$$\mathbb{R}^s = \mathbf{v}_1\mathbb{R} \oplus \cdots \oplus \mathbf{v}_{q-1}\mathbb{R} \oplus [\mathbf{v}_q \ldots \mathbf{v}_k]\mathbb{R}^{k-q} \ .$$

Further, any surface point $\mathbf{s}(\mathbf{x})$ lies in the convex hull of all control points $\mathbf{c}_\mathbf{i}^m$, where m is fixed and $B(m\mathbf{x} - \mathbf{i}) > 0$. We show that the diameters of these convex hulls shrink to zero.

Since $[\mathbf{v}_1 \ldots \mathbf{v}_k]\mathbb{Z}^k = \mathbb{Z}^s$ and since span $[\mathbf{v}_1 \ldots \mathbf{v}_r^* \ldots \mathbf{v}_k] \neq \mathbb{R}^s$ for $r = 1, \ldots, q-1$, it follows from the fact $B(m\mathbf{x} - \mathbf{i}) > 0$ that $B(m\mathbf{x} - \mathbf{k}) = 0$ for all $\mathbf{k} \in \mathbb{Z}^s$, with $\mathbf{i} - \mathbf{k} \notin \text{span}[\mathbf{v}_q \ldots \mathbf{v}_k]$.

Consequently, any two points $\mathbf{c}_\mathbf{i}^m$ and $\mathbf{c}_\mathbf{k}^m$ that $\mathbf{s}(\mathbf{x})$ depends on satisfy $\mathbf{i} - \mathbf{k} =: \mathbf{v} \in \text{span}[\mathbf{v}_q \ldots \mathbf{v}_k]$, and $+\mathbf{v}$ or $-\mathbf{v}$ lie in supp$B(\mathbf{x})$.

Since $[\mathbf{v}_1 \ldots \mathbf{v}_k]\mathbb{Z}^k = \mathbb{Z}^s$ and since $\mathbb{Z}^s$ is the direct sum

$$\mathbf{v}_1\mathbb{Z} \oplus \cdots \oplus \mathbf{v}_{q-1}\mathbb{Z} \oplus [\mathbf{v}_{q-1} \ldots \mathbf{v}_k]\mathbb{Z}^{k-q} \ ,$$

any $\mathbf{v} \in \mathbb{Z}^s \cap \text{span}[\mathbf{v}_q \ldots \mathbf{v}_k]$ can be written as a sum of possibly repeated vectors from the set $\{\mathbf{v}_q, -\mathbf{v}_q, \ldots, \mathbf{v}_k, -\mathbf{v}_k\}$. Therefore, any difference $\nabla_\mathbf{v}\mathbf{c}_\mathbf{i}^m$ can be expressed as a sum of differences $\nabla_{\mathbf{v}_r}\mathbf{c}_\mathbf{i}^m$, $r = q, \ldots, k$. Thus, bounding the differences $\nabla_{\mathbf{v}_r}\mathbf{c}_\mathbf{i}^m$ is all we need to do to finish the proof.

For $r \geq q$, we can drop the rth averaging step of the subdivision algorithm 17.6 and obtain, up to the factor m, the $\nabla_{\mathbf{v}_r}$-difference scheme, see 15.5 and 15.6. This means that, if we apply the subdivision algorithm 17.6 without the rth averaging step to the differences $\nabla_{\mathbf{v}_r}\mathbf{c}_\mathbf{i}^1$, then we obtain the divided differences $m\nabla_{\mathbf{v}_r}\mathbf{c}_\mathbf{i}^m$. Hence, it follows from (7) that

$$\sup_{\mathbf{i}\in\mathbb{Z}^s} \|\nabla_{\mathbf{v}_r}\mathbf{c}_\mathbf{i}^m\| \leq hM/m \ ,$$

which concludes the proof. ◇

On the other hand, if $[\mathbf{v}_1 \ldots \mathbf{v}_k]\mathbb{Z}^k \neq \mathbb{Z}^s$, then convergence cannot be proved as we show for $s(\mathbf{x}) = B(\mathbf{x})$:

If some grid point $\mathbf{i} \in \mathbb{Z}^s$ exists that does not lie in $[\mathbf{v}_1 \ldots \mathbf{v}_k]\mathbb{Z}^k$, then none of the grid points $\mathbf{j} \in J := \mathbf{i} + [\mathbf{v}_1 \ldots \mathbf{v}_s]\mathbb{Z}^s$ lies in $[\mathbf{v}_1 \ldots \mathbf{v}_k]\mathbb{Z}^k$ also. Since $\sum_{\mathbf{j}\in J} B(\mathbf{x} - \mathbf{j}) > 0$, see (4) in 17.6, there exists, for every $\mathbf{x} \in \mathbb{R}^s$, some $\mathbf{j} \in J$ such that $B(\mathbf{x} - \mathbf{j}) > 0$.

Subdividing the single box spline $B(\mathbf{x})$ by the algorithm in 17.5 produces control points $c_\mathbf{i}^m$ such that

$$B(\mathbf{x}) = \sum_{\mathbf{i}\in\mathbb{Z}^s} c_\mathbf{i}^m B(m\mathbf{x} - \mathbf{i}) \ .$$

The geometric derivation of the algorithm in 17.6 shows that $c_{\mathbf{i}}^m \neq 0$ if and only if $\mathbf{i} \in [\mathbf{v}_1 \dots \mathbf{v}_k]\mathbb{Z}_m^k$.

Thus, if $[\mathbf{v}_1 \dots \mathbf{v}_k]\mathbb{Z}^k \neq \mathbb{Z}^s$ or $J \neq \emptyset$, there is for every $\mathbf{x} \in \mathbb{R}^s$ some zero control point $c_{\mathbf{j}}^m = 0$ with $B(m\mathbf{x} - \mathbf{j}) > 0$. For all $\mathbf{x}$, where $B(\mathbf{x}) > 0$, these zero control points do not converge to $B(\mathbf{x})$, as m tends to infinity.

17.8 Half-box splines

Half-box splines are defined over the triangular grid spanned by $\mathbf{e}_1 = [1\ 0]^t$, $\mathbf{e}_2 = [0\ 1]^t$ and $\mathbf{e}_3 = [1\ 1]^t$. In 15.10, they come up in the context of the subdivision algorithm for hexagonal nets. Here, we provide the inductive definition, which is due to Sabin [Sabin '77], and its geometric interpretation.

Splitting the unit square along its diagonal in direction $\mathbf{e}_3$ produces the two triangles

$$\Delta = \{\mathbf{x} | 0 \leq x \leq y < 1\} \quad \text{and} \quad \nabla = \{\mathbf{x} | 0 \leq y < x < 1\} \ .$$

These are the supports of the two piecewise constant half-box splines

$$H_\Delta(\mathbf{x}|) = \begin{cases} 1 & \text{if} \quad \mathbf{x} \in \Delta \\ 0 & \text{otherwise} \end{cases}$$

and

$$H_\nabla(\mathbf{x}|) = \begin{cases} 1 & \text{if} \quad \mathbf{x} \in \nabla \\ 0 & \text{otherwise} \end{cases} \ .$$

As with box splines, we obtain half-box splines of higher order by successive convolutions, given by

$$H_\Delta(\mathbf{x}|\mathbf{v}_3 \dots \mathbf{v}_k) = \int_0^1 H_\Delta(\mathbf{x} - t\mathbf{v}_k|\mathbf{v}_3 \dots \mathbf{v}_{k-1})dt$$

and

$$H_\nabla(\mathbf{x}|\mathbf{v}_3 \dots \mathbf{v}_k) = \int_0^1 H_\nabla(\mathbf{x} - t\mathbf{v}_k|\mathbf{v}_3 \dots \mathbf{v}_{k-1})dt \ ,$$

where $k \geq 3$. We assume that the directions $\mathbf{v}_i$ are the unit directions,

$$\mathbf{v}_3, \dots, \mathbf{v}_k \in \{\mathbf{e}_1, \mathbf{e}_2, \mathbf{e}_3\} \ .$$

Any two half-box splines $H_\Delta(\mathbf{x}|\mathbf{v}_3 \dots \mathbf{v}_k)$ and $H_\nabla(\mathbf{x}|\mathbf{v}_3 \dots \mathbf{v}_k)$ sum to the box spline $B(\mathbf{x}|\mathbf{e}_1\mathbf{e}_2\mathbf{v}_3 \dots \mathbf{v}_k)$ even for $k = 2$. The two piecewise cubic C^1 half-box splines $H_{\dots}(\mathbf{x}|\mathbf{e}_1\mathbf{e}_2\mathbf{e})$ are shown in Figure 17.6.

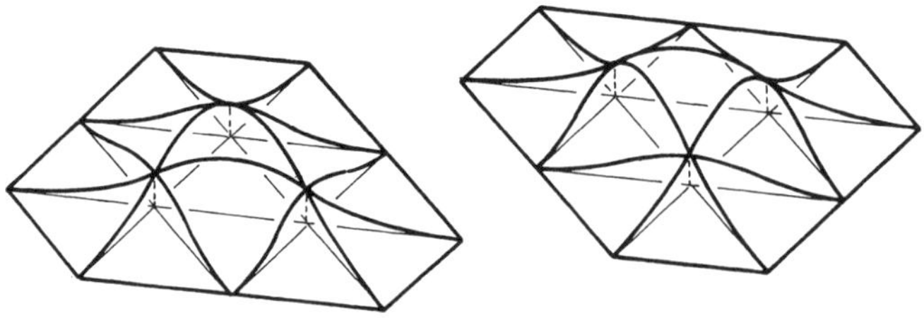

Figure 17.6: The two piecewise cubic C^1half-box splines.

As done for box splines in 17.1, we can derive the analogous properties for half-box splines. Because of symmetry, it suffices to list these properties for $H(\mathbf{x}) = H_\Delta(\mathbf{x}|\mathbf{v}_3 \ldots \mathbf{v}_k)$.

The half-box splines are **normalized** such that

$$\int_{\mathbb{R}^2} H(\mathbf{x}) d\mathbf{x} = 1/2 \ .$$

Any k independent directions $\mathbf{u}_1, \ldots, \mathbf{u}_k \in \mathbb{R}^k$ define a half-box

$$\vartheta = \{\sum \mathbf{u}_i \alpha_i | 0 \le \alpha_1 \le \alpha_2 \quad \text{and} \quad \alpha_2, \ldots, \alpha_k \in [0,1]\} \ .$$

The **density of a shadow** of this half-box represents a half-box spline: If π denotes the orthogonal projection from $\mathbb{R}^k$ onto $\mathbb{R}^2$ mapping $\mathbf{u}_1, \ldots, \mathbf{u}_k$ to $\mathbf{e}_1, \mathbf{e}_2, \mathbf{v}_3, \ldots, \mathbf{v}_k$, then

$$H(\mathbf{x}|\mathbf{v}_1 \ \ldots \ \mathbf{v}_k) = \frac{1}{2 \operatorname{vol}_k \vartheta} \operatorname{vol}_{k-2}(\pi^{-1}\mathbf{x} \cap \vartheta) \ .$$

From this geometric construction it follows that $H(\mathbf{x})$

- *does not depend on the* **ordering** *of* $\mathbf{v}_3 \ \ldots \ \mathbf{v}_k$,
- *is* **positive** *over the convex set* $\Delta + [\mathbf{v}_3 \ldots \mathbf{v}_k](0,1)^{k-2}$,
- *has the* **support** closure$(\Delta) + [\mathbf{v}_3 \ldots \mathbf{v}_k][0,1]^{k-2}$,
- *has the* **directional derivative**

  $$D_{\mathbf{v}_r} H(\mathbf{x}) = H(\mathbf{x}|\mathbf{v}_3 \ \ldots \ \mathbf{v}_r^* \ \ldots \ \mathbf{v}_k) - H(\mathbf{x} - \mathbf{v}_r|\mathbf{v}_3 \ \ldots \ \mathbf{v}_r^* \ \ldots \ \mathbf{v}_k)$$

  with respect to $\mathbf{v}_r$, $r \ge 3$,
- *is* r *times* **continuously differentiable**, *provided that all subsets of* $\{\mathbf{v}_1, \ \ldots \ , \mathbf{v}_k\}$ *obtained by deleting* $r+1$ *vectors* $\mathbf{v}_i$ *span* $\mathbb{R}^2$,

- *is polynomial of total degree* $\leq k-2$ *over each triangle* $\mathbf{i}+\Delta$ *and* $\mathbf{i}+\nabla, \mathbf{i} \in \mathbb{Z}^2$.

Remark 10: The half-box splines $H_\Delta(\mathbf{x}|\mathbf{e}_1 \overset{k}{\ldots} \mathbf{e}_1\mathbf{e}_2 \overset{k}{\ldots} \mathbf{e}_2\mathbf{e}_3 \overset{k}{\ldots} \mathbf{e}_3)$ are $2k-1$ times continuously differentiable and of polynomial degree $\leq 3k$.

17.9 Half-box spline surfaces

Any pair of half-box splines

$$\begin{aligned} H_\Delta(\mathbf{x}) &= H(\mathbf{x}|\mathbf{e}_3\mathbf{v}_3 \ldots \mathbf{v}_k) \ , \\ H_\nabla(\mathbf{x}) &= H(\mathbf{x}|\mathbf{e}_3\mathbf{v}_3 \ldots \mathbf{v}_k) \end{aligned}$$

sums to the box spline $B(\mathbf{x}|\mathbf{e}_1\mathbf{e}_2\mathbf{e}_3\mathbf{v}_3 \ldots \mathbf{v}_k)$. Hence, the translates $H_\Delta(\mathbf{x}-\mathbf{i})$ and $H_\nabla(\mathbf{x}-\mathbf{i})$, $\mathbf{i} \in \mathbb{Z}^2$, form a **partition of unity**.

Consequently, any **half-box spline surface**

$$\mathbf{s}(\mathbf{x}) = \sum_{\mathbf{i} \in \mathbb{Z}^2} \left(\mathbf{c}_\mathbf{i}^\Delta H_\Delta(\mathbf{x}-\mathbf{i}) + \mathbf{c}_\mathbf{i}^\nabla H_\nabla(\mathbf{x}-\mathbf{i})\right)$$

is an affine combination of its **control points** and this representation is **affinely invariant**, i.e., under any affine map the images of the control points control the image of $\mathbf{s}(\mathbf{x})$.

Since the half-box splines are non-negative, $\mathbf{s}(\mathbf{x})$ is even a convex combination of its control points and lies in their **convex hull**.

If we connect control points $\mathbf{c}_\mathbf{i}^\Delta$ and $\mathbf{c}_\mathbf{j}^\nabla$ whose associated triangles $\mathbf{i}+\Delta$ and $\mathbf{j}+\nabla$ have a common edge, then we obtain a hexagonal net, the control net of $\mathbf{s}$. An example is illustrated schematically in Figure 17.7

The **directional derivative** of $\mathbf{s}$ with respect to $\mathbf{v}_r$ can be computed by the derivative formula given in 17.8 for a single half-box spline. It is given by

$$D_{\mathbf{v}_r}\mathbf{s}(\mathbf{x}) = \\ \textstyle\sum_{\mathbf{i} \in \mathbb{Z}^s} \left(\nabla_{\mathbf{v}_r}\mathbf{c}_\mathbf{i}^\Delta H_\Delta(\mathbf{x}|\mathbf{v}_3 \ldots \mathbf{v}_r^* \ldots \mathbf{v}_k) + \nabla_{\mathbf{v}_r}\mathbf{c}_\mathbf{i}^\nabla H_\nabla(\mathbf{x}|\mathbf{v}_3 \ldots \mathbf{v}_r^* \ldots \mathbf{v}_k)\right) \ ,$$

where $\nabla_\mathbf{v}\mathbf{c}_\mathbf{i} = \mathbf{c}_\mathbf{i} - \mathbf{c}_{\mathbf{i}-\mathbf{v}}$.

If $H_\Delta(\mathbf{x})$ is continuous or, equivalently, if there exist two independent directions among $\mathbf{v}_3, \ldots, \mathbf{v}_k$, then all directional derivatives of the sum of all shifts, $\sum H_\Delta(\mathbf{x}-\mathbf{i})$, are zero. Therefore, this sum is a constant function. Because of symmetry, and since the shifts of both half-box splines H_Δ and

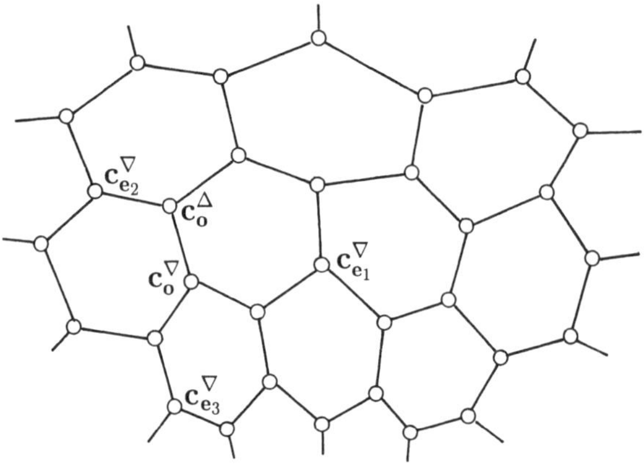

Figure 17.7: A hexagonal control net.

H_∇ form a partition of unity, we obtain

$$\sum_{i\in\mathbb{Z}^2} H_\Delta(\mathbf{x}-\mathbf{i}) = \sum_{i\in\mathbb{Z}^2} H_\nabla(\mathbf{x}-\mathbf{i}) = 1/2 \ . \tag{8}$$

In particular, this implies that the shifts of H_Δ and H_∇ are **linearly dependent**.

Further, if the box spline

$$B(\mathbf{x}) = B(\mathbf{x}|\mathbf{e}_1\mathbf{e}_2\mathbf{v}_3 \ldots \mathbf{v}_k) = H_\Delta(\mathbf{x}) + H_\nabla(\mathbf{x})$$

is continuous, then we recall from (6) in 17.5 that

$$\sum_{i\in\mathbb{Z}^2} \mathbf{m}_\mathbf{i}\,(H_\Delta(\mathbf{x}-\mathbf{i}) + H_\nabla(\mathbf{x}-\mathbf{i})) = \mathbf{x} \ ,$$

where $\mathbf{m}_\mathbf{i}$ is the center of supp$B(\mathbf{x}-\mathbf{i})$. If H_Δ is continuous, we can use (8) and obtain for any $\mathbf{v} \in \mathbb{R}^2$

$$\sum_{i\in\mathbb{Z}^2} ((\mathbf{m}_\mathbf{i}+\mathbf{v})H_\Delta(\mathbf{x}-\mathbf{i}) + (\mathbf{m}_\mathbf{i}-\mathbf{v})H_\nabla(\mathbf{x}-\mathbf{i})) = \mathbf{x} \ .$$

For example, if $\mathbf{v} = (\mathbf{e}_2 - \mathbf{e}_1)/6$, then the points $\mathbf{m}_\mathbf{i}^\Delta = \mathbf{m}_\mathbf{i} + \mathbf{v}$ and $\mathbf{m}_\mathbf{i}^\nabla = \mathbf{m}_\mathbf{i} - \mathbf{v}$ form a regular hexagonal grid, as illustrated in Figure 17.8.

Since the half-box spline representation is affinely invariant, we obtain for

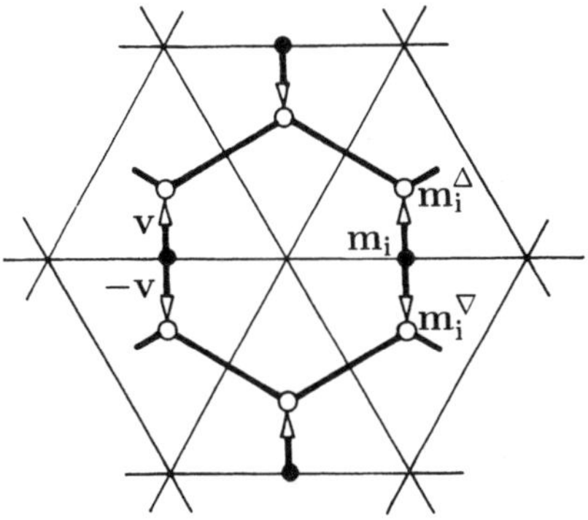

Figure 17.8: The hexagonal grid of the "centers" $\mathbf{m}_i^\Delta$ and $\mathbf{m}_i^\nabla$.

any linear polynomial $l(\mathbf{x})$ the half-box spline representation

$$l(\mathbf{x}) = \sum_{\mathbf{i}\in\mathbb{Z}^2} \left(l(\mathbf{m}_i^\Delta) H_\Delta(\mathbf{x}-\mathbf{i}) + l(\mathbf{m}_i^\nabla) H_\nabla(\mathbf{x}-\mathbf{i}) \right) .$$

This property is referred to as the **linear precision** of the half-box spline representation.

Remark 11: Any half-box spline surface

$$s(\mathbf{x}) = \sum_{\mathbf{i}\in\mathbb{Z}^2} \left(\mathbf{c}_i^\Delta H_\Delta(\mathbf{x}-\mathbf{i}) + \mathbf{c}_i^\nabla H_\nabla(\mathbf{x}-\mathbf{i}) \right)$$

has also a "finer" representation

$$s(\mathbf{x}) = \sum_{\mathbf{i}\in\mathbb{Z}^2} \left(\mathbf{d}_i^\Delta H_\Delta(m\mathbf{x}-\mathbf{i}) + \mathbf{d}_i^\nabla H_\nabla(m\mathbf{x}-\mathbf{i}) \right)$$

for any $m \in \mathbb{N}$. In particular, for $m = 2^k, k \in \mathbb{N}$, the new control points $\mathbf{d}_i^\Delta$ and $\mathbf{d}_i^\nabla$, which depend on m, can be computed by k repeated applications of the subdivision algorithm 15.9. Similar to the subdivision algorithm for box splines as described in Section 17.6, this algorithm has an obvious generalization that generates the points $\mathbf{d}_i^\Delta$ and $\mathbf{d}_i^\nabla$ for any arbitrary $m \in \mathbb{N}$. We leave it as an exercise to work out this generalization.

17.10 Problems

1 Let the directions $\mathbf{v}_1, \ldots, \mathbf{v}_k \in \mathbb{Z}^s$ span $\mathbb{R}^s$, and assume that the associated box spline $B(\mathbf{x}) = B(\mathbf{x}|\mathbf{v}_1 \ldots \mathbf{v}_k)$ is r times continuously differentiable. Use the derivative formula (5) given in 17.5 to show, by induction over k, that for any polynomial $c(\mathbf{x})$ of total degree $d \leq r+1$ the spline

$$s(\mathbf{x}) = \sum_{\mathbf{i} \in \mathbb{Z}^s} c(\mathbf{i}) B(\mathbf{x} - \mathbf{i})$$

is also a polynomial of degree d.

2 Continuing with Problem 1, show that the map $c(\mathbf{x}) \mapsto \sum c(\mathbf{i}) B(\mathbf{x} - \mathbf{i})$ is a regular linear map on the space of all polynomials of degree $\leq r+1$.

3 Show by induction that the box splines $B(\mathbf{x}-\mathbf{i}|\mathbf{v}_1 \ldots \mathbf{v}_k)$, for $\mathbf{i} \in \mathbb{Z}^s$, are linearly independent if $[\mathbf{v}_1 \ldots \mathbf{v}_k]$ is unimodular.

4 Verify the recursion formula for box splines

$$\begin{aligned} B(\mathbf{x}|\mathbf{v}_1 \ldots \mathbf{v}_k) &= \frac{1}{k-s} \sum_{r=1}^{k} \big(\alpha_r B(\mathbf{x}|\mathbf{v}_1 \ldots \mathbf{v}_r^* \ldots \mathbf{v}_k) \\ &\quad + (1-\alpha_r) B(\mathbf{x} - \mathbf{v}_r|\mathbf{v}_1 \ldots \mathbf{v}_r^* \ldots \mathbf{v}_k) \big) , \end{aligned}$$

where $\mathbf{x} = \sum_{r=1}^{k} \alpha_r \mathbf{v}_r$. This formula is due to de Boor and Höllig [de Boor & Höllig '82]. A geometric proof and a box spline evaluation algorithm similar to de Boor's algorithm can be found in [Boehm '84a].

5 Given the hexagonal control net of a cubic C^1 half-box spline surface, one can compute its Bézier points by the three masks shown in Figure 17.9 and symmetric versions of these masks. The figure shows schematically the relevant part of the hexagonal net and one Bézier triangle where the solid dot depicts the Bézier point computed by the mask. Use these masks to compute the Bézier representation of a single cubic half-box spline and verify its correctness.

6 Given a quartic C^2-box spline surface $\sum_{\mathbf{i} \in \mathbb{Z}^2} \mathbf{c_i} B(\mathbf{x} - \mathbf{i}|\mathbf{e}_1\mathbf{e}_1\mathbf{e}_2\mathbf{e}_2\mathbf{e}_3\mathbf{e}_3)$ over $\mathbb{R}^2$, one can compute its Bézier net as follows. First, the control net is subdivided twice by the refinement operator $\mathcal{R}$, see Remark 10 in 15.7. Second, the mask shown in Figure 17.10 is used to compute the Bézier points from the subdivided net [Boehm '83, Prautzsch & Boehm '02].

7 Show that the control points $\mathbf{c}_\mathbf{i}^m$ computed by the subdivision algorithm 17.6, Remark 5, are convex combinations of the initial control points $\mathbf{c}_\mathbf{i}^1$ when

$$[\mathbf{v}_1 \ldots \mathbf{v}_k]\mathbb{Z}^k = \mathbb{Z}^s .$$

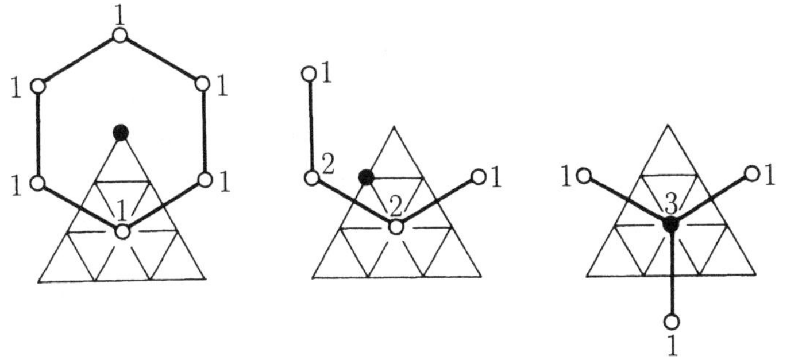

Figure 17.9: Masks to compute Bézier points.

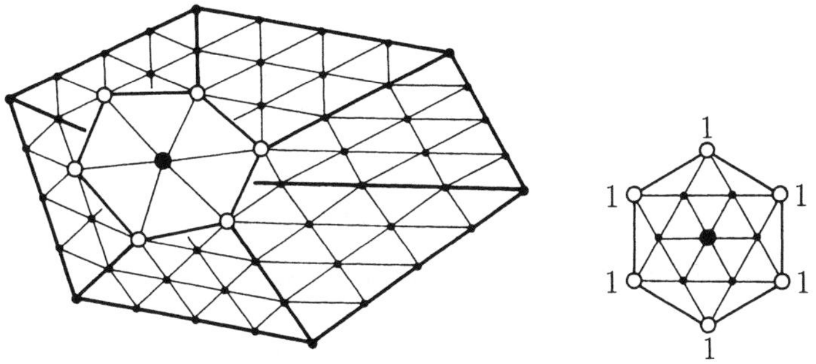

Figure 17.10: Computing the Bézier points of a quartic C^2-box spline surface.

8 Let $\mathbf{v}_1, \ldots, \mathbf{v}_k$ be directions in $\mathbb{R}^2$ such that the angles between $\mathbf{v}_1$ and $\mathbf{v}_r$ increase with $r = 1, \ldots, k$ and are all less than 180°. Show that the boundary of the support of the box spline $B(\mathbf{x}|\mathbf{v}_1 \ldots \mathbf{v}_k)$ is formed by the points

$$\mathbf{x} = \mathbf{v}_1 + \cdots + \mathbf{v}_{r-1} + \mathbf{v}_r \alpha$$

and

$$\mathbf{x} = \mathbf{v}_k + \cdots + \mathbf{v}_{r+1} + \mathbf{v}_r \alpha \ ,$$

where $r = 1, \ldots, k$ and $\alpha \in [0, 1]$, see Figure 17.2.

9 Let S be the support of any, for simplicity, bivariate box spline. Show that the intersection of S with any translate of S forms the support of some box spline.

10 Let $k = 4$ and $[\mathbf{v}_1 \ldots \mathbf{v}_k] = \begin{bmatrix} 1 & 0 & 1 & -1 \\ 0 & 1 & 1 & 1 \end{bmatrix}$. Show that the points $\mathbf{c}^2_\mathbf{i}$

obtained from the subdivision algorithm in 17.6 can also be computed by the following simple averaging scheme illustrated in Figure 17.11.

1. Compute the midpoints of all edges $\mathbf{c}_\mathbf{i}^1\,\mathbf{c}_{\mathbf{i}-\mathbf{v}_1}^1$ and $\mathbf{c}_\mathbf{i}^1\,\mathbf{c}_{\mathbf{i}-\mathbf{v}_2}^1$.
2. Connect the midpoints of any two edges with a common end point by a "new" edge.
3. The midpoints of the new edges are the points $\mathbf{c}_\mathbf{i}^2$.

This scheme is due to Sabin [Sabin '86] and can also be applied to arbitrary control nets, see [Peters & Reif '97].

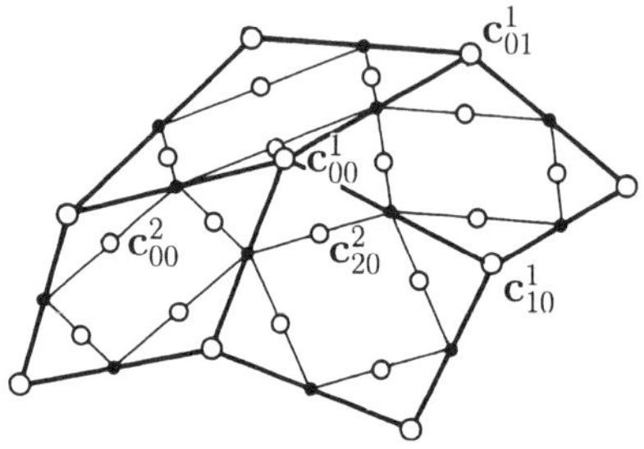

Figure 17.11: The simplest subdivision scheme.

11 Derive, geometrically or analytically, the subdivision algorithm for half-box spline surfaces, see Remark 11 in 17.9, for arbitrary m.

18 Simplex splines

The B-splines introduced in 5 can also be constructed by projecting simplices onto the real line. The density function of a simplex shadow is a B-spline. Schoenberg found this geometric interpretation in 1965, and de Boor used it in 1976 to define multivariate B-splines, see [de Boor '76b]. This beautiful geometric construction of B-splines allows us to immediately see or to derive in a straightforward manner properties, such as smoothness and recursion, the knot insertion and degree elevation formulas.

18.1 Shadows of simplices

To construct multivariate splines, we use the projection π from $\mathbb{R}^k$ onto $\mathbb{R}^s$ given by

$$\pi[x_1 \dots x_k]^t = [x_1 \dots x_s]^t \ .$$

The **fibers** of π, which are the sets of points mapped to single points $\mathbf{x}$, i.e.,

$$\pi^{-1}\mathbf{x} = \{\mathbf{z} \in \mathbb{R}^k | \pi\mathbf{z} = \mathbf{x}\}, \quad \mathbf{x} \in \mathbb{R}^s \ ,$$

form $k-s$-dimensional parallel subspaces of $\mathbb{R}^k$.

Let $\mathbf{p}_0, \dots, \mathbf{p}_k$ be $k+1$ independent points in $\mathbb{R}^k$. Their convex hull forms a k-dimensional **simplex** σ. The shadow of σ in $\mathbb{R}^s$ with respect to π defines an s-variate **simplex spline**, given by

$$M_\sigma(\mathbf{x}) = \mathrm{vol}_{k-s}(\sigma \cap \pi^{-1}\mathbf{x}) \ .$$

For $k = 3$ this geometric construction is illustrated in Figures 18.1 and 18.2

with $s = 1$ and $s = 2$, respectively.

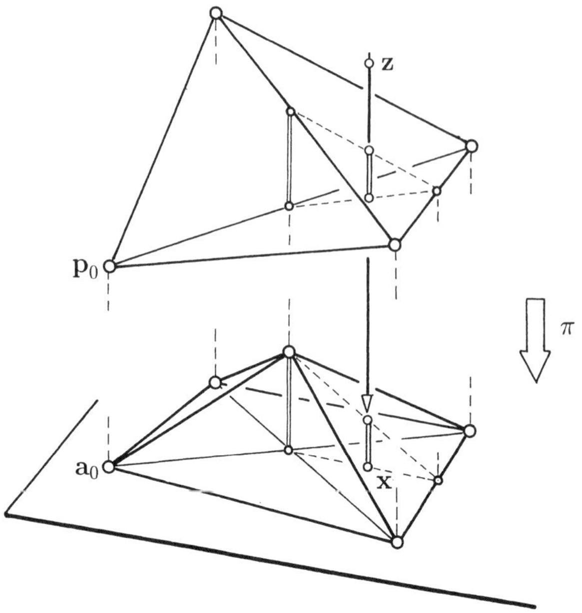

Figure 18.1: The shadow of a tetrahedron over a plane.

Thus, $M_\sigma(\mathbf{x})$ is the $k-s$-dimensional volume of a slice of σ. In particular, if $k = s$, then M_σ is the **characteristic function** of σ, which is defined as

$$M_\sigma(\mathbf{x}) = \chi_\sigma(\mathbf{x}) := \begin{cases} 1 & \text{if} \quad \mathbf{x} \in \sigma \\ 0 & \text{otherwise} \end{cases} \quad .$$

If $k = s + 1$, then M_σ is piecewise linear.

18.2 Properties of simplex splines

Obviously, any simplex spline M_σ is non-negative, and the projection $\pi\sigma$ of σ is its **support**. Further, M_σ is continuous over its support and, if it is zero on the boundary of its support, also over $\mathbb{R}^s$.

The vertices $\mathbf{p}_i$ of σ are projected to the points $\mathbf{a}_i = \pi\mathbf{p}_i$, which are called the **knots** of M_σ. We call the convex hull of any m knots an **m-knot**. In particular, $\pi\sigma$ is a $(k+1)$-knot, and if there is an $s-1$-dimensional k-knot, it lies on the boundary of $\pi\sigma$. Further,

M_σ is continuous everywhere over $\mathbb{R}^s$, except at an $s-1$-dimensional

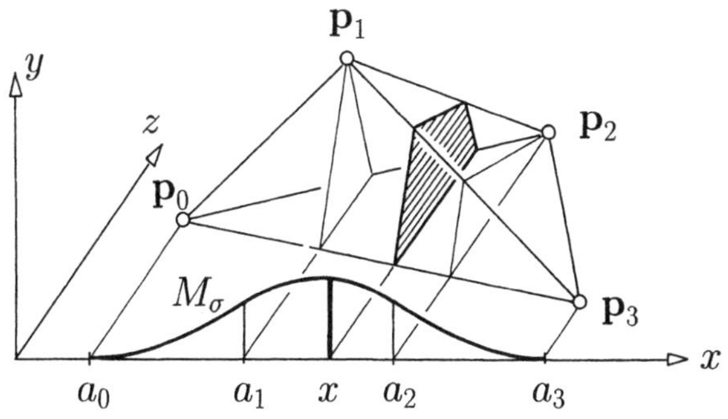

Figure 18.2: The shadow of a tetrahedron over a line.

k-knot.

For a proof, let $\mathbf{a}_1, \ldots, \mathbf{a}_m$ determine an $s-1$-dimensional m-knot on the boundary of $\pi\sigma = \text{supp}M_\sigma$, and suppose that the affine hull of this m-knot does not contain any other knot. Let ρ be the corresponding simplex $\mathbf{p}_1 \ldots \mathbf{p}_m$. Since $\pi\rho$ lies on the boundary of $\pi\sigma$, it follows for all $\mathbf{x} \in \pi\rho$ that

$$\pi^{-1}\mathbf{x} \cap \rho = \pi^{-1}\mathbf{x} \cap \sigma \ .$$

Since $\pi^{-1}\mathbf{x} \cap \rho$ is $(m-1)-(s-1)$- or $m-s$-dimensional, $M_\sigma(\mathbf{x})$ is non-zero over $\pi\rho$ if and only if $m = k$. $\diamond$

The (non-degenerate) s-knots partition the support of M_σ. Examples are shown in Figure 18.3 for $k = 4$ and $s = 2$.

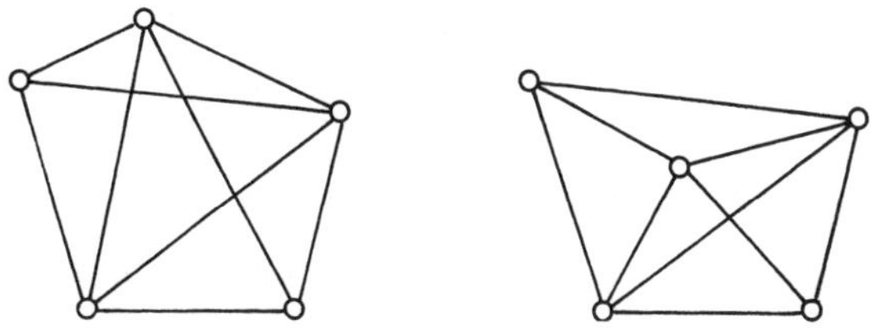

Figure 18.3: Support knots and 2-knots of bivariate quadratic simplex splines.

We will see that M_σ, when restricted to any tile of this partition, is a polynomial of total degree $\leq k-s$ and that M_σ is $k-m-1$ times continuously differentiable over any s-knot whose affine hull contains m knots. Thus, if the knots are in general position, then M_σ is $k-s-1$ times differentiable.

Remark 1: For all continuous functions f on $\mathbb{R}^s$, the equation

$$\int_{\mathbb{R}^s} M_\sigma(\mathbf{x}) f(\mathbf{x}) d\mathbf{x} = \int_\sigma f(\pi \mathbf{y}) d\mathbf{y}$$

holds. Thus, M_σ can also be defined by requiring that this equation hold for all f, see [Micchelli '80] and Remark 1 in 17.2.

18.3 Normalized simplex splines

An s-variate simplex spline M_σ depends on a k-dimensional simplex σ. However, we would rather get rid of this dependency and specify an s-variate spline only by its knots in $\mathbb{R}^s$. Fortunately, it is possible to do this.

If we consider a second simplex ρ in $\mathbb{R}^k$ whose vertices are projected onto the same knots $\mathbf{a}_i$ as the vertices $\mathbf{p}_i$ of σ, then the affine map φ mapping σ onto ρ satisfies $\pi \circ \varphi = \pi$. Hence, φ maps any slice $\sigma_\mathbf{x} = \sigma \cap \pi^{-1}\mathbf{x}$ to the slice $\rho_\mathbf{x} = \rho \cap \pi^{-1}\mathbf{x}$.

Any two slices $\sigma_\mathbf{x}$ and $\sigma_\mathbf{y}$ are parallel. Consequently, the ratio of their volumes is preserved under φ, which implies that the ratios

$$M_\sigma(\mathbf{x}) : M_\sigma(\mathbf{y}) = M_\rho(\mathbf{x}) : M_\rho(\mathbf{y})$$

and

$$M_\sigma(\mathbf{x}) : M_\rho(\mathbf{x}) = M_\sigma(\mathbf{y}) : M_\rho(\mathbf{y})$$

are constant for all $\mathbf{x}$.

Since

$$\int_{\mathbb{R}^s} M_\sigma(\mathbf{x}) d\mathbf{x} = \mathrm{vol}_k \sigma \ ,$$

we finally conclude that the **normalized simplex spline**

$$M(\mathbf{x}|\mathbf{a}_0 \dots \mathbf{a}_k) = M_\sigma(\mathbf{x}) / \mathrm{vol}_k \sigma$$

depends only on the knots $\mathbf{a}_0, \dots, \mathbf{a}_k$, but not on σ.

Remark 2: The integral of a normalized simplex spline is one, i.e.,

$$\int_{\mathbb{R}^s} M(\mathbf{x}|\mathbf{a}_0 \dots \mathbf{a}_k) d\mathbf{x} = 1 \ .$$

Remark 3: The smoothness properties mentioned in 18.2 characterize a univariate simplex spline $M(x|a_0 \dots a_k)$ as being some multiple of the ordinary B-spline $N(x|a_0 \dots a_k)$ with the same knots, see 5.4. Hence, we conclude, from Problem 2 in 5.12, that

$$M(x|a_0 \ldots a_k) = \frac{k}{a_k - a_0} N(x|a_0 \ldots a_k) \ ,$$

where we assume that $[a_0, a_k]$ is the support.

18.4 Knot insertion

The knot insertion formula for univariate B-splines discussed in 6.1 corresponds to a simple subdivision of the generating simplex and can be generalized.

Let σ be the simplex $\mathbf{p}_0 \ldots \mathbf{p}_k$ and let $\mathbf{a}_i = \pi \mathbf{p}_i$ be the corresponding knots in $\mathbb{R}^s$. Further, let

$$\mathbf{a}_{k+1} = \sum_{i=0}^{k} \alpha_i \mathbf{a}_i \ , \quad 1 = \sum_{i=0}^{k} \alpha_i \ ,$$

be any "new" knot. We view it as the projection of the point

$$\mathbf{p}_{k+1} = \sum \alpha_i \mathbf{p}_i$$

in $\mathbb{R}^k$. The weights α_i are not uniquely defined by $\mathbf{a}_{k+1}$ if $k > s$. Finally, let σ_i denote the simplex $\mathbf{p}_0 \ldots \mathbf{p}_i^* \ldots \mathbf{p}_{k+1}$, where the asterisk denotes the absence of this term. In particular, this means $\sigma_{k+1} = \sigma$. The other simplices $\sigma_0, \ldots, \sigma_k$ form a partition of the simplex σ in such a way that the signed simplex splines M_{σ_i} sum to the simplex spline M_σ, i.e.,

$$M_\sigma = \sum_{i=0}^{k} (\operatorname{sign} \alpha_i) M_{\sigma_i} \ .$$

Two examples are shown in Figure 18.4 for $k = 2$, where $M_\sigma = M_{\sigma_1} + M_{\sigma_2} - M_{\sigma_0}$, and for $k = 3$, where $M_\sigma = M_{\sigma_0} + M_{\sigma_1} + M_{\sigma_2} + M_{\sigma_3}$.

For a proof, consider a point $\mathbf{y} = \sum \eta_i \mathbf{p}_i$ and let $\eta_{k+1} = 0$ and $\alpha_{k+1} = -1$. Then, for any $\lambda \in \mathbb{R}$ we obtain the equation

$$\mathbf{y} = \sum_{i=0}^{k+1} (\eta_i - \lambda \alpha_i) \mathbf{p}_i \ .$$

In particular, for $\lambda = \lambda_j := \eta_j / \alpha_j$, the weights $\eta_i - \lambda_j \alpha_i$ are the barycentric coordinates of $\mathbf{y}$ with respect to σ_j and $\mathbf{y}$ lies in the interior of σ_j if and only if all weights $\eta_i - \lambda_j \alpha_i, j \neq i$, are positive. If $\lambda_j = \lambda_k$, then $\mathbf{y}$ lies on the

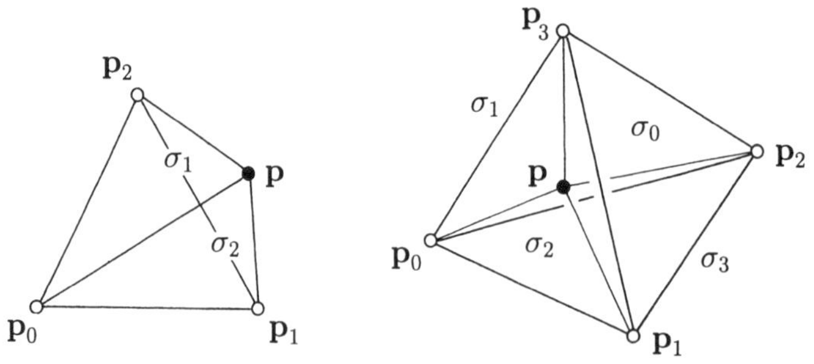

Figure 18.4: Subdividing a triangle and a tetrahedron.

common boundary of σ_j and σ_k. We assume that this is not the case.

Since the weights $\eta_i - \lambda\alpha_i$ are monotone in λ and since there are positive and negative α_i, there is at most one non-empty interval (λ_j, λ_k) over which all weights $\eta_i - \lambda\alpha_i$ are positive. Hence, $\mathbf{y}$ lies in none or in exactly two simplices σ_j and σ_k. In the later case α_j and α_k have opposite signs, see Figure 18.5.

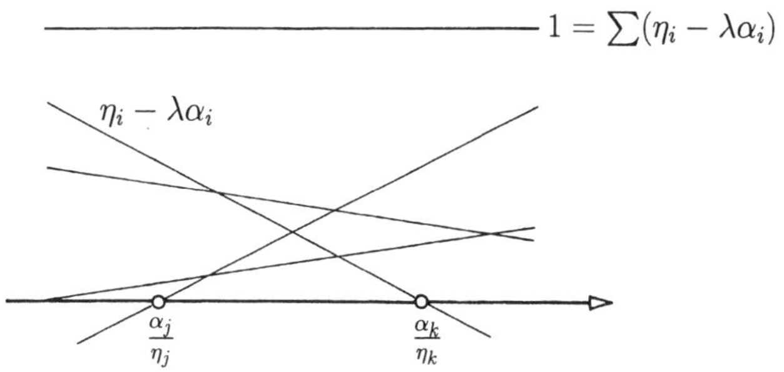

Figure 18.5: The weights $\eta_i - \lambda\alpha_i$.

Thus, the characteristic maps of the simplices satisfy

$$0 = (\text{sign}\alpha_0)\chi_{\sigma_0} + \cdots + (\text{sign}\alpha_k)\chi_{\sigma_k} - \chi_{\sigma_{k+1}}$$

over $\mathbb{R}^k$ without the lower dimensional faces of the simplices σ_i.

Hence, it follows that $M_\sigma = \text{sign}\alpha_0 M_{\sigma_0} + \cdots + \text{sign}\alpha_k M_{\sigma_k}$, which concludes the proof. $\diamond$

After normalization, this sum becomes Micchelli's **knot insertion formula**

(1) $$M(\mathbf{x}|\mathbf{a}_0 \ldots \mathbf{a}_k) = \sum_{i=0}^{k} \alpha_i M(\mathbf{x}|\mathbf{a}_0 \ldots \mathbf{a}_i^* \ldots \mathbf{a}_{k+1})$$

for all $\mathbf{x}$ for which the splines involved are continuous, see [Micchelli '80] and also Problems 2 and 3. Note that the right hand side of the knot insertion formula is an affine combination.

18.5 A recurrence relation

If the new knot $\mathbf{a}_{k+1}$ coincides with $\mathbf{x}$, then the knot insertion formula (1) from 18.4 represents a simplex spline at $\mathbf{x}$ as an affine combination of simplex splines with $\mathbf{x}$ as a knot, i.e.,

$$M(\mathbf{x}|\mathbf{a}_0 \ldots \mathbf{a}_k) = \sum_{i=0}^{k} \xi_i M(\mathbf{x}|\mathbf{x}\,\mathbf{a}_0 \ldots \mathbf{a}_i^* \ldots \mathbf{a}_k) \ ,$$

where $\mathbf{x} = \sum \xi_i \mathbf{a}_i$ and $1 = \sum \xi_i$.

This leads to a recursive formula since the simplex splines on the right are simplex splines of lower degree. For example, let σ_0 be the simplex $\mathbf{p}\,\mathbf{p}_1 \ldots \mathbf{p}_k$ in $\mathbb{R}^k$ with shadow

$$M_{\sigma_0}(\mathbf{x}) = M(\mathbf{x}|\mathbf{x}\,\mathbf{a}_1 \ldots \mathbf{a}_k) \ ,$$

where $\pi\mathbf{p}\pi\mathbf{p}_i, \ldots, \pi\mathbf{p}_k$ are the knots $\mathbf{x}\mathbf{x}_1, \ldots, \mathbf{a}_k$. We assume that the "base simplex" ρ with the vertices $\mathbf{p}_1, \ldots, \mathbf{p}_k$ lies in a hyperplane orthogonal to the fibers of the projection π.

Hence, if h denotes the Euclidean distance between $\mathbf{p}$ and ρ, we obtain

$$\mathrm{vol}_k \sigma_0 = \frac{1}{k} h \cdot \mathrm{vol}_{k-1} \rho$$

and

$$\mathrm{vol}_{k-s}(\sigma_0 \cap \pi^{-1}\mathbf{x}) = \frac{1}{k-s} h \cdot \mathrm{vol}_{k-s-1}(\rho \cap \pi^{-1}\mathbf{x})$$

as illustrated in Figure 18.6 for $k = 3$ and $s = 1$.

Dividing the second by the first equation, we obtain

$$M(\mathbf{x}|\mathbf{x}\,\mathbf{a}_1 \ldots \mathbf{a}_k) = \frac{k}{k-s} M(\mathbf{x}|\mathbf{a}_1 \ldots \mathbf{a}_k) \ .$$

Hence, we can transform the knot insertion formula (1) into **Micchelli's recurrence relation**

$$(2) \qquad M(\mathbf{x}|\mathbf{a}_0 \ldots \mathbf{a}_k) = \frac{k}{k-s} \sum_{i=0}^{k} \xi_i M(\mathbf{x}|\mathbf{a}_0 \ldots \mathbf{a}_i^* \ldots \mathbf{a}_k) \ ,$$

which represents a simplex spline with $k+1$ knots as a linear combination of simplex splines with k knots, see [Micchelli '80]. Since the weights ξ_i depend

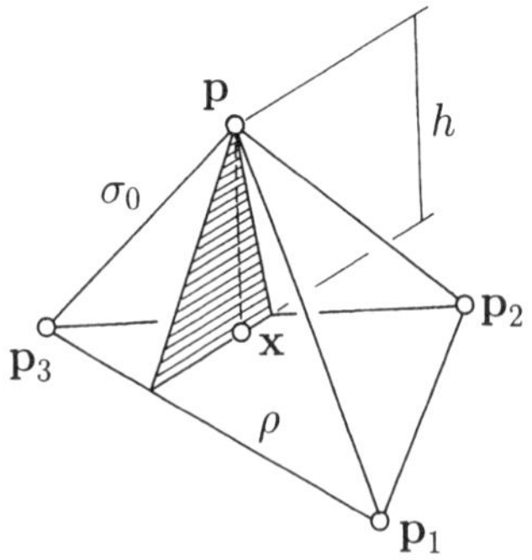

Figure 18.6: Computing the volumes of σ_0 and $\sigma_0 \cap \pi^{-1}\mathbf{x}$.

linearly on $\mathbf{x}$, repeated application of this recurrence relation shows that

an s-variate simplex spline with $k+1$ knots is piecewise polynomial of total degree $\le k-s$.

Remark 4: Comparing the recurrence relations for simplex splines and Bernstein polynomials, see 10.1, we see that an s-variate simplex spline with only $s+1$ distinct knots is a Bernstein polynomial, i.e.,

$$M(\mathbf{x}|\mathbf{a}_0 \overset{i_0+1}{\dots} \mathbf{a}_0 \dots \mathbf{a}_s \overset{i_s+1}{\dots} \mathbf{a}_s) = \frac{\binom{k}{s}}{\text{vol}_s\Delta} B_{\mathbb{i}}^{k-s}(\mathbb{u}) \ ,$$

where $\mathbb{i} = (i_0 \dots i_s), k = i_0 + \dots + i_s + s, \Delta$ denotes the simplex $\mathbf{a}_0 \dots \mathbf{a}_s$ and $\mathbb{u}$ the barycentric coordinate column of $\mathbf{x}$ with respect to Δ, see Figure 18.7.

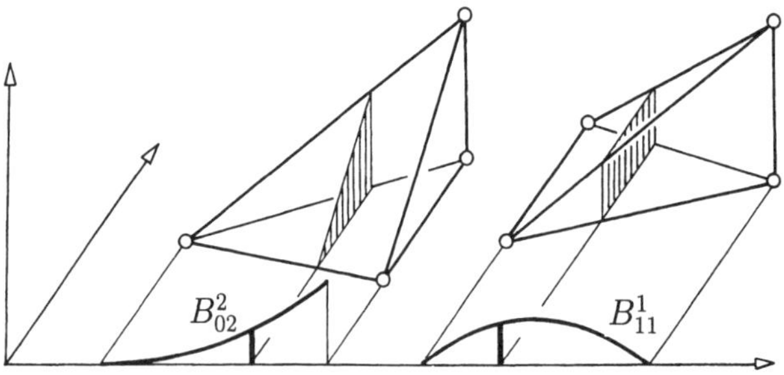

Figure 18.7: Quadratic Bernstein polynomials.

Remark 5: Repeated application of the recursion formula(2) also shows that each polynomial segment of a multivariate B-spline can be written as a product of k weights ξ, which represent $\mathbf{x}$ as certain affine combinations of all or some knots $\mathbf{a}_i$. Since these weights depend continuously on the knots $\mathbf{a}_i$ (or on the vertices $\mathbf{p}_i$ of some associated simplex), it follows that each polynomial segment depends continuously on the knots $\mathbf{a}_i$.

18.6 Derivatives

Let $\mathbf{x}$ vary with t on a line, i.e., let

$$\mathbf{x} = \mathbf{x}_0 + t\mathbf{v} \ ,$$

where $\mathbf{x}_0 = \sum \alpha_i \mathbf{p}_i, 1 = \sum \alpha_i$, represents a point and $\mathbf{v} = \sum \nu_i \mathbf{p}_i, 0 = \sum \nu_i$, a vector in $\mathbb{R}^s$. Differentiating an s-variate B-spline $M(\mathbf{x}|\mathbf{a}_0 \ldots \mathbf{a}_k)$ with respect to t results in its **directional derivative** at $\mathbf{x}$ with respect to $\mathbf{v}$, which we denote by $\dot{M}$. It can be expressed in terms of the lower degree B-splines $M_i = M(\mathbf{x}|\,\mathbf{a}_0 \ldots \mathbf{a}_i^* \ldots \mathbf{a}_k)$:

$$\dot{M} = k \sum \nu_i M_i \ . \tag{3}$$

For a proof, first, let $k = s + 1$, first. Then, we obtain relation (3) by differentiating Micchelli's recurrence relation in 18.5.

Second, if k is greater than $s + 1$, we view $\mathbf{a}_0, \ldots, \mathbf{a}_k$ as projections under $\pi : [x_1 \ldots x_{k-1}]^t \mapsto [x_1 \ldots x_s]^t$ of some points $\bar{\mathbf{a}}_0, \ldots, \bar{\mathbf{a}}_k$ in $\mathbb{R}^{k-1}$ and obtain

$$M(\mathbf{x}|\mathbf{a}_0 \ldots \mathbf{a}_k) = \int_{\mathbb{R}} \overset{k-s-1}{\cdots} \int_{\mathbb{R}} M(\begin{bmatrix} \mathbf{x} \\ \mathbf{y} \end{bmatrix} |\bar{\mathbf{a}}_0 \ldots \bar{\mathbf{a}}_k) d\mathbf{y} \ . \tag{4}$$

Since $M(\mathbf{x}|\mathbf{a}_0 \ldots \mathbf{a}_k)$ has continuous left- and right-hand side derivatives, we can differentiate both sides of this identity and interchange the integral and the derivative. This leads to

$$\frac{d}{dt} M(\mathbf{x}|\mathbf{a}_0 \ldots \mathbf{a}_k) = \int_{\mathbb{R}} \overset{k-s-1}{\cdots} \int_{\mathbb{R}} \frac{d}{dt} M(\begin{bmatrix} \mathbf{x} \\ \mathbf{y} \end{bmatrix} |\bar{\mathbf{a}}_0 \ldots \bar{\mathbf{a}}_k) d\mathbf{y} \ .$$

With the aid of formula (3), applied to the integrand, and (4), we can transform this equation to formula (3) for $\dot{M}$. ◇

Repeated differentiation shows that the rth derivative of an s-variate B-spline $M(\mathbf{x}|\mathbf{a}_0 \ldots \mathbf{a}_k)$ can be written as a linear combination of B-splines with $k + 1 - r$ knots $\mathbf{a}_i$. These splines are continuous over any s-knot whose affine hull contains no more than $k - r - 1$ knots. Hence, $M(\mathbf{x}|\mathbf{a}_0 \ldots \mathbf{a}_k)$ is $k - m - 1$ times continuously differentiable over any s-knot whose affine hull contains no more than m knots.

Remark 6: Differentiating Micchelli's recursion formula(2) in 18.5 leads to

$$\dot{M} = \frac{k}{k-s}\sum_{i=0}^{k}(\nu_i M_i + \xi_i \dot{M}_i) \ .$$

Because of formula (3), this is equivalent to the equation

$$\dot{M} = \frac{1}{k-s}\dot{M} + \frac{k}{k-s}\sum_{i=0}^{k}\xi_i \dot{M}_i \ ,$$

which can be transformed to the **derivative recursion** [Micchelli '80]

$$\frac{d}{dt}M(\mathbf{x}|\mathbf{a}_0 \ldots \mathbf{a}_k) = \frac{k}{k-s-1}\sum_{i=0}^{k}\xi_i \frac{d}{dt}M(\mathbf{x}|\mathbf{a}_0 \ldots \mathbf{a}_i^* \ldots \mathbf{a}_k) \ .$$

18.7 Problems

1 The volume of a simplex $\mathbf{q}_0 \ldots \mathbf{q}_{k-s}$ in $\mathbb{R}^{k-s}$ is given by

$$\mathrm{vol}_{k-s}[\mathbf{q}_0 \ \cdots \ \mathbf{q}_{k-s}] = \left| \frac{1}{(k-s)!}\det\begin{bmatrix} \mathbf{q}_0 & \cdots & \mathbf{q}_{k-s} \\ 1 & \ldots & 1 \end{bmatrix} \right| \ .$$

Use this formula to observe directly from the definition in 18.1 that an s-variate simplex spline with $k+1$ knots is piecewise polynomial of degree $k-s$.

2 Show that the knot insertion formula from 18.4 is also valid on the boundary of the convex hull $[\mathbf{a}\,\mathbf{a}_0 \ldots \mathbf{a}_k]$ in case that some of the splines involved are discontinuous there.

3 Derive the knot insertion formula for univariate B-splines from the general formula (1) in 18.4.

4 Derive the recurrence relation for univariate B-splines from the general formula (2) in 18.5.

5 A simplex $\mathbf{p}_1 \ldots \mathbf{p}_k$ in $\mathbb{R}^k$ and a direction $\mathbf{v}$ define a **prism**. The prism is the convex hull of its $2k$ vertices $\mathbf{p}_i$ and $\mathbf{q}_i = \mathbf{p}_i + \mathbf{v}$, $i = 1, \ldots, k$. Show that the k simplices $\mathbf{q}_1 \ldots \mathbf{q}_i \mathbf{p}_i \ldots \mathbf{p}_k$, for $i = 1, \ldots, k$, have equal volume and form a partition of the prism, see Figure 18.8.

6 Use the statement of Problem 5 to prove Micchelli's **degree elevation**

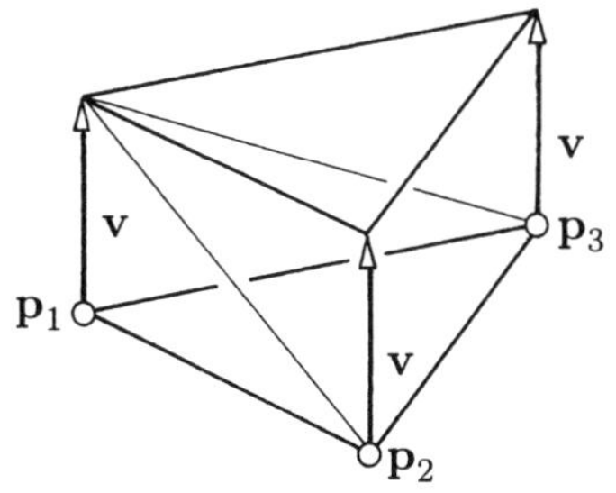

Figure 18.8: Triangulation of a prism.

formula,

$$M(\mathbf{x}|\mathbf{a}_0 \ldots \mathbf{a}_k) = \frac{1}{k+1} \sum_{i=0}^{k} M(\mathbf{x}|\mathbf{a}_0\,\mathbf{a}_1 \ldots \mathbf{a}_i\,\mathbf{a}_i \ldots \mathbf{a}_k) \ ,$$

for simplex splines [Micchelli '79].

19 Multivariate splines

De Boor's algorithm is a generalization of de Casteljau's and univariate B-splines are a generalization of Bernstein polynomials. Analogously, it is possible to generalize de Casteljau's algorithm for multivariate polynomials to multivariate splines. The underlying basis functions of this generalized algorithm are simplex splines forming a partition of unity. These simplex splines are proper generalizations of the univariate B-splines and many properties of the univariate B-splines also hold for these multivariate simplex splines.

19.1 Generalizing de Casteljau's algorithm

Univariate Bernstein polynomials restricted to a simplex can be viewed as simplex splines with multiple knots, see Remark 4 in 18.5. As a consequence, de Casteljau's algorithm is a special instance of de Boor's algorithm for univariate splines.

Moreover, it is easily possible to generalize, in an analogous fashion, de Casteljau's algorithm for multivariate polynomials in Bézier representation, see [Seidel '91]. The idea is illustrated in Figures 19.1 and 19.2 for a quadratic polynomial $\mathbf{c}(\mathbf{x})$. The points on the associated polar form $\mathbf{c}[\mathbf{x}\,\mathbf{y}]$ are denoted just by the argument $\mathbf{x}\,\mathbf{y}$ as in Figure 11.1.

Figure 19.1 shows de Casteljau's algorithm with respect to the triangle $\mathbf{012}$. From the six Bézier points $\mathbf{c}[\mathbf{i}\,\mathbf{j}]$, we compute first the three points $\mathbf{c}[\mathbf{j}\,\mathbf{x}]$ and then compute from these three points the point $\mathbf{c}(\mathbf{x}) = \mathbf{c}[\mathbf{x}\,\mathbf{x}]$.

Figure 19.2 shows the generalization. Instead of three knots, there are three knot sequences $\mathbf{0}\bar{\mathbf{0}}, \mathbf{1}\bar{\mathbf{1}}, \mathbf{2}\bar{\mathbf{2}}$. From the six control points $\mathbf{c}[\mathbf{0}\bar{\mathbf{0}}]$, $\mathbf{c}[\bar{\mathbf{0}}\bar{\mathbf{1}}]$, $\mathbf{c}[\bar{\mathbf{1}}\mathbf{1}]$,

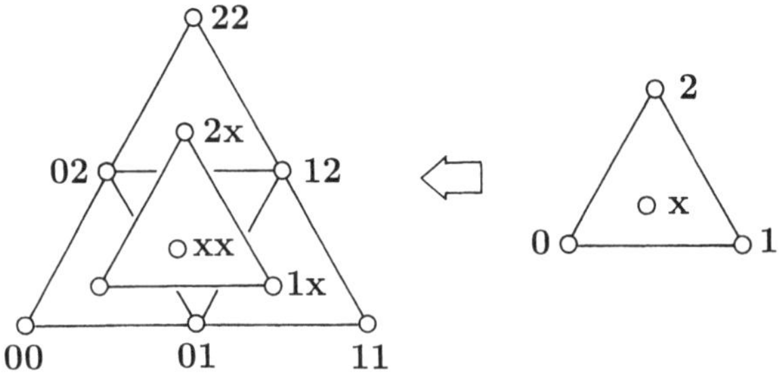

Figure 19.1: De Casteljau's algorithm for a quadratic.

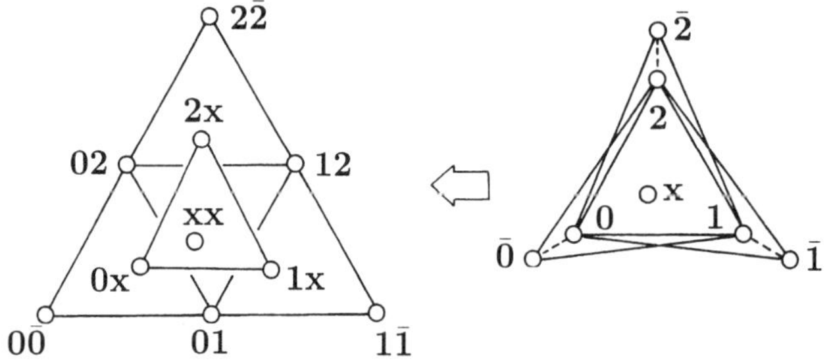

Figure 19.2: Seidel's generalization of de Casteljau's algorithm for a quadratic.

$\mathbf{c}[\bar{\mathbf{1}}\bar{\mathbf{2}}]$, $\mathbf{c}[\bar{\mathbf{2}}\mathbf{2}]$ and $\mathbf{c}[\bar{\mathbf{2}}\bar{\mathbf{0}}]$, we compute the three points $\mathbf{c}[\mathbf{i}\,\mathbf{x}]$ and then compute from these three points the point $\mathbf{c}(\mathbf{x}) = \mathbf{c}[\mathbf{x}\,\mathbf{x}]$.

To present the generalization of de Casteljau's algorithm, let $\mathbf{a}_i^0, \ldots, \mathbf{a}_i^{n-1}$ denote a sequence, also called **chain** or **cloud of knots**, for $i = 0, \ldots, s$. If these knots are in general position, then any s-variate polynomial surface $\mathbf{c}(\mathbf{x})$ with polar form $\mathbf{c}[\mathbf{x}_1 \ldots \mathbf{x}_n]$ is completely defined by its **B – points**

$$\mathbf{c}_{\mathbb{i}} = \mathbf{c}[\mathbf{a}_0^0 \ldots \mathbf{a}_0^{i_0-1} \ldots \mathbf{a}_s^0 \ldots \mathbf{a}_s^{i_s-1}] \ ,$$

where $\mathbb{i} = [i_0 \ldots i_s] \in \Delta_n$ and

$$\Delta_n = \{\mathbb{i} | \mathbb{i} \in \mathbb{Z}^{s+1}, \mathbb{o} \le \mathbb{i}, i_0 + \cdots + i_s = n\} \ .$$

Namely, $\mathbf{c}(\mathbf{x})$ can be computed by means of the recursion formula

$$\mathbf{c}_{\mathbb{j}} = \xi_0 \mathbf{c}_{\mathbb{j}+\mathbb{e}_0} + \cdots + \xi_s \mathbf{c}_{\mathbb{j}+\mathbb{e}_s} \quad \mathbb{j} \in \Delta := \Delta_{n-1} \cup \cdots \cup \Delta_0 \ ,$$

where the ξ_k are the barycentric coordinates of $\mathbf{x}$ with respect to the simplex

$$S_{\mathbb{j}} = [\mathbf{a}_0^{j_0} \ldots \mathbf{a}_s^{j_s}]$$

and $\mathbb{e}_k$ denotes the kth unit vector in $\mathbb{R}^{s+1}$ as before. This means that

$$\mathbf{c}_{\mathbb{j}} = \mathbf{c}[\mathbf{a}_0^0 \ldots \mathbf{a}_0^{j_0-1} \ldots \mathbf{a}_s^0 \ldots \mathbf{a}_s^{j_s-1} \mathbf{x} \ldots \mathbf{x}]$$

and that the knots are in **general position** if all simplices $S_{\mathbb{j}}$ are non-degenerate.

Remark 1: If $\mathbf{c}(\mathbf{x})$ is quadratic, then its Bézier points $\mathbf{b}_{\mathbb{i}}$ over the simplex $S_{\mathbb{o}}$ satisfy

$$\mathbf{b}_{\mathbb{e}_i+\mathbb{e}_j} = \mathbf{c}_{\mathbb{e}_i+\mathbb{e}_j} \quad \text{for all} \quad i \neq j \ ,$$

and $\mathbf{b}_{\mathbb{e}_i+\mathbb{e}_i}$ lies in the plane spanned by the points $\mathbf{c}_{\mathbb{e}_i+\mathbb{e}_j}$, $j = 0, \ldots, s$. An example is shown in Figure 19.5.

19.2 B-polynomials and B-patches

We can apply the generalized de Casteljau algorithm from Section 19.1 to any arbitrary net of control points $\mathbf{c}_{\mathbb{i}}$, $|\mathbb{i}| \in \Delta_n$. If we start with the control points

$$c_{\mathbb{i}} = \begin{cases} 1 & \text{if} \ \ \mathbb{i} = \mathbb{k} \\ 0 & \text{otherwise} \end{cases} \quad , \quad \mathbb{i} \in \Delta_n \ ,$$

then the resulting scalar valued polynomials $C_{\mathbb{k}}^n(\mathbf{x})$ form a basis for the space of all nth degree polynomials. Namely, since the generalized de Casteljau algorithm in 19.1 is linear in the control points, any polynomial $\mathbf{c}(\mathbf{x})$ of degree $\leq n$ can be written as

$$\mathbf{c}(\mathbf{x}) = \sum_{\mathbb{i} \in \Delta_n} \mathbf{c}_{\mathbb{i}} C_{\mathbb{i}}^n(\mathbf{x}) \ ,$$

where $\mathbf{c}_{\mathbb{i}}$ is defined as in 19.1. Further, the $C_{\mathbb{i}}^n$ are linearly independent since their number $\binom{n+s}{n}$ equals the dimension of the space of all polynomials of total degree $\leq n$.

The polynomials $C_{\mathbb{i}}^n$ are called the **B-polynomials** and the representation of a polynomial as a linear combination of B-polynomials is referred to as a **B-patch representation**. From their construction above, we obtain the following properties, which are similar to the properties of Bernstein polynomials, given in Section 10.1. The s-variate B-polynomials of degree n

- *form a* **basis** *for all s-variate polynomials of total degree $\leq n$,*
- *form a* **partition of unity**, *i.e.,*

$$\sum_{\mathbb{i} \in \Delta_n} C_{\mathbb{i}}^n(\mathbf{x}) = 1 \ ,$$

- *are* **positive** *for all* $\mathbf{x}$ *in the interior of the intersection* Γ *of all simplices* $S_{\mathfrak{j}}, |\mathfrak{j}| \le n-1$,
- *and satisfy the* **recursion**

$$C_{\mathfrak{i}}^{n}(\mathbf{x}) = \sum_{k=0}^{s} \xi_k C_{\mathfrak{i}-\mathbf{e}_k}^{n-1}(\mathbf{x}) \ ,$$

where ξ_k is the kth barycentric coordinate of $\mathbf{x}$ with respect to $S_{\mathfrak{i}-\mathbf{e}_k}$ and $C_{\mathfrak{j}}^{n} = 0$ if $\mathfrak{j}$ has a negative coordinate.

The intersection Γ of the simplices $S_{\mathfrak{j}}, \mathfrak{j} \in \Delta_n$, is illustrated in Figure 19.3 for $n = 2$ and $s = 2$. It can be empty, depending on the knot positions.

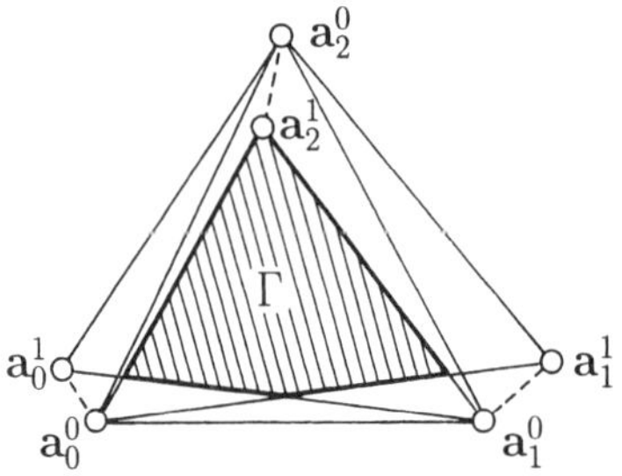

Figure 19.3: The region Γ over which all B-polynomials are positive for $n = s = 2$.

Since the B-polynomials form a partition of unity, any polynomial surface $\mathbf{c}(\mathbf{x})$ is an affine combination of its B-points $\mathbf{c}_{\mathfrak{i}}$. Consequently, a B-patch representation is affinely invariant. Furthermore, the patch $\mathbf{c}(\mathbf{x}), \mathbf{x} \in \Gamma$, lies in the convex hull of the B-points $\mathbf{c}_{\mathfrak{i}}$ since the B-polynomials sum to one and are non-negative over Γ.

Remark 2: If all knots of each knot chain are equal, i.e., $S_{\mathfrak{o}} = S_{\mathfrak{j}}$ for all $\mathfrak{j}$, then the B-polynomials $C_{\mathfrak{i}}^{n}$ are the Bernstein polynomials $B_{\mathfrak{i}}^{n}$ over the simplex $S_{\mathfrak{o}}$ since both sets of polynomials satisfy the same recursion.

19.3 Linear precision

The linear polynomial $\mathbf{x}$, which is the identity map on $\mathbb{R}^s$, has a B-patch representation of degree n with respect to the knots $\mathbf{a}_i^j$ introduced in 19.1. Since $\mathbf{x}$ has the polar form

$$\mathbf{x}[\mathbf{x}_1 \dots \mathbf{x}_n] = \frac{1}{n}(\mathbf{x}_1 + \cdots + \mathbf{x}_n) \ ,$$

its B-points are

$$\mathbf{x}_{\mathfrak{i}} := (\mathbf{a}_0^0 + \cdots + \mathbf{a}_0^{i_0-1} + \cdots + \mathbf{a}_s^0 + \cdots + \mathbf{a}_s^{i_s-1}) \ ,$$

see 19.1 and Figure 19.4 for an illustration, where $n = s = 2$.

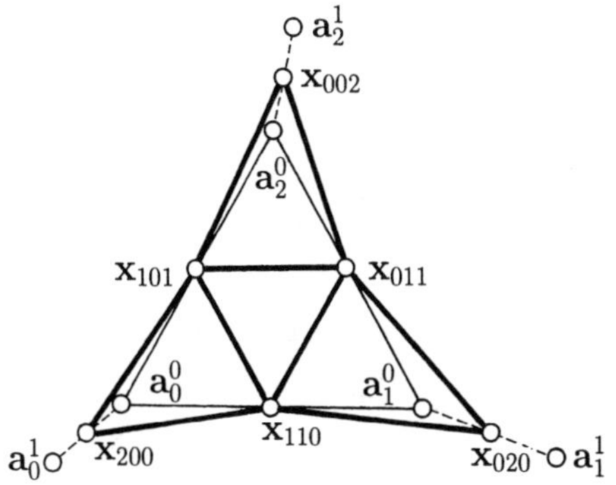

Figure 19.4: The quadratic B-patch representation of the identity on $\mathbb{R}^2$.

Similarly, we obtain the B-points of any other linear polynomial $\mathbf{c}(\mathbf{x})$. Since $\mathbf{c}(\mathbf{x})$ is an affine map, its n-affine polar form is given by

$$\mathbf{c}[\mathbf{x}_1 \ldots \mathbf{x}_n] = \mathbf{c}(\mathbf{x}[\mathbf{x}_1 \ldots \mathbf{x}_n]) \ .$$

Hence, $\mathbf{c}(\mathbf{x})$ has the B-points $\mathbf{c}_{\mathfrak{i}} = \mathbf{c}(\mathbf{x}_{\mathfrak{i}})$. This property is referred to as the **linear precision** of the B-patch representation.

Remark 3: As a consequence of the linear precision, any functional surface

$$\mathbf{c}(\mathbf{x}) = \begin{bmatrix} \mathbf{x} \\ c(\mathbf{x}) \end{bmatrix} , \quad \text{where} \quad c(\mathbf{x}) = \sum c_{\mathfrak{i}} C_{\mathfrak{i}}^n(\mathbf{x}) \ ,$$

of degree $\le n$ has the B-points $\mathbf{c}_{\mathfrak{i}} = [\mathbf{x}_{\mathfrak{i}}^t \, c_{\mathfrak{i}}]^t$, as illustrated in Figure 19.5 for $n = s = 2$.

19.4 Derivatives of a B-patch

Any directional derivative of a polynomial

$$\mathbf{c}(\mathbf{x}) = \sum \mathbf{c}_{\mathfrak{i}} C_{\mathfrak{i}}^n(\mathbf{x})$$

given in B-patch form with respect to a knot sequence $A_{\mathfrak{n}-\mathfrak{e}}$ has a B-patch representation of degree $n-1$ with respect to the knot sequence $A_{\mathfrak{n}-2\mathfrak{e}}$. This representation depends in a simple manner on the points $\mathbf{c}_{\mathfrak{i}}$.

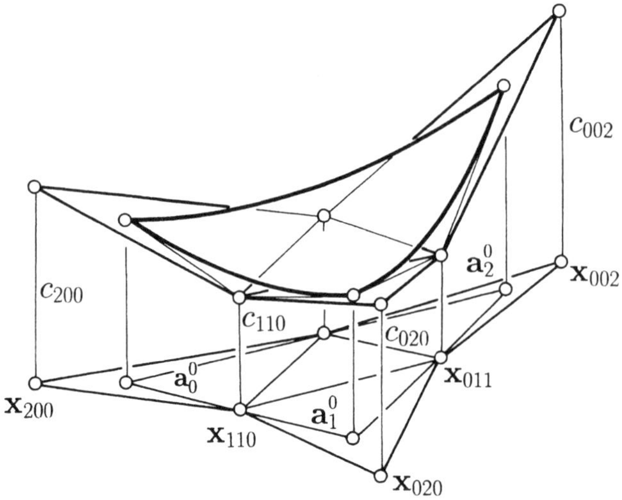

Figure 19.5: A quadratic polynomial with its B-patch and Bézier ordinates.

Let $\Delta\mathbf{x}$ be any direction in $\mathbb{R}^s$ and let

$$\dot{\mathbf{c}}(\mathbf{x}) = \frac{d}{dt}\mathbf{c}(\mathbf{x} + t\Delta\mathbf{x})|_{t=0}$$

denote the directional derivative of $\mathbf{c}$ with respect to $\Delta\mathbf{x}$ at $\mathbf{x}$. Furthermore, let $\mathbf{c}[\mathbf{x}_1 \ldots \mathbf{x}_n]$ be the symmetric polynomial of $\mathbf{c}(\mathbf{x})$. Then, the symmetric polynomial of $\dot{\mathbf{c}}(\mathbf{x})$ is given by

$$\dot{\mathbf{c}}[\mathbf{x}_2 \ldots \mathbf{x}_n] = n\mathbf{c}[\Delta\mathbf{x}\, \mathbf{x}_2 \ldots \mathbf{x}_n] \ ,$$

as explained in 11.6. Hence, the B-points of $\dot{\mathbf{c}}(\mathbf{x})$ are given by

$$n\dot{\mathbf{c}}_{\mathfrak{j}} = n\mathbf{c}[\Delta\mathbf{x}\, \mathbf{a}_0^0 \ldots \mathbf{a}_0^{i_0-1} \ldots \mathbf{a}_s^0 \ldots \mathbf{a}_s^{i_s-1}] \ , \quad \mathfrak{j} \in \Delta_{n-1} \ .$$

Expressing $\Delta\mathbf{x}$ with respect to $S_{\mathfrak{j}}$, i.e., writing $\Delta\mathbf{x}$ as

$$\Delta\mathbf{x} = \sum_{k=0}^{s} \nu_k \mathbf{a}_k^{j_k}, \quad 0 = \sum_{k=0}^{s} \nu_k \ ,$$

and using the fact that polar forms are multiaffine, we obtain for the control points $\dot{\mathbf{c}}_{\mathfrak{j}}$ of the derivative

$$\dot{\mathbf{c}}_{\mathfrak{j}} \ = \ \sum_{k=0}^{s} \nu_k \mathbf{c}_{\mathfrak{j}+\mathbf{e}_k} \ .$$

This result is illustrated in Figure 19.6 for $s = n = 2$.

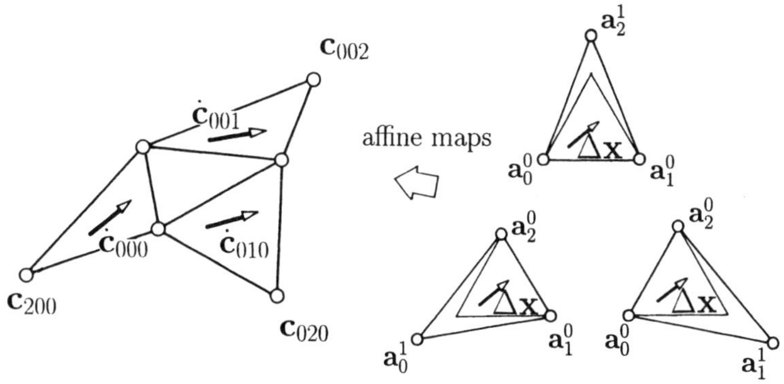

Figure 19.6: The differences $\dot{\mathbf{c}}_{\mathfrak{j}}$.

Remark 4: Let $\mathbf{x}_{\mathfrak{i}}$ be the B-points of the identity polynomial $\mathbf{c}(\mathbf{x}) = \mathbf{x}$ with respect to the knots $\mathbf{a}_k^i$ and view the B-net of $\mathbf{c}(\mathbf{x})$ as the collection of the linear patches

$$\mathbf{b}_{\mathfrak{j}}(\mathbf{x}) := \sum_{k=0}^{s} \gamma_k \mathbf{c}_{\mathfrak{j}+\mathbf{e}_k} \;, \quad \mathfrak{j} \in \Delta_{n-1} \;,$$

over the simplices $\mathbf{x}_{\mathfrak{j}+\mathbf{e}_0} \ldots \mathbf{x}_{\mathfrak{j}+\mathbf{e}_s}$, where $\gamma_0, \ldots, \gamma_s$ are the barycentric coordinates of $\mathbf{x}$ with respect to $\mathbf{x}_{\mathfrak{j}+\mathbf{e}_0} \ldots \mathbf{x}_{\mathfrak{j}+\mathbf{e}_s}$.

Since $\dot{\gamma}_k = n\nu_k$, the directional derivative of $\mathbf{c}_{\mathfrak{j}}(\mathbf{x})$ with respect to $\Delta\mathbf{x}$ is $n\dot{\mathbf{c}}_{\mathfrak{j}}$. Hence,

> *the derivative of the B-net consists of the B-points of the derivative* $\dot{\mathbf{c}}(\mathbf{x})$.

This is illustrated in Figure 19.7 for a functional bivariate quadratic surface.

19.5 Multivariate B-splines

The B-polynomials $C_{\mathfrak{i}}$, $\mathfrak{i} \in \Delta_n$, are defined with respect to the knots

$$\mathbf{a}_0^0, \ldots, \mathbf{a}_0^{n-1}, \ldots, \mathbf{a}_s^0, \ldots, \mathbf{a}_s^{n-1} \;.$$

Here, we add $s+1$ knots $\mathbf{a}_0^n, \ldots, \mathbf{a}_s^n$ and show that

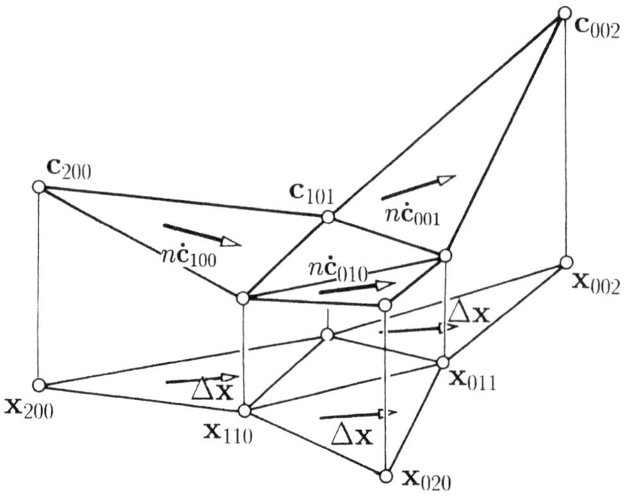

Figure 19.7: The directional derivative of a B-net for $n = s = 2$.

the normalized simplex splines

$$N_{\mathfrak{i}} := \frac{\text{vol } S_{\mathfrak{i}}}{\binom{n+s}{s}} M(\mathbf{x}|A_{\mathfrak{i}}) \ , \quad \mathfrak{i} \in \Delta_n \ ,$$

with the knot sequences $A_{\mathfrak{i}} = \mathbf{a}_0^0 \ldots \mathbf{a}_0^{i_0} \ldots \mathbf{a}_s^0 \ldots \mathbf{a}_s^{i_s}$ *coincide with the B-polynomials* $C_{\mathfrak{i}}(\mathbf{x})$ *for all*

$$\mathbf{x} \in \Omega := \text{interior } \Big(\bigcap_{\mathfrak{j} \in \Delta_0 \cup \cdots \cup \Delta_n} S_{\mathfrak{j}} \Big) \ .$$

This result is due to [Dahmen et al. '92].

The splines $N_{\mathfrak{i}}$ are multivariate **B-splines** with properties similar to those of univariate B-splines. In fact, for $s = 1$ the B-splines $N_{\mathfrak{i}}(x)$ are the univariate B-splines defined in 5.3.

Proof: From the definition of constant simplex splines in 18.1 and from the recurrence formula (2) in 18.5 for simplex splines, we obtain for $\mathbf{x} \in \mathbb{R}^s$ the recurrence relation

$$N_{\mathfrak{o}}^0(\mathbf{x}) = \begin{cases} 1 & \text{if } \ \mathbf{x} \in S_{\mathfrak{o}} \\ 0 & \text{otherwise} \end{cases} \ ,$$

$$N_{\mathfrak{j}}(\mathbf{x}) = \sum_{k=0}^{s} \frac{\text{vol } S_{\mathfrak{j}}}{\binom{j+s}{s}} \ \frac{j+s}{s} \ \frac{\sigma_{\mathfrak{j}}^k \text{vol } S_{\mathfrak{j}}^k}{\sigma_{\mathfrak{j}} \text{vol } S_{\mathfrak{j}}} \ M(\mathbf{x}|A_{\mathfrak{j}-\mathbf{e}_k}) \ , \quad j := |\mathfrak{j}| \ ,$$

where $S_{\mathfrak{j}}^k$ is obtained from $S_{\mathfrak{j}}$ by replacing $\mathbf{a}_{j_k}$ by $\mathbf{x}$. The orientations -1 or $+1$ of the sequences $S_{\mathfrak{j}}$ and $S_{\mathfrak{j}}^k$ are denoted by $\sigma_{\mathfrak{j}}$ and $\sigma_{\mathfrak{j}}^k$, respectively.

Since both simplices $S_{\mathfrak{j}}$ and $S_{\mathfrak{j}-\mathbf{e}_k}$ contain Ω, their orientations are equal and the equation above is equivalent to

$$\begin{aligned} N_{\mathfrak{j}}(\mathbf{x}) &= \sum_{k=0}^{s} \frac{\sigma_{\mathfrak{j}}^k}{\sigma_{\mathfrak{j}-\mathbf{e}_k} \operatorname{vol} S_{\mathfrak{j}}^k} \operatorname{vol} S_{\mathfrak{j}-\mathbf{e}_k} N_{\mathfrak{j}-\mathbf{e}_k}(\mathbf{x}) \\ &= \sum_{k=0}^{s} \xi_{\mathfrak{j}-\mathbf{e}_k}^k N_{\mathfrak{j}-\mathbf{e}_k}(\mathbf{x}) \ , \end{aligned} \tag{1}$$

where $\xi_{\mathfrak{j}-\mathbf{e}_k}^k$ is the kth barycentric coordinate of $\mathbf{x}$ with respect to $S_{\mathfrak{j}-\mathbf{e}_k}$. To have all terms well-defined, we define, for $j_k = 0$,

$$\operatorname{vol} S_{\mathfrak{j}-\mathbf{e}_k} = 1 \quad \text{and} \quad N_{\mathfrak{j}-\mathbf{e}_k}(\mathbf{x}) = \frac{1}{\binom{j+s-1}{s}} M(\mathbf{x}|A_{\mathfrak{j}-\mathbf{e}_k}) \ . \tag{2}$$

If $j_k = 0$, then $N_{\mathfrak{j}-\mathbf{e}_k}$ is a simplex spline over the knot sequence $A_{\mathfrak{j}-\mathbf{e}_k}$, which does not contain any knot from the kth knot cloud $\mathbf{a}_k^0 \dots \mathbf{a}_k^n$. Its support and Ω are disjoint, which we prove next.

The support of $N_{\mathfrak{j}-\mathbf{e}_k}$ is the convex hull of all knots in $A_{\mathfrak{j}-\mathbf{e}_k}$. Hence, every point in this support is a convex combination of some points $\mathbf{a}_0, \dots, \mathbf{a}_k^*, \dots, \mathbf{a}_s$, where $\mathbf{a}_i$ lies in the convex hull of the first $j_i + 1$ knots $\mathbf{a}_i^0, \dots, \mathbf{a}_i^{j_i}$ of the ith knot chain.

We need to show that any such simplex $\mathbf{a}_0 \dots \mathbf{a}_k^* \dots \mathbf{a}_s$ does not intersect Ω. Since Ω is contained in all simplices $S_{\mathfrak{k}}$, where $\mathbb{0} \le \mathfrak{k} \le \mathfrak{j}$, it is also contained in any simplex $\mathbf{a}_0\, \mathbf{a}_1^0 \dots \mathbf{a}_s^0$. Applying this argument repeatedly, we find that the open set Ω is contained in the interior of the simplex $\mathbf{a}_0 \dots \mathbf{a}_s$, which implies that Ω and the support of $N_{\mathfrak{j}-\mathbf{e}_k}$ are disjoint.

So, for all $\mathbf{x}$ in Ω, the recurrence formula of the B-splines $N_{\mathfrak{j}}$ coincides with the recurrence formula of the B-polynomials $C_{\mathfrak{j}}$. Hence, it follows that $N_{\mathfrak{i}}(\mathbf{x}) = C_{\mathfrak{i}}(\mathbf{x})$ for all $\mathbf{x} \in \Omega$. ◇

19.6 Linear combinations of B-splines

We have discussed multivariate B-splines over one complex of $s+1$ knot chains. However, the true value of these B-splines is that they define useful spline spaces over the entire $\mathbb{R}^s$ partitioned into complexes of such knot chains.

Let $\mathbf{a}_k^0$, $k \in \mathbb{Z}$, be the vertices of a triangulation of $\mathbb{R}^s$, and let $K \subset \mathbb{Z}^{s+1}$

represent the simplices

$$\mathbf{a}^0_{k_0} \ldots \mathbf{a}^0_{k_s} \ , \quad \Bbbk = [k_0 \ldots k_s] \in K \ ,$$

of this triangulation. Further, let each vertex belong to a chain $\mathbf{a}^0_k \ldots \mathbf{a}^n_k$, let $S_{\Bbbk\mathbb{j}}$ denote the simplex $\mathbf{a}^{j_0}_{k_0} \ldots \mathbf{a}^{j_s}_{k_s}$, and let $A_{\Bbbk\mathbb{j}}$ denote the knot sequence

$$\mathbf{a}^0_{k_0} \ldots \mathbf{a}^{j_0}_{k_0} \ldots \mathbf{a}^0_{k_s} \ldots \mathbf{a}^{j_s}_{k_s}, \quad \text{where} \quad \Bbbk \in K, \mathbb{j} \in \Delta_0 \cup \ldots \Delta_{n-1} \ .$$

We also assume that for all $\Bbbk \in K$ the intersection

$$\Omega_\Bbbk = \text{interior} \bigcap_{\mathbb{j} \in \Delta_0 \cup \cdots \cup \Delta_n} S_{\Bbbk\mathbb{j}}$$

is non-empty, and we denote the multivariate B-spline over the knot sequence $A_{\Bbbk\mathbb{i}}$, defined in 19.5, by $N_{\Bbbk\mathbb{i}}$.

Any linear combination

$$\mathbf{s}(\mathbf{x}) = \sum_{\Bbbk \in K} \sum_{\mathbb{i} \in \Delta_n} \mathbf{c}_{\Bbbk\mathbb{i}} N_{\Bbbk\mathbb{i}}(\mathbf{x})$$

of the B-splines $N_{\Bbbk\mathbb{i}}$ forms a piecewise polynomial C^{n-1} surface of polynomial degree $\leq n$ if the knots are in general positions. The coefficients $\mathbf{c}_{\Bbbk\mathbb{i}}$ form the **control net** of $\mathbf{s}(\mathbf{x})$, as illustrated in Figure 19.8 for $n = s = 2$. The control net is not connected, in general, but decomposes into smaller nets, one for each $\Bbbk \in K$. These nets are the B-nets of the polynomial segments $\mathbf{s}(\mathbf{x})$, $\mathbf{x} \in \Omega_\Bbbk$.

We say that the control points

$$\mathbf{c}_{\Bbbk\mathbb{i}} \ , \quad \Bbbk \in K \ , \quad \mathbb{i} \in \Delta_n \ ,$$

form a **connected net** when for any two adjacent simplices $S_{\Bbbk\mathbb{o}}$ and $S_{\bar{\Bbbk}\mathbb{o}}$ and suitable index transformations such that

$$k_0 \neq \overline{k}_0 \quad \text{and} \quad [k_1 \ldots k_s] = [\overline{k}_1 \ldots \overline{k}_s] \ ,$$

the control points $\mathbf{c}_{\Bbbk\mathbb{i}}$ and $\mathbf{c}_{\bar{\Bbbk}\mathbb{i}}$ are equal for all $\mathbb{i}$, where $i_0 = 0$.

19.7 A recurrence relation

If the spline

$$\mathbf{s}(\mathbf{x}) = \sum_{\Bbbk} \sum_{\mathbb{i}} \mathbf{c}_{\Bbbk\mathbb{i}} N_{\Bbbk\mathbb{i}}(\mathbf{x})$$

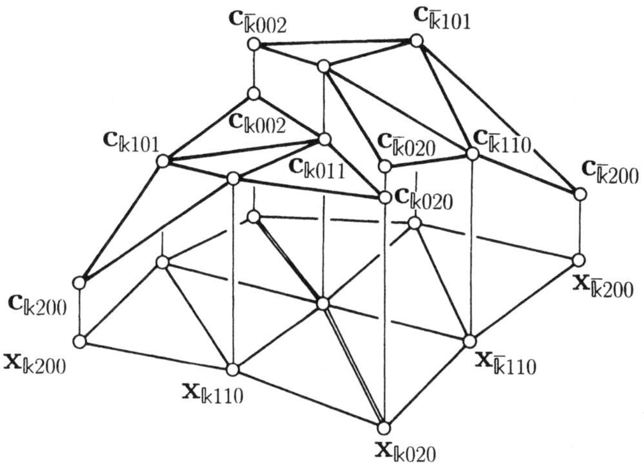

Figure 19.8: The control net of a bivariate quadratic spline over two triangles.

has a connected control net, then the recurrence relation (1) in 19.5 shows that $\mathbf{s}(\mathbf{x})$ can be written as a linear combination of the B-splines $N_{\mathbb{k}\mathbb{j}}$ of degree $|\mathbb{j}| = n-1$, i.e.,

$$(3) \qquad \mathbf{s}(\mathbf{x}) = \sum_{\mathbb{k}\in K}\sum_{\mathbb{j}\in\Delta_{n-1}} \mathbf{d}_{\mathbb{k}\mathbb{j}} N_{\mathbb{k}\mathbb{j}}(\mathbf{x}) \ ,$$

where

$$\mathbf{d}_{\mathbb{k}\mathbb{j}} = \sum_{l=0}^{s} \xi^{l}_{\mathbb{k}\mathbb{j}} \mathbf{c}_{\mathbb{k},\mathbb{j}+\mathbb{e}_l}$$

and $\xi^0_{\mathbb{k}\mathbb{j}}, \ldots, \xi^s_{\mathbb{k}\mathbb{j}}$ are the barycentric coordinates of $\mathbf{x}$ with respect to $S_{\mathbb{k}\mathbb{j}}$.

For a proof, we apply the recurrence relation (1) in 19.5 to all B-splines $N_{\mathbb{k}\mathbb{i}}$ and obtain

$$\mathbf{s}(\mathbf{x}) = \sum_{\mathbb{k}}\sum_{\mathbb{i}}\sum_{l=0}^{s} \xi^{l}_{\mathbb{k},\mathbb{i}-\mathbb{e}_l} \mathbf{c}_{\mathbb{k}\mathbb{i}} N_{\mathbb{k},\mathbb{i}-\mathbb{e}_l} \ .$$

This sum involves terms

$$\mathbf{c}_{\mathbb{k},\mathbb{j}+\mathbb{e}_l} N_{\mathbb{k}\mathbb{j}} \ ,$$

with $j_l = -1$. We will show that all these terms sum to zero, which means that the sum of all the other terms takes on the simple form we are about to prove.

Because of index transformations, it suffices to consider an index $\mathbb{k}\mathbb{j}$, where $j_0 = -1$ and $\mathbb{k} = [0 \ldots s]$. Further, we may assume that $\bar{\mathbb{k}} = [-1\, 1 \ldots s]$ is also in K. Hence, $S_{\Bbbk\mathbb{o}}$ and $S_{\bar{\Bbbk}\mathbb{o}}$ have opposite orientations, i.e.,

$$\sigma_{\Bbbk\mathbb{o}} = -\sigma_{\bar{\Bbbk}\mathbb{o}} \ .$$

In turn, it follows by successive applications of the argument used to derive equation (1) that

$$\sigma_{\Bbbk,\mathbb{j}} = -\sigma_{\bar{\Bbbk},\mathbb{j}}$$

and, consequently,

$$\xi^0_{\Bbbk\mathbb{j}} = -\xi^0_{\bar{\Bbbk}\mathbb{j}} \ .$$

Since the control net is connected and $N_{\Bbbk\mathbb{j}} = N_{\bar{\Bbbk}\mathbb{j}}$, see (2), we obtain, therefore,

$$\xi^0_{\Bbbk\mathbb{j}}\mathbf{c}_{\Bbbk,\mathbb{j}+\mathbb{e}_0}N_{\Bbbk\mathbb{j}} + \xi^0_{\bar{\Bbbk}\mathbb{j}}\mathbf{c}_{\bar{\Bbbk},\mathbb{j}+\mathbb{e}_0}N_{\bar{\Bbbk}\mathbb{j}} = 0 \ ,$$

which concludes the proof. ◇

19.8 Derivatives of a spline

The recurrence relation (2) in 18.4 and the derivative formula (3) in 18.5 for simplex splines are similar. One obtains the derivative formula up to a constant factor just by differentiating the weights in the recurrence formula although one might expect further terms due to the product rule.

Because of this, we can transform the recurrence (1) in 19.5 for multivariate splines to a derivative formula. Differentiating the barycentric coordinates ξ in the proof of the recurrence formula in 19.7, we obtain the directional derivative with respect to a direction $\Delta\mathbf{x}$ of a spline

$$\mathbf{s}(\mathbf{x}) = \sum_{\Bbbk\in K}\sum_{\mathbb{i}} \mathbf{c}_{\Bbbk\mathbb{i}}N_{\Bbbk\mathbb{i}}(\mathbf{x})$$

of degree $n = |\mathbb{i}|$ with connected control net is given by

$$\frac{d}{dt}\mathbf{s}(\mathbf{x}+t\Delta\mathbf{x})|_{t=0} = n\sum_{\Bbbk\in K}\sum_{\mathbb{j}\in\Delta} \dot{\mathbf{c}}_{\Bbbk\mathbb{j}}N_{\Bbbk\mathbb{j}}(\mathbf{x}) \ ,$$

where

$$\dot{\mathbf{c}}_{\Bbbk\mathbb{j}} = \sum_{l=0}^{s} \nu_l\mathbf{c}_{\Bbbk,\mathbb{j}+\mathbb{e}_l}$$

and $\nu_0,\ldots,\nu_s$ are the barycentric coordinates of the direction $\Delta\mathbf{x}$ with respect to the simplex $S_{\Bbbk\mathbb{j}}$, see also 19.4.

Remark 5: As already observed in 19.4, the directions $\dot{\mathbf{c}}_{\Bbbk\mathbb{j}}$ are part of the directional derivatives with respect to $\Delta\mathbf{x}$ of the control nets $\mathbf{c}_{\Bbbk\mathbb{j}}(\mathbf{x})$ of $\mathbf{s}(\mathbf{x})$ over the simplices $\mathbf{x}_{\Bbbk,\mathbb{j}+\mathbb{e}_0}\ \ldots\ \mathbf{x}_{\Bbbk,\mathbb{j}+\mathbb{e}_s}$.

19.9 The main theorem

We use the notation from 19.6 and assume that the first $r+1$ knots $\mathbf{a}_k^0, \ldots, \mathbf{a}_k^r$ of every knot cluster coincide. Still, the simplices $S_{\mathbb{k}\mathbb{o}}$ form a triangulation of $\mathbb{R}^s$. We show that any C^{n-r} piecewise polynomial function over this triangulation is a linear combination of the B-splines $N_{\mathbb{k}\mathbb{i}}$ and relate its control points to the polar forms of its polynomial segments.

Let $\mathbf{s}(\mathbf{x})$ be any $(n-r)$ times differentiable spline which, over each simplex $S_{\mathbb{k}\mathbb{o}}$, is identical with some polynomial $\mathbf{s}_{\mathbb{k}}(\mathbf{x})$ of degree $\leq n$. Then, the multivariate version of the **main theorem** 5.5 is as follows [Seidel '92].

The spline $\mathbf{s}(\mathbf{x})$ *can be written as*

$$\mathbf{s}(\mathbf{x}) = \sum_{\mathbb{k}\in K} \sum_{\mathbb{i}\in\Delta_n} \mathbf{s}_{\mathbb{k}}[A_{\mathbb{k},\mathbb{i}-\mathbb{e}}] N_{\mathbb{k}\mathbb{i}}(\mathbf{x}) \ , \tag{4}$$

where $\mathbf{s}_{\mathbb{k}}[\mathbf{x}_1 \ldots \mathbf{x}_n]$ *is the polar form of* $\mathbf{s}_{\mathbb{k}}$.

We prove this by induction over $n-r$. If $n-r=-1$, then the B-splines $N_{\mathbb{k}\mathbb{i}}$ are the Bernstein polynomials over $S_{\mathbb{k}\mathbb{o}}$, see 19.5 and Remark 2 in 19.2. In this case identity (4) follows from the main theorem 11.2. Now let $n-r \geq 0$ and let $D\mathbf{s}(\mathbf{x})$ be the directional derivative of $\mathbf{s}(\mathbf{x})$ with respect to some non-zero vector $\Delta\mathbf{x}$ and assume that

$$D\mathbf{s}(\mathbf{x}) = \sum_{\mathbb{k}\in K} \sum_{\mathbb{j}\in\Delta_{n-1}} D\mathbf{s}_{\mathbb{k}}[A_{\mathbb{k},\mathbb{j}-\mathbb{e}}] N_{\mathbb{k}\mathbb{j}}(\mathbf{x}) \ .$$

According to Section 11.6, the polar form of $D\mathbf{s}_{\mathbb{k}}$ can be expressed by the polar form of $\mathbf{s}_{\mathbb{k}}$, i.e.

$$\begin{aligned} D\mathbf{s}_{\mathbb{k}}[A_{\mathbb{k}\mathbb{j}-\mathbb{e}}] &= n\mathbf{s}_{\mathbb{k}}[\Delta\mathbf{x}\ A_{\mathbb{k},\mathbb{j}-\mathbb{e}}] \\ &= n\sum_{l=0}^{s} \nu_l \mathbf{s}_{\mathbb{k}}[A_{\mathbb{k},\mathbb{j}-\mathbb{e}+\mathbb{e}_l}] \ , \end{aligned}$$

where $\nu_0, \ldots, \nu_s$ are the barycentric coordinates of $\Delta\mathbf{x}$ with respect to $S_{\mathbb{k}\mathbb{j}}$.

Because of Remark 7 in 11.7, the control points $\mathbf{s}_{\mathbb{k}}[A_{\mathbb{k},\mathbb{i}-\mathbb{e}}]$ form a connected net. Hence, it follows from 19.8 that

$$D\sum_{\mathbb{k}}\sum_{\mathbb{i}} \mathbf{s}_{\mathbb{k}}[A_{\mathbb{k},\mathbb{i}-\mathbb{e}}] N_{\mathbb{k}\mathbb{i}} = D\mathbf{s}$$

Therefore, (4) is true up to some constant term. Since (4) holds over every $\Omega_{\mathbb{k}}$, see 19.4 and 19.5, this constant is zero, which concludes the proof. ◇

In particular, it follows that the B-splines from a **partition of unity**, i.e.,

$$\sum_{\mathbb{k}\in K}\sum_{\mathbb{i}\in\Delta_n} N_{\mathbb{k}\mathbb{i}} = 1 \ .$$

Further, the identity $\mathbf{p}(\mathbf{x}) = \mathbf{x}$ has the control points

$$\mathbf{x}_{\mathbb{k}\mathbb{i}} = \frac{1}{n} A_{\mathbb{k}\mathbb{i}}\mathbf{e} \ ,$$

see also 19.3 and Figure 19.8 for an illustration, where $n = s = 2$.

19.10 Problems

1 Derive a degree elevation algorithm for B-patches. Given a sequence of knot chains $A_{\mathbb{n}} = \mathbf{a}_0^0 \dots \mathbf{a}_s^n$ in $\mathbb{R}^s$ and a polynomial $\mathbf{p}(\mathbf{x})$ of degree n, consider the two B-patch representations

$$\mathbf{p}(\mathbf{x}) = \sum_{\mathbb{i}\in\Delta_n} \mathbf{p}_{\mathbb{i}} C_{\mathbb{i}}^n(\mathbf{x})$$

and

$$\mathbf{p}(\mathbf{x}) = \sum_{\mathbb{k}\in\Delta_{n+1}} \mathbf{p}_{\mathbb{k}} C_{\mathbb{k}}^{n+1}(\mathbf{x})$$

with respect to $A_{\mathbb{n}-\mathbb{e}}$ and $A_{\mathbb{n}}$, respectively. Express the control points $\mathbf{p}_{\mathbb{k}}$ as affine combinations of the points $\mathbf{p}_{\mathbb{i}}$.

2 Come up with an algorithm to rewrite a spline over the triangulation $S_{\mathbb{k},\mathbb{o}}, \mathbb{k} \in K$, defined in 19.9 with r-fold knots $\mathbf{a}_k^0 = \dots = \mathbf{a}_k^{r-1}$, as a linear combination of B-splines with $(r+1)$-fold knots $\mathbf{a}_k^0 = \dots = \mathbf{a}_k^r$.

3 Derive a degree elevation algorithm for multivariate splines over a triangulation of $\mathbb{R}^s$ in B-patch form.

4 Come up with an algorithm to compute the Bézier representation of a spline over the triangulation $S_{\mathbb{k}\mathbb{o}}, \mathbb{k} \in K$, defined in 19.9.

5 Let $A_{\mathbb{n}} = [\mathbf{a}_0^0 \dots \mathbf{a}_s^n]$ be the matrix of $s+1$ knot chains with $n+1$ knots each, and let $\mathbf{x}_{\mathbb{i}} = \frac{1}{n} A_{\mathbb{i}}\mathbf{e}$ be the corresponding **Greville abscissae**, see Remark 3 in 19.3. The points $\mathbf{x}_{\mathbb{j}}$ are **well-positioned** if all simplices $\mathbf{x}_{\mathbb{j}+\mathbb{e}_0} \dots \mathbf{x}_{\mathbb{j}+\mathbb{e}_s}$, for $\mathbb{j} \in \Delta_{n-1}$, have disjoint interiors. Show that the $\mathbf{x}_{\mathbb{i}}$ are well-positioned if any knot $\mathbf{a}_{\mu\nu}$ can be written as

$$\mathbf{a}_{\mu\nu} = \mathbf{a}_{\mu 0} + \sum_{k=0}^{s} \alpha_k \mathbf{a}_{k0} \ ,$$

where $0 = \alpha_0 + \cdots + \alpha_s$ and $|\alpha_0|, \ldots, |\alpha_s| < 1/2$, see also [Prautzsch '97, Thm. 4.2].

6 Let $A_{\mathbb{n}}, n_0 = n_1 = n_2 = 2$, be a knot sequence in $\mathbb{R}^2$ with well-positioned Greville abscissae. Show that any bivariate scalar-valued quadratic polynomial is convex over $\mathbb{R}^2$ if its B-net with respect to $A_{\mathbb{n}}$ is convex, see Section 10.6 and Figures 10.6 and 19.5.

7 Let $A_{\mathbb{n}}$ be a knot sequence in $\mathbb{R}^2$ with well-positioned Greville abscissae. Show that any bivariate scalar-valued polynomial is convex over the domain Ω defined in 19.5 if its B-net with respect to $A_{\mathbb{n}}$ is convex.

References

C. Bangert and H. Prautzsch (1999). A geometric criterion for the convexity of Powell-Sabin interpolants and its multivariate extension. *Computer Aided Geometric Design*, 16(6):529–538.

R.E. Barnhill and R.F. Riesenfeld, editors (1974). *Computer Aided Geometric Design.* Academic Press, New York, 1974.

W. Boehm (1977). Über die Konstruktion von B-Spline-Kurven. *Computing*, 18:161–166.

W. Boehm (1980). Inserting new knots into B-Spline-Curves. *Computer-Aided Design*, 12(4):199–201.

W. Boehm (1983). Generating the Bézier points of triangular splines. In R.E. Barnhill and W. Boehm, editors (1983), *Surfaces in Computer Aided Geometric Design*, pages 77–91. Elsevier Science Publishers B.V. (North-Holland), 1983.

W. Boehm (1984a). Calculating with box splines. *Computer Aided Geometric Design*, 1:149–162.

W. Boehm (1984b). Efficient evaluation of splines. *Computing*, 33:171–177.

W. Boehm (1985). Curvature continuous curves and surfaces. *Computer Aided Geometric Design*, 2:313–323.

W. Boehm (1987). Smooth curves and surfaces. In G. Farin, editor (1987), *Geometric Modeling: Algorithms and New Trends*, pages 175–184. SIAM, 1987.

W. Boehm, G. Farin, and J. Kahmann (1984). A survey of curve and surface methods in CAGD. *Computer Aided Geometric Design*, 1(1):1–60.

W. Boehm and H. Prautzsch (1993). *Numerical Methods.* Vieweg, Wiesbaden and AK Peters, Wellesley, Mass.

W. Boehm, H. Prautzsch, and P. Arner (1987). On triangular splines. *Constructive Approximation*, 3:157–167.

C. DE BOOR (1972). On calculating with B-splines. *Journal of Approximation Theory*, 6:50–62.

C. DE BOOR (1976). Total positivity of the spline collocation matrix. *Indiana University Journal Math.*, 25:541–551.

C. DE BOOR (1976a). Splines as linear combinations of B-splines. In SCHUMAKER, LORENTZ, CHUI, editor (1976a), *Approximation Theory II*, pages 1–47. Academic Press, New York, 1976.

C. DE BOOR (1978). *A Practical Guide to Splines.* Springer, New York.

C. DE BOOR (1984). Splinefunktionen. Research Report 84-05, Seminar für angewandte Mathematik, Eidgenössische Technische Hochschule, Zürich, Switzerland, 1984.

C. DE BOOR (1987). B-form basics. In G. FARIN, editor (1987), *Geometric Modeling: Algorithms and New Trends*, pages 131–148. SIAM, Philadelphia, 1987.

C. DE BOOR AND K. HÖLLIG (1982). Recurrence relations for multivariate B-splines. *Proceedings of the American Mathematical Society*, 85(3).

C. DE BOOR, K. HÖLLIG, AND S. RIEMENSCHNEIDER (1993). *Box Splines.* Applied Mathematical Sciences 98. Springer, New York.

M. DO CARMO (1976). Differential Geometry of Curves and Surfaces. Prentice Hall, Englewood Cliffs, New Jersey, 1976.

J.M. CARNICER AND W. DAHMEN (1992). Convexity preserving interpolation and Powell-Sabin elements. *Computer Aided Geometric Design*, 9(4):279–290.

P. DE FAGET DE CASTELJAU (1959). Outillage méthodes calcul. *Enveloppe Soleau 40.040, Institute National de la Proprieté Industrielle, Paris.*

P. DE FAGET DE CASTELJAU (1985). Formes a poles. In *Mathematiques et CAO 2.* Hermes Publishing, Paris, 1985.

E. CATMULL AND J. CLARK (1978). Recursively generated B-spline surfaces on arbitrary topological meshes. *Computer-Aided Design*, 10(6):350–355.

A.S. CAVARETTA, W. DAHMEN, AND C.A. MICCHELLI (1991). Stationary subdivision. *Memoirs of the American Mathematical Society*, 93(453):1–186.

G.M. CHAIKIN (1974). An algorithm for high-speed curve generation. *Computer Graphics and Image Processing*, 3:346–349.

R.W. Clough and J.L. Tocher (1965). Finite element stiffness matrices for the analysis of plate bending. In *Proceedings of the 1st Conference on Matrix Methods in Structural Mechanics*, Wright-Patterson AFB AFFDL TR, pages 515–545, 1965.

E. Cohen, T. Lyche, and R.F. Riesenfeld (1980). Discrete B-splines and subdivision techniques in computer-aided geometric design and computer graphics. *Computer Graphics and Image Processing*, 14:87–111.

E. Cohen, T. Lyche, and R.F. Riesenfeld (1984). Discrete box splines and refinement algorithms. *Computer Aided Geometric Design*, 1:131–148.

E. Cohen, T. Lyche, and L.L. Schumaker (1985). Algorithms for degree raising of splines. *ACM Transactions on Graphics*, 4(3):171–181.

M.G. Cox (1972). The numerical evaluation of B-splines. *J. Inst. Maths. Applics.*, 10:134–149.

H.B. Curry and I.J. Schoenberg (1966). On Pólya frequency functions IV: The fundamental spline functions and their limits. *J. d'Analyse Math.*, 17:71–107.

W. Dahmen and C.A. Micchelli (1983). Translates of multivariate splines. *Linear Algebra and its Applications*, 217-234.

W. Dahmen and C.A. Micchelli (1984). Subdivision algorithms for the generation of box spline surfaces. *Computer Aided Geometric Design*, 1:115–129.

W. Dahmen and C.A. Micchelli (1985). On the local linear independence of translates of a box spline. *Studia Math.*, 82:243-262.

W. Dahmen, C.A. Micchelli, and H.-P. Seidel (1992). Blossoming begets B-splines built better by B-patches. *Mathematics of computation 59*, 199:97–115.

W. Degen (1988). Some remarks on Bézier curves. *Computer Aided Geometric Design*, 5:259–268.

A.D. DeRose, R.N. Goldmann, H. Hagen, and S. Mann (1993). Functional composition algorithms via blossoming. *ACM Transactions on Graphics*, 12:113–135.

D.W.H. Doo and M.A. Sabin (1978). Behavior of recursive division surfaces near extraordinary points. *Computer-Aided Design*, 10(6):356–360.

N. DYN, J. GREGORY, AND D. LEVIN (1990). A butterfly subdivision scheme for surface interpolation with tension control. *ACM Transactions on Graphics*, 9:160–169.

N. DYN, J.A. GREGORY, AND D. LEVIN (1991). Analysis of uniform binary subdivision schemes for curve design. *Constructive Approximation*, 7:127–147.

N. DYN, D. LEVIN, AND J.A. GREGORY (1987). A 4-point interpolatory subdivision scheme for curve design. *Computer Aided Geometric Design*, 4:257–268.

N. DYN AND C.A. MICCHELLI (1988). Piecewise polynomial spaces and geometric continuity of curves. *Numerical Mathematics*, 54:319–337.

M. ECK (1993). Degree reduction of Bézier curves. *Computer-Aided Design*, 10:237–251.

M. ECK (1995). Least squares degree reduction of Bézier curves. *Computer-Aided Design*, 27 (11):845–853.

G. FARIN (1979). *Subsplines über Dreiecken.* Diss., TU Braunschweig, Germany, 1979.

G. FARIN (1982). Visually C^2 cubic splines. *Computer-Aided Design*, 14:137–139.

G. FARIN (1986). Triangular Bernstein-Bézier patches. *Computer Aided Geometric Design*, 3(2):83–127.

G. FARIN (2002). *Curves and Surfaces for Computer Aided Geometric Design.* Morgan Kaufmann Publishers, 5th edition.

D. FILIP, R. MAGDSON, AND R. MARKOT (1986). Surface algorithms using bounds on derivatives. *Computer Aided Geometric Design*, 3(4):295–311.

M.S. FLOATER (1997). A counterexample to a theorem about the convexity of Powell-Sabin elements. *Computer Aided Geometric Design*, 14:383–385.

T.A. FOLEY AND G.M. NIELSON (1989). Knot selection for parametric spline interpolation. In T. LYCHE AND L.L. SCHUMAKER, editors (1989), *Mathematical Methods in Computer Aided Geometric Design*, pages 261–271. Academic Press, Boston, 1989.

W. GORDON AND R.E. RIESENFELD (1974). B-spline curves and surfaces. In *[Barnhill & Riesenfeld '74]*, pages 95–125, 1974.

T.A. GRANDINE (1989). On convexity of piecewise polynomial functions on triangulations. *Computer Aided Geometric Design*, 6(3):181–187.

J. GRAVESEN (1997). Adaptive subdivision and the length of Bézier curves. *Computational Geometry*, 8:13–31.

J.A. GREGORY (1991). An introduction to bivariate uniform subdivision. In D.F. GRIFFITHS AND G.A. WATSON, editors (1991), *Numerical Analysis*, pages 103–117. Pitman Research Notes in Mathematics, Longman Scientific and Technical, 1991.

T.N.E. GREVILLE (1967). On the normalization of the B-splines and the location of the nodes for the case of unequally spaced knots. In O. SHISKA, editor (1967), *Inequalities*. Academic Press, New York, 1967.

J. HOSCHEK AND D. LASSER (1992). *Grundlagen der geometrischen Datenverarbeitung*. B.G. Teubner, Stuttgart, second edition. English translation: Fundamentals of Computer Aided Geometric Design, AK Peters, Wellesley, 1993.

R.-Q. JIA (1983). Linear independence of translates of a box spline. *Journal of Approximation Theory*, 40:158-160.

R.-Q. JIA (1985). Local linear independence of the translates of a box spline. *Constructive Approximation*, 1:175-182.

S. KARLIN (1968). *Total Positivity*. Stanford University Press, 1968.

W. KLINGENBERG (1978). *A Course in Differential Geometry*. Springer, New York.

L. KOBBELT (1994). *Iterative Erzeugung glatter Interpolanten*. Diss., Universität Karlsruhe, 1994.

L. KOBBELT (2000). $\sqrt{3}$-Subdivision. *Proceedings of SIGGRAPH 2000*, Academic Press, New York, pages 103–112.

L. KOBBELT AND H. PRAUTZSCH (1995). Approximating the length of a spline by its control polygon. In T. LYCHE M. DÆHLEN AND L.L. SCHUMAKER, editors (1995), *Mathematical Methods for Curves and Surfaces*, pages 291–292, 1995.

M.-J. LAI (1991). A characterization theorem of multivariate splines in blossoming form. *Computer Aided Geometric Design*, 8:513–521.

J.M. LANE AND R.F. RIESENFELD (1980). A theoretical development for the computer generation and display of piecewise polynomial surfaces. *IEEE Transactions on Pattern Analysis and Machine Intelligence*, 2(1):35–46.

E.T.Y. Lee (1982). A simplified B-spline computation routine. *Computing*, 29:365–373.

E.T.Y. Lee (1989). Choosing nodes in parametric curve interpolation. *Computer-Aided Design*, 21:363–370.

E.T.Y. Lee (1994). Remarks on an identity related to degree elevation. *Computer Aided Geometric Design*, 11:109–111.

W. Liu (1997). A simple, efficient degree raising algorithm for B-spline curves. *Computer Aided Geometric Design*, 14:693–698.

C.T. Loop (1987) Smooth subdivision surfaces based on triangles. Master thesis, University of Utah, 1987.

D. Lutterkort, J. Peters and U. Reif (1999). Polynomial degree reduction in the L_2-norm equals best Euclidean approximation of Bézier coefficients. *Computer Aided Geometric Design*, 16:607–612.

M.-L. Mazure and H. Pottmann (1996). Tchebycheff curves. In M. Gasca and C.A. Micchelli, editors (1996), *Total Positivity and its Applications*, pages 187–218. Kluwer, 1996.

C.A. Micchelli (1979). On a numerically efficient method for computing multivariate B-splines. In W. Schempp and K. Zeller, editors (1979), *Multivariate Approximation Theory*, pages 211–248. Birkhäuser, Basel, 1979.

C.A. Micchelli (1980). A constructive approach to Kergin interpolation in R^k: multivariate B-splines and Lagrange interpolation. *Rocky Mountain Journal of Mathematics*, 10(3):485–497.

C.A. Micchelli and H. Prautzsch (1987). Refinement and subdivision for spaces of integer translates of a compactly supported function. In D.F. Griffiths and G.A. Watson, editors (1987), *Numerical Analysis*, pages 192–222. Longman, London, 1987.

C.A. Micchelli and H. Prautzsch (1989). Computing curves invariant under subdivision. *Computer Aided Geometric Design*, 4:321–328.

D. Nairn, J. Peters, and D. Lutterkort (1999). Sharp, quantitative bounds on the distance between a polynomial piece and its Bézier control polygon. *Computer Aided Geometric Design*, 16(7):613–633.

G. Nielson (1974). Some piecewise polynomial alternatives to splines under tension. In R.E. Barnhill and R.F. Riesenfeld, editors (1974), *Computer Aided Geometric Design*, pages 209–235. Academic Press, 1974.

J. PETERS (1991). Smooth interpolation of a mesh of curves. *Constructive Approximation*, 7:221–246.

J. PETERS (1994). Evaluation and approximate evaluation of the multivariate Bernstein-Bézier form on a regularly partitioned simplex. *ACM Transactions on Mathematical Software*, 20(4):460–480.

J. PETERS AND U. REIF (1997). The simplest subdivision scheme for smoothing polyhedra. *ACM Transaction on Graphics*, 16(4):420–431.

L. PIEGL AND W. TILLER (1994). Software-engineering approach to degree elevation of B-spline curves. *Computer-Aided Design*, 26(1):17–28.

B. PIPER (1987). Visually smooth interpolation with triangular Bézier patches. In G. FARIN, editor (1987), *Geometric Modeling: Algorithms and New Trends*, Philadelphia, pages 221–233. SIAM, 1987.

H. POTTMANN (1993). The geometry of Tchebycheffian splines. *Computer Aided Geometric Design*, 10:181–210.

H. POTTMANN AND A.D. DEROSE (1991). Classification using normal curves. *Curves and Surfaces in Computer Vision and Graphics 2*, 1610:217–228.

M.J.D. POWELL (1981). Approximation theory and methods. *Cambridge University Press.*

M.J.D. POWELL AND M.A. SABIN (1977). Piecewise quadratic approximations on triangles. *ACM Transactions on Mathematical Software*, 3(4):316–325.

H. PRAUTZSCH (1984a). Degree elevation of B-spline curves. *Computer Aided Geometric Design*, 1:193–198.

H. PRAUTZSCH (1984b). Unterteilungsalgorithmen für multivariate Splines Ein geometrischer Zugang. Diss., TU Braunschweig, 1984.

H. PRAUTZSCH (1989). A round trip to B-splines via de Casteljau. *ACM Transactions on Graphics*, 8(3):243–254.

H. PRAUTZSCH (1997). Freeform splines. *Computer Aided Geometric Design*, 14:201–206.

H. PRAUTZSCH (1998). Smoothness of subdivision surfaces at extraordinary points. *Advances in Computational Mathematics*, 9:377–389.

H. PRAUTZSCH (2002). B-splines with arbitrary connection matrices. Preprint, 2002.

H. PRAUTZSCH AND W. BOEHM (2002). Box Splines. In KIM ET AL, editor (2002), *Handbook of Computer Aided Design*, Elsevier, 2002.

H. PRAUTZSCH AND B. PIPER (1991). A fast algorithm to raise the degree of spline curves. *Computer Aided Geometric Design*, 8:253–265.

H. PRAUTZSCH AND U. REIF (1999). Degree estimates for C^k-piecewise polynomial subdivision surfaces. *Advances in Computational Mathematics*, 10:209–217.

L. RAMSHAW (1987). Blossoming: a Connect-the-Dots Approach to Splines. Technical report, Digital Systems Research Center, Palo Alto, California, June 21, 1987.

U. REIF (1993). *Neue Aspekte in der Theorie der Freiformflächen beliebiger Topologie.* Diss., Mathematisches Institut A der Universität Stuttgart, 1993.

U. REIF (1995a). A note on degenerate triangular Bézier patches. *Computer Aided Geometric Design*, 12:547–550.

U. REIF (1995b). A unified approach to subdivision algorithms near extraordinary vertices. *Computer Aided Geometric Design*, 12:153 174.

U. REIF (1996). A degree estimate for subdivision surfaces of higher regularity. *Proc. of the AMS*, 124(7):2167–2174.

U. REIF (1998). TURBS - Topologically unrestricted rational B-splines. *Constructive Approximation*, 14(1):57–78.

U. REIF (2000). Best bounds on the approximation of polynomials and splines by their control structure. *Computer Aided Geometric Design*, 17:579–589.

G. DE RHAM (1947). Un peu de mathématiques à propos d'une courbe plane. *Elemente der Mathematik*, 2(4):89–104.

V.L. RVACHEV (1990). Compactly supported solutions of functional differential equations and their applications. *Russian Math. Surveys*, 45:87–120.

M.A. SABIN (1977). *The use of piecewise forms for the numerical representation of shape.* PhD thesis, Hungarian Academy of Science, Budapest, 1977.

M.A. SABIN (1986). Recursive subdivision. In GREGORY, editor (1986), *The Mathematics of Surfaces*, Claredon Press, Oxford, England, pages 269-282.

P. SABLONNIERE (1978). Spline and Bézier polygons associated with a polynomial spline curve. *Computer-Aided Design*, 10:257–261.

R. SCHABACK (1993). Error estimates for approximations from control nets. *Computer Aided Geometric Design*, 10:57–66.

I.J. SCHOENBERG (1967). On spline functions. In O. SISCHA, editor (1967), *Inequalities*, pages 255–291. Academic Press, New York, 1967.

I.J. SCHOENBERG AND A. WHITNEY (1953). On Pólya frequency functions, 3: The positivity of translation determinants with applications to the interpolation problem by spline curves. *TAMS*, 74:246–259.

L.L. SCHUMAKER (1973). Constructive aspects of discrete polynomial spline functions. In G.G. LORENTZ ET AL, editor (1973), *Approximation Theory*, pages 469–476. Academic Press, New York, 1973.

L.L. SCHUMAKER (1981). *Spline functions – Basic theory.* John Wiley & Sons, New York.

H.-P. SEIDEL (1989). Computing B-spline control points. In W. STRASSER AND H.-P. SEIDEL, editors (1989), *Theory and Practice of geometric modeling*, pages 17–32. Springer, Berlin, 1989.

H.-P. SEIDEL (1991). Symmetric recursive algorithms for surfaces: B-patches and the de Boor algorithm for polynomials over triangles. *Constructive Approximation*, 7:257–279.

H.-P. SEIDEL (1992). New algorithms and techniques for computing with geometrically continuous spline curves of arbitrary degree. *Mathematical Modelling and Numerical Analysis*, 26(1):149–176.

P. SHENKMAN, D. DYN, AND D. LEVIN (1999). Normals of the butterfly subdivision scheme surfaces and their application. *Journal of Computational and Applied Mathematics*, 102(1):157–180.

E. STÄRK (1976). *Mehrfach differenzierbare Bézier-Kurven und Bézier-Flächen.* Diss., TU Braunschweig, 1976.

W. TRUMP (2001). *Neue Algorithmen zur Graderhöung und Umparametrisierung von Bézier- und B-Spline-Darstellungen.* Diss., Universität Karlsruhe, 2001.

W. TRUMP AND H. PRAUTZSCH (1996). Arbitrary degree elevation for simplicial Bézier polynomials. *Computer Aided Geometric Design*, 13:387–398.

Index

C

D

E

F

G

H

I

J

K

L

M

N

O

P

Q

R

S

T

U

V

Z

Druck:	Strauss Offsetdruck, Mörlenbach
Verarbeitung:	Schäffer, Grünstadt